# Power System Capacitors

# POWER ENGINEERING

*Series Editor*
## H. Lee Willis
*ABB Inc.*
*Raleigh, North Carolina*

*Advisory Editor*
## Muhammad H. Rashid
*University of West Florida*
*Pensacola, Florida*

# Power System Capacitors

**Ramasamy Natarajan**

CRC Press
Taylor & Francis Group
Boca Raton London New York

CRC Press is an imprint of the
Taylor & Francis Group, an **informa** business
A TAYLOR & FRANCIS BOOK

CRC Press
Taylor & Francis Group
6000 Broken Sound Parkway NW, Suite 300
Boca Raton, FL 33487-2742

First issued in paperback 2019

ISBN-13: 978-1-57444-710-1 (hbk)
ISBN-13: 978-0-367-39316-8 (pbk)

**Library of Congress Cataloging-in-Publication Data**

Catalog record is available from the Library of Congress

**Visit the Taylor & Francis Web site at
http://www.taylorandfrancis.com**

**and the CRC Press Web site at
http://www.crcpress.com**

# *Dedicated to*

*My father, K. Ramasamy Gounder; my mother, Palaniammal;*
*Midappadi Village, India*

# Series Introduction

Power engineering is the oldest and most traditional of the various areas of electrical engineering, yet no other facet of modern technology continues to undergo a more persistent evolution in technology and industry structure. The use of capacitors in both shunt and series applications demonstrates this quite well. Since the earliest applications of AC systems, engineers have known that tight control of VAR flow and power factor, using capacitors in many cases, is necessary to attain good operating efficiency and stable voltage. But whereas power factor engineering was once based on simple rules-of-thumb, with the capacitors literally sized and assembled plate by plate in the field, today's capacitor engineering is an intricate analytical process that can select from a diverse range of types of capacitor technologies, including power electronic devices that simulate their action.

*Power System Capacitors* is a well-organized and comprehensive view of the theory behind capacitors and their practical applications to modern electric power systems. At both the introductory and advanced levels, this book provides a solid foundation of theory, fact, nomenclature, and formula, and offers sound insight into the philosophies of power factor engineering techniques and capacitor selection and sizing. Dr. Natarajan's book will be useful as a day-to-day reference but will also make an excellent self-paced tutorial: the book begins with a thorough review of the basics and builds upon that foundation in a comprehensive and very broad presentation of all aspects of modern theory and engineering, including the latest analysis and modeling techniques.

Like all the books in Marcel Dekker's Power Engineering Series, *Power System Capacitors* puts modern technology in a context of practical application, making this a useful reference book as well as a self-study tool and a textbook for advanced classroom use. This series includes books covering the entire field of power engineering, its specialties and subgenres, all aimed at providing practicing power engineers with the knowledge and techniques they need to meet the electric industry's challenges in the 21st century.

*H. Lee Willis*

# Preface

In power systems, the reactive power compensation is provided locally at all voltage levels using fixed capacitors, switched capacitors, substation capacitor banks, or static VAR compensators. Whatever the nature of the compensation, capacitors are the common elements in all the devices. The power factor correction approach using shunt capacitors has been employed for the past several decades. The capacitor banks used for power factor correction include fuses, circuit breakers, protective relaying, surge arresters, and various mounting approaches. Technological changes have been tremendous in the past decade, and these developments have been reported in technical papers in various journals, but no book has been published on this subject in the U.S. since the 1950s. This book contributes to this new direction.

Power system shunt capacitors are important in power factor correction. Series capacitors are vital to improving the performance of long-distance transmission. Chapters 1 through 9 examine the fundamentals of capacitors, mainly related to power factor correction. Topics include power factor concepts, industry standards, capacitor specifications, testing of capacitors, location of shunt capacitors, power factor improvement, and system benefits. Chapters 10 through 14 deal with series capacitors, surge capacitors, capacitors for motor applications, capacitors for other applications, and static VAR compensation. Chapters 15 through 18 discuss the protection of shunt capacitors, overcurrent protection, circuit breakers, and surge protection of shunt capacitor banks. Chapter 19 explains shunt capacitor bank maintenance. Chapters 20 through 23 explore system impact issues such as harmonics, switching transients, and induced voltages on the control cables. Chapter 24 discusses economic analysis applicable to the power factor compensation. The chapters are ordered from simple concepts to more complex items. The complexity of capacitor bank installation is related to protection and system impact. The reader is expected to have a basic knowledge of protection equipment such as fuses, circuit breakers, surge arresters, and unbalance protection, as well as an understanding of the basic concepts of time-domain analysis in order to follow the transient effects of capacitor switching. It should be noted that transient-related failures occur momentarily and sometimes in remote locations.

This book describes the fundamentals, capacitor applications, protection issues, and system impact based on practical installations in a step-by-step, easily understandable manner. The subject may seem complicated because capacitor failures are related to equipment voltage ratings and power system transients and are very difficult to correlate even if time-domain measurements are available. For several years I was involved in system studies for industrial power systems and utilities. I hope that this book will be a useful tool for power system engineers in industry, utilities, and consulting, as well as for professionals involved in the evaluation of practical power systems. I wish to make it clear that this book has been my spare-time activity, and I am responsible for its contents, including any errors that may inadvertently appear in it.

I wish to thank the software manufacturers for permission to use the copyrighted material in this book, including the EMTP program from Dr. H. W. Dommel, University of British Columbia, Canada, and SuperHarm and TOP, the output processor from Electrotek Concepts, Knoxville, Tennessee. The reprint permission granted by various publishers and organizations such as IEEE is greatly appreciated. John Jenkins' contribution of the photograph of Leyden jars from the Spark Museum is greatly appreciated.

I also thank the following organizations for providing photographs of various devices associated with the capacitor bank equipment: Aerovox Division, PPC, New Bedford, Massachusetts; Alpes Technologies Capacitors, Annecy-Le-Vieux, France; B&K Precision, Melrose, California; Capacitor Industries Inc., Chicago, Illinois; Capacitor Technologies, Victoria, Australia; Duke Energy, Charlotte, North Carolina; FLIR Systems, North Billerica, Massachusetts; Gilbert Electrical Systems and Products, Beckley, West Virginia; High Energy Corporation, Parkesburg, Pennsylvania; Maxwell Technologies, San Diego, California; Maysteel LLC, Menomonee

Falls, Wisconsin; Mitsubishi Electric Products Inc., Warrendale, Pennsylvania; National Oceanic and Atmospheric Administration (NOAA), Washington, D.C.; Ohio Brass (Hubbell Power Systems), Aiken, South Carolina; and Toshiba Power Systems & Services, Sydney, Australia.

Finally, I wish to thank the many engineers who, over the past several years, have discussed the technical problems presented in this book. I acknowledge the contributions of technical reviewers who read the manuscript and provided valuable comments. I also express my sincere thanks to Nora Konopka, Christine Andreasen, and Preethi Cholmondeley of Taylor & Francis for coordinating editorial activities. The efforts of Dan Collacott of Keyword Publishing are greatly appreciated.

I also thank H. Lee Willis, Vice President for Utility Systems and Asset Management, KEMA T&D Consulting, for his help with my book projects and other power system projects. I also appreciate his contributions to the power system area as the editor of the Marcel Dekker Power Engineering Series.

*Ramasamy Natarajan*

# About the author

Ramasamy Natarajan, Ph.D., is the President of Practical Power Associates, Raleigh, North Carolina. He received his B.S (1970) and M.Sc. (1973) degrees from the University of Madras, India; his M.S degree (1982) from the University of Saskatchewan, Canada; and his Ph.D. degree (1986) in electrical power engineering from the University of Washington, Seattle. Dr. Natarajan is the author or co-author of more than 60 professional publications and the book *Computer-Aided Power System Analysis* (Marcel Dekker, 2002). He is a senior member of the Institute of Electrical and Electronic Engineers. His fields of interest include power system analysis, electromagnetic transients, traction power, harmonics, power system measurements, power electronics and rotating electrical machines.

# Contents

# 1

# INTRODUCTION

## 1.1  HISTORY OF CAPACITORS

William Gilbert, who experimented with magnetism and electricity in the 1600s, was the first to apply the term "electric" (Greek electron, "amber") to the force realized upon rubbing certain substances together. He called the force between the two objects charged by friction "electric." Gilbert also identified the difference between magnetism and electricity. In 1729, Stephen Gray discovered the electrical conductor. Charles Dufay first expressed the idea of a repulsive electric force in 1733.

In 1745, Pieter van Musschenbroek, a professor of physics and mathematics at the University of Leyden, the Netherlands, invented the electrostatic charge. He experimented with a glass jar, which is now known as the Leyden jar. A photograph of Leyden jars is shown in Figure 1.1 [1]. Other accounts claim that Ewald Georg von Kleist, a German inventor, discovered the first capacitor at around the same time. The first Leyden jar was a glass jar, partially filled with water and closed with a cork, with a wire that dipped into the water inserted through its center. The wire is brought into contact with a static electricity producer and the jar becomes charged.

**Figure 1.1** Leyden jars. (Courtesy of John Jenkins, www.sparkmuseum.com.)

Any conductor that comes into contact with or in close proximity to the wire will discharge the Leyden jar.

The American statesman Benjamin Franklin conducted experiments with the Leyden jar and illustrated that the charge could also be caused by thunder and lightning. In 1752, Franklin proposed the "electrical fluid theory" to prove that lightning is an electrical phenomenon. It was also Franklin who used "+" and "−" for the two types of electric forces, attraction (positive) and repulsion (negative). His discoveries using the Leyden jar led him to show that a flat piece of glass would work just as well as a glass container for making a capacitor. These are called flat capacitors or "Franklin squares." Michael Faraday's experimental research on disk type static electricity led him to invent a means of measuring the capacitance. The plate type electrical capacitor that was used by Faraday was 50 in. in diameter with two sets of rubbers; its prime conductor consisted of two brass cylinders connected by a third, for a total length of 12 ft. The surface area in contact with air was about 1422 in.$^2$. Under conditions of good excitation, one revolution of the plate will give 10 or 12 sparks from the conductors, each 1 in. in length.

**TABLE 1.1** Historical Developments Related to Capacitance

| | |
|---|---|
| 1600s | William Gilbert experimented on magnetism and electricity. |
| 1729 | Stephen Gray discovered the electrical conductor. |
| 1745 | Pieter van Musschenbroek invented the capacitor (Leyden jar). |
| 1745 | Ewald Georg von Kleist invented the capacitor at the same time. |
| 1752 | Benjamin Franklin tested his theory on electrical fluid theory to prove that lightning is an electrical phenomenon. He invented the flat capacitor. |
| 1785 | Charles Coulomb demonstrated the manner in which charges repel one another. |
| 1831 | Michael Faraday's induction principles. |
| 1820 | Hans C. Oersted and André M. Ampère discovered that an electric field produces a magnetic field. |
| 1915 | Karly Willy Wagner and George Campell invented the LC ladder network for electronic filtering. |
| 1938 | Polypropylene was discovered, later used in capacitor manufacture. |
| 1970 | All-film design of shunt capacitors. |
| 1970s | PCBs (polychlorinated biphenyls) were found to be harmful, prompting the switchover to biodegradable fluids in power capacitors. |
| 1970s | The vacuum and SF6 circuit breakers were developed. Useful in capacitor switching. |

Sparks from 10 to 14 in. in length may easily be drawn from the conductors.

It was Faraday who discovered that moving a magnet inside a coil creates an electric current. This concept led to the development of rotating machines such as the dynamo, generator, and electric motor. To honor his contributions to electrical engineering, the unit of capacitance, the farad (F), is named after Faraday. Some of the historical events related to the invention of the capacitor are summarized in Table 1.1. Although many of the discoveries seem to be isolated, the technology was developed over several hundred years. These and other unreported discoveries are closely related to the overall design of capacitors.

### 1.1.1 Other Capacitors

Capacitors are also known as condensers. They find extensive application in the power industry in power factor correction. In the electronics industry, numerous capacitor elements are used in circuit design for control applications. These capacitors come in different shapes and sizes. The materials used in the production of these capacitors include aluminum foil, polypropylene, polyester, polycarbonate, mica, Epoxy resins, Teflon, and Tantalum. The capacitor may vary in size from a fraction of a picofarad (pF) to large ones of the order of several microfarads (μF). Electronic circuits are usually designed to operate at DC voltages of the order of 5 or 10 V. The smaller versions of the components can be soldered easily to electronic circuit boards and mounted in protected enclosures. Electrolytic capacitors are used in certain DC and motor starting applications.

## 1.1.2  Power Capacitors

References [2–4] provide detailed information on shunt capacitors. Reference [5] is a collection of standards that provide guidelines for the application of shunt capacitors. The first capacitor units were constructed using Kraft paper as the active dielectric and Askarel as the liquid impregnant. The unit, with sizes up to 100 kVAR, was commonly known as a paper capacitor. With the introduction of electrical grade plastic film, also known as polypropylene, the capacitor design was changed, and larger kVAR ratings were introduced using a dielectric composed of a combination of Kraft paper and plastic film. PCBs (polychlorinated biphenyls) were found to present problems due to their nonbiodegradability and persistence in the environment. As a result, all capacitor units are now manufactured with a non-PCB biodegradable dielectric fluid.

The capacitor units with all-film design are low loss units, with increased safety due to reduced probability of tank rupture and increased reliability due to lower operating temperatures. Figure 1.2 shows the concept of the dielectrics in paper and all-film capacitor units. The corresponding average loss/kVAR for all three designs [4] are:

| | |
|---|---|
| Paper, oil impregnated | (2.0–2.5) W/kVAR |
| Paper, PCB impregnated | (3.0–3.5) W/kVAR |

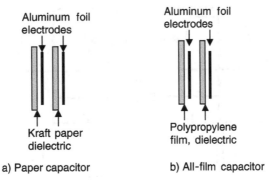

a) Paper capacitor     b) All-film capacitor

**Figure 1.2** The basic components of (a) a paper capacitor and (b) an all-film capacitor.

| | |
|---|---|
| Plastic file/paper PCB impregnated | (0.5–1.0) W/kVAR |
| Plastic film, oil impregnated | (0.5–1.0) W/kVAR |
| Metalized film | Under 0.5 W/kVAR |
| Plastic film | Under 0.2 W/kVAR |

## 1.2 CAPACITORS FOR THE POWER INDUSTRY

Power factor correction is the main application for shunt capacitor units in the power system. The advantage of improved power factor is reduced line and transformer losses, improved voltage profile, reduced maximum demand, and improved power quality. The capacitors are installed in a distribution system on pole-mounted racks, substation banks, and high voltage (HV) or extra-high voltage (EHV) units for bulk power applications. In industrial systems, the power factor correction capacitor units are utilized for group or individual loads.

The capacitors are constructed in an enclosure (tank) and are designed to provide maintenance-free service. The major components of a capacitor unit are shown in Figure 1.3. A cut view of a typical two-terminal high voltage capacitor unit is shown in Figure 1.4. The capacitor units are designed and tested according to industry standards. The ratings are expressed in kVAR, voltage, and frequency of operation

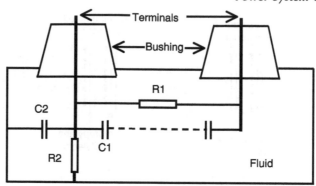

**Figure 1.3**  Components of a capacitor.

**Figure 1.4**  Cut-view of a two-terminal, all-film capacitor. 1. Terminal for connection; 2. Porcelain terminal; 3. Fixing lug; 4. Stainless steel case; 5. Active part. (Reproduced from the website of Alpes Capacitors, Annecy-Le-Vieux, France; Reference [6].)

within an ambient temperature range of −40 to +46°C. Single-phase units are available in 50, 100, 200, 300, and 400 kVAR sizes. The voltage ratings can be in the range 2.4–23 kV. A basic capacitor is made up of electrodes, a dielectric, a case, dielectric liquid, bushings, and a discharge resistor. Brief descriptions of the various components are given below.

### 1.2.1 The Electrodes

Typically, the electrodes are thin sheets, of the order of 6 mm of pure self-annealed aluminum foil. The electrodes are separated by sheets of dielectric and wound into a roll according to design specifications. Electrical connections are soldered or welded to the electrodes and terminations are made. Usually two electrodes are used in every single-phase capacitor.

### 1.2.2 The Dielectric

The dielectric or insulation material is used in the capacitors to separate the electrodes. The various dielectric materials used in the capacitor industry are listed in Table 2.1 (Chapter 2). These materials have high dielectric strength, high dielectric constant, and low dielectric loss. Kraft paper was used in earlier designs, but polypropylene film is used in present-day capacitor designs. The paper used for capacitor applications can be of low density (density $= 0.8 \, g/cm^3$), medium density (density $= 1.0 \, g/cm^3$), or high density (density $= 1.2 \, g/cm^3$). The permittivity of the paper used depends on the type of impregnants used on the paper.

Polypropylene film was invented in the 1930s and was introduced in the capacitor industry in the 1960s. This material has a very low loss and relatively high permittivity, of the order of 2.25 in the range 50 Hz–1 MHz. The film thickness is very low, of the order of 10 μm, and hence the overall volume of the capacitor for the given kVAR will be small. Some characteristics of the commercially available polypropylene film for capacitor production are listed in Table 1.2 [4].

### 1.2.3 The Case

The rolled assembly of the aluminum electrodes and dielectric material is placed in a metallic case and insulation is provided in between. A dielectric liquid is used to fill the case. The case is then sealed and the electrodes are terminated through

TABLE 1.2    Characteristics of Plastic Films Used in Capacitor
Production

| Plastic Film | Permittivity | Loss Factor | Dielectric Strength MV/m |
|---|---|---|---|
| Polypropylene | 2.25 | 0.0005 | >32 |
| Polycarbonate | (2.7–3.1) | 0.0005 | 120 |
| Polyethylene terephthalate | (3.0–3.2) | 0.03 | 100/160 |

bushings. In high voltage capacitor units, the case may be
made of stainless steel, as shown in Figure 1.4.

### 1.2.4   The Bushings

In low voltage capacitors, the electrodes are terminated
through ordinary terminals. In high voltage capacitors, the
termination is through porcelain bushings. In the terminal
arrangement, the bushings are provided with metal inserts.

Liquids are used in capacitors to fill the voids in the dielectric material. This is important, because the voids may result
in electrical discharges between the electrodes through the
dielectric, leading to failures. In pre-1975 designs, PCB was
used as the main impregnant. It was then claimed that PCB
was not biodegradable, thus constituting both a health risk and
an environmental hazard. Capacitor manufacturers were
therefore compelled to develop alternative impregnants for
the entire range of capacitors. Some of the characteristics of
the impregnants that were developed are listed in Table 1.3 [4].
The metalized film designs were produced with a non-PCB
insulating fluid. This was followed by completely dry metalized
film designs during the late 1970s. PCB is a fire-resistant fluid
with superior qualities. None of the replacement liquids has
the same characteristics.

### 1.2.5   Discharge Resistors

As part of the capacitor construction, discharge resistors are
connected to the electrodes to discharge the capacitor when

**TABLE 1.3** Liquid Impregnants for Capacitors

| Impregnant | Permittivity | Dielectric Strength MV/m | Pour Point (°C) | Flash Point (°C) | Fire Point (°C) |
|---|---|---|---|---|---|
| Dioctyl phthalate | 5.26 | 10.6 | −45 | 225 | 251 |
| Diisononyl phthalate | 4.68 | 11.8 | −48 | 221 | 257 |
| Isopropyl biphenyl | 2.83 | 56 | −55 | 155 | 175 |
| Benzyl neocaprate | 3.8 | 76 | −60 | 155 | 165 |
| PCB | 6 | 74 | −23 | – | – |

disconnected from the source. A typical discharge resistor arrangement is shown in Figure 1.4.

## 1.3 ORGANIZATION OF THE BOOK

In this book, the fundamentals of capacitance and details about the applications of both shunt and series capacitor units are presented in a step-by-step manner. The benefits of capacitors in the power system are described in both qualitative and quantitative manners. The complex aspects of capacitor applications in low and high voltage systems are also addressed.

In Chapter 2, the fundamental concepts of capacitance, the parallel plate capacitor, the spherical capacitor, the capacitor in composite medium, and the cylindrical capacitor are presented. The energy stored in the capacitor units, the charging and discharging of capacitors, the time constraint, and related concepts are also discussed.

The concepts of real current, reactive current, and total current are discussed in Chapter 3. The general concepts of power factor and the power factor correction are explained with numerical examples.

In Chapter 4, the organizations responsible for the development and maintenance of the necessary industry standards are identified. The industry standards for the design, manufacture, and application of capacitor equipment are identified. These standards include guidelines for specification of capacitors, protection, approaches used in various applications, grounding, unbalanced detection, selection methodology for circuit breakers in capacitor switching, surge arresters for protection from switching and lightning, fuses for protection against overcurrents, and other similar application details.

The capacitor units are manufactured, tested, and used in power factor correction applications. Proper specification of the applicable parameters is important for safe and efficient operation. Some of the specifications applicable to these units are voltage, frequency, insulation class, momentary ratings, nominal kVAR rating, and allowable operating service conditions. These specifications are presented in Chapter 5.

In Chapter 6, the various tests performed on the capacitor units are described. These include capacitor design tests, routine tests, and field tests. Routine tests are performed by the manufacturer on all capacitor units and field tests are performed after assembly at the installed location. These tests are usually performed according to industry standards.

The shunt capacitor banks can be installed at distribution systems, loads, feeders, high voltage systems, or extra-high voltage systems. Further, the capacitor banks can be applied on individual loads, branch locations, or at the group load. Furthermore, the banks can be fixed or switched. There are various methods available for switching these capacitor banks. The shunt capacitor installations are used in distribution and high voltage systems, where significant reactive power is supplied to the power system. In such installations, the maximum and minimum sizes from the voltage regulation point of view are presented. Also, if a capacitor unit fails, then the voltage of the good capacitor units increases. These voltages have to be kept within acceptable limits. All the above-mentioned issues are discussed in Chapter 7.

In Chapter 8, the basics of power factor improvement are discussed. The typical power factors of various loads are presented. The concept of selecting fixed shunt compensation and switched capacitors is discussed using a typical maximum kVAR demand curve. A look-up table is provided based on the existing power factor, known kW, and desired power factor. Using the constant from the table, the required shunt capacitor size can be calculated. The released capacity due to the power factor correction is shown using an example.

Chapter 9 deals with the system benefits that result from the installation of power system capacitors. The improvement in the voltage profile, voltage rise due to the addition of shunt capacities on unloaded lines, and related calculations are examined. The release of power system capacity in the transformer and generator circuits are discussed with suitable numerical examples. Finally, the cost savings from reduced energy losses in various circuits due to the shunt capacitors are presented.

Series capacitors are used in long lines and distribution lines to compensate and transmit more power. The protection of these series capacitors is an important aspect of these applications. Various approaches used to protect the series capacitor installations are shown. The limitations and the self-excitation problems associated with the induction motor-series capacitor combination are explained. Typical examples for the series capacitor installations for both distribution and transmission lines are also illustrated in Chapter 10.

In Chapter 11, surge capacitors suitable for rotating machines are presented. The acceptable insulation withstand voltages are available in the form of a curve. The necessary surge protection for rotating machines includes a surge arrester and surge capacitor installed at the terminals of the motor. The surge arrester limits the overvoltage magnitude, and the surge capacitor limits the rate of rise of the surge voltage. Example calculations are used to explain the selection of suitable surge protection for a motor or generator.

In Chapter 12, the fundamental concepts of the application of shunt capacitors at motor terminals are discussed. The reactive power requirements for various induction motor designs are identified. In order to compensate for the reactive power, the shunt capacitors are used at the motor terminals. The associated technical problems due to self-excitation and transient torques are also discussed with a view to understanding how to contain the problems. The motor starting application using shunt capacitors is presented. The location of the shunt capacitors can be chosen based on the given operating conditions. The motor starting transients are illustrated using practical measurements for cases without and with terminal capacitors. The effect of shunt capacitors on the voltage and current waveforms as obtained from laboratory measurements is shown. Also discussed in Chapter 12 are the motor starting capacitor and surge capacitor for large motors.

There are many other applications in the industry where capacitors are used for power factor correction or energy storage. These include arc furnaces, resistance welding equipment, single-phase capacitor start, capacitor run motors, and lighting applications. In lighting applications, the capacitor is used for power factor correction of fluorescent lamps, mercury vapor lamps, sodium vapor lamps, and other similar lamps of various designs. The features of these miscellaneous capacitor applications are discussed in Chapter 13.

In Chapter 14, the basic principles of static VAR compensators (SVC) are presented. A brief description of the SVC is given along with the various types in use. The effects of SVC on the bulk power system in terms of damping, power transfer, reactive power support, and voltage support are also discussed.

The overall protection of shunt capacitors requires protection against bus faults, system surge voltages, inrush currents, discharge current from parallel banks, overcurrent protection, and rack failures. Chapter 15 presents an extensive analysis of unbalanced protection for grounded and ungrounded capacitor banks.

In Chapter 16, the overcurrent protection of shunt capacitors using fuses is discussed. The expulsion type K- and T-link fuses are selected for capacitor protection based on the steady-state currents. The T-link fuses are used for slow speed applications, while K-link fuses are used for fast speed requirements. Whenever the fault current levels are excessive, the current limiting fuses are used. The other considerations in selecting the fuse links, such as transient overcurrents, fault currents, and back-to-back switching currents, are also discussed in this chapter.

The switching devices for capacitor circuits require careful consideration because of the high frequency oscillations. Various types of circuit breakers available for capacitor switching applications are discussed. The important characteristics of the circuit breakers needed for capacitor switching and the performance of these devices are outlined in Chapter 17.

In Chapter 18, the surge protection issues of the shunt capacitor banks are presented. The type of lightning surges responsible for the overvoltages in the power systems are identified. These include direct strokes, back flashover, and multiple strokes. The characteristics of the metal oxide varistor type of surge arrester are discussed. The distribution class, intermediate class, and station type surge arresters, suitable for the protection of various voltage levels, are identified. The nature of overvoltages due to the various switching operations such as energization, de-energization, fault clearing, backup fault clearing, reclosing, restriking, current chopping, sustained overvoltages, ferroresonance, and ground faults are identified from the surge protection point of view. The selection procedure for a suitable surge arrester for a typical shunt capacitor installation is presented.

In Chapter 19, the maintenance issues of the shunt capacitor bank are presented. The possible failure mechanisms are identified for the shunt capacitor units. Shunt capacitor failures leave very few symptoms of the possible cause. The various approaches to tracing the origin of failure are presented in a logical manner. The possible remedial approaches are also discussed. Typical measurement devices required for shunt capacitor installations are identified.

In Chapter 20, the harmonic currents from converters, inverters, pulse width modulation converters, cycloconverters, static VAR converters, and switched mode power supplies are discussed. The use of industry standards to design suitable harmonic filters, including the voltage distortion and current distortion limits, is discussed. The typical harmonic filters are outlined and the use of multiple filters is discussed briefly. An example of a harmonic filter design is shown.

In certain applications, the transformer and the shunt capacitors are switched together. An arc furnace is one such load where the furnace transformer and the shunt capacitors are energized or de-energized together. Sometimes such a configuration is found to produce harmonic resonance. In EHV and HV systems, the shunt capacitors are installed at the substation and the lines are terminated through step-down transformers. The energization of the shunt capacitors produces significant transients at the transformer location. Two specific issues and their remedial measures are discussed in Chapter 21.

The capacitor switching produces overvoltages, and the related issues are discussed in Chapter 22. Specifically, the effects of energization, de-energization, fault clearing, backup fault clearing, reclosing, restriking, prestrike, back-to-back switching, voltage magnification, outrush currents due to close-in faults, and sustained overvoltages are discussed. The insulation coordination principles for the capacitor-switching transients are shown.

The induced voltages in substation control cables can produce unwanted tripping of the relay or circuit breaker. In the presence of the shunt capacitor equipment, the switching transients are higher and hence the induced voltages in control cables have to be minimized. The calculation approaches and mitigation techniques for induced voltages in the substations equipped with shunt capacitors are presented in Chapter 23.

The economic analysis suitable for power factor correction projects is presented in Chapter 24. The fundamental concepts of present worth and future worth of money are discussed. The cost components of a power factor correction project are identified along with the benefits due to the savings from all possible tariff structures. The economic analysis is presented using the present worth of money, tax, depreciation, and inflation effects. The economic analysis for this type of project can be performed using the payback period, cost-benefit analysis, rate of return, and break-even analysis. Some of the important references on shunt capacitors, harmonic filters, and series capacitors are also presented.

## PROBLEMS

1.1. Give a brief account of the historical development of capacitors.

1.2. What are the basic components used in the manufacture of a high voltage capacitor?

1.3. What is the difference between a low voltage capacitor and a high voltage capacitor?

1.4. Draw the cross-sectional view of a power capacitor and identify the various components used in its construction.

1.5. In order to reconstruct the ancient capacitor, prepare two pieces of insulation and two pieces of electrodes. Cut 1 in. × 40 in. aluminum foil for the electrode. Cut 1.5 in. × 45 in. paper for insulation. Use two small pieces of copper wire for the termination. Place paper as the first layer, aluminum foil as the second layer, paper as the third layer, and aluminum foil as the fourth layer. Roll the entire assembly around a 1.5 in. cardboard cylinder. Attach the copper wire on the two aluminum foils as the terminals. Use a 6 V DC battery and charge the assembly. This is a basic capacitor. Remove the battery and use a multimeter to measure the voltage across the terminals. What is the value of the retained charge in volts? Short circuit the terminals and see if a spark occurs. Change the terminal position and see if there is a difference in the retained charge.

## REFERENCES

1. Jenkins, J., www.sparkmuseum.com.

2. Bloomquist, W. C. (1950). *Capacitors for Industry*, John Wiley & Sons, New York.

3. Marbury, R. E. (1949). *Power Capacitors*, McGraw-Hill, New York.

4. Longland, T., Hunt, T. W., Brecknell, W. A. (1984). *Power Capacitor Handbook*, Butterworth-Heinemann, London.

5. Power Factor Capacitor, *IEEE Standards Code* (1994). Collection of IEEE Standards.

6. www.alpestechnologies.com, Alpes Technologies, Annecy-Le-Vieux, France, accessed on March 3, 2004.

# 2

# CAPACITOR FUNDAMENTALS

A capacitor consists of two conducting plates separated by a layer of insulating material called the dielectric. The conducting surfaces can be in the form of circular or rectangular plates. The purpose of the capacitor is to store electric energy. In this chapter, the definitions of capacitance for parallel plates, isolated sphere, hollow sphere, parallel plate with a dielectric, as well as composite medium, are presented [1,2].

## 2.1  CAPACITANCE OF PARALLEL PLATES AND SPHERES

A typical parallel plate capacitor is shown in Figure 2.1. One plate is connected to the positive end of the DC supply, the other to the negative end, and it is grounded. When such a capacitor is connected across a battery, there is a momentary flow of electrons from X to Y. As negatively charged electrons are withdrawn from X, it becomes positive, and as these electrons collect on Y, it becomes negative. Hence, a potential difference is established between plates X and Y. The transient flow of electrons gives rise to a charging current. The strength of the charging current is maximum when the two plates are uncharged, then decreases and finally ceases when the charges across the plates become equal and opposite to the

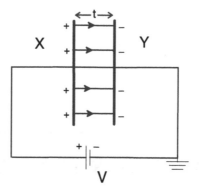

**Figure 2.1** A parallel plate capacitor.

electromotive force of the battery. The capacitance of a capacitor is defined as the amount of charge required for creating a unit potential difference between the plates. Suppose $Q$ coulombs of charge is given to one of the two plates of a capacitor and if a potential difference of $V$ volts is established between the plates, then the capacitance is:

$$C = \frac{Q}{V} = \frac{\text{Charge}}{\text{Potential Difference}} \qquad (2.1)$$

Therefore, the capacitance is the charge required per unit potential difference. Pieter van Musschenbroek discovered the concept of capacitance and Michael Faraday developed the method for measurement of capacitance. The unit of capacitance is the coulomb/volt, which is also called the farad, in honor of Faraday:

$$1 \text{ farad} = 1 \text{ coulomb/volt}$$

One farad is defined as the capacitance of a capacitor, which requires a charge of 1 coulomb to establish a potential difference of 1 volt between the plates. One farad is actually too large for practical applications, hence much smaller units like the microfarad and picofarad are used.

$$1 \text{ Microfarad } (1\,\mu F) = 10^{-6}\,F$$
$$1 \text{ Nanofarad } (1\,nF) = 10^{-9}\,F$$
$$1 \text{ Picofarad } (1\,pF) = 10^{-12}\,F$$

## Parallel Plate Capacitor

Consider a parallel plate capacitor consisting of two plates X and Y, each of area $A$ square meters separated by a thickness of $t$ meters of a medium of relative permittivity $\varepsilon_r$ as shown in Figure 2.1. If a charge of $Q$ coulombs is given to plate X, then the flux passing through the medium produces a flux density of $B$.

$$\text{The flux density, } B = \frac{Q}{A} \qquad (2.2)$$

$$\text{The electric field intensity, } E = \frac{V}{t} \qquad (2.3)$$

$$B = \varepsilon E \qquad (2.4)$$

$$\frac{Q}{A} = \frac{\varepsilon V}{t} \qquad (2.5)$$

$$C = \frac{Q}{V} = \frac{\varepsilon A}{t} = \frac{\varepsilon_0 \varepsilon_r A}{t} \qquad (2.6)$$

where $\varepsilon_0 = 8.854 \times 10^{-12}$ F/m; $\varepsilon_r =$ relative permittivity.

For measuring relative permittivity, vacuum or free space is chosen as the reference medium. The relative permittivity of vacuum with reference to itself is unity. The relative permittivity and dielectric strength in kV/mm for various materials are shown in Table 2.1.

TABLE 2.1   Dielectric Constant and Dielectric Strength

| Insulating Material | Relative Permittivity | Dielectric Strength, kV/mm |
|---|---|---|
| Air | 1.0 | 3.2 |
| Glass | (5–12) | (12.0–20) |
| Wood | (4–6) | (20–60) |
| Micanite | (2.5–6.7) | — |
| Paper | (4.6–6.0) | (25–35) |
| Wax | (1.8–2.6) | 30 |
| Porcelain | (1.7–2.3) | 15 |
| Quartz | (4.5–4.7) | 8 |

## Capacitance in a Composite Medium

Consider a parallel plate capacitor with three different mediums with thickness $t_1 + t_2 + t_3$ as shown in Figure 2.2. If $V$ is the potential difference across the plates and $V_1$, $V_2$, and $V_3$ are the potential differences across the three dielectric plates, then:

$$V = V_1 + V_2 + V_3 = E_1\, t_1 + E_2\, t_2 + E_3\, t_3$$

$$= \frac{B}{\varepsilon_0 \varepsilon_{r1}} \cdot t_1 + \frac{B}{\varepsilon_0 \varepsilon_{r2}} \cdot t_2 + \frac{B}{\varepsilon_0 \varepsilon_{r3}} \cdot t_3$$

$$= \frac{Q}{\varepsilon_0 A} \left( \frac{t_1}{\varepsilon_{r1}} + \frac{t_2}{\varepsilon_{r2}} + \frac{t_3}{\varepsilon_{r3}} \right) \tag{2.7}$$

$$C = \frac{Q}{V} = \frac{\varepsilon_0 A}{\left( \dfrac{t_1}{\varepsilon_{r1}} + \dfrac{t_2}{\varepsilon_{r2}} + \dfrac{t_3}{\varepsilon_{r3}} \right)} \tag{2.8}$$

where $\varepsilon_{r1}$, $\varepsilon_{r2}$, and $\varepsilon_{r3}$ are the permittivity of the various plates.

## Example 2.1

A capacitor has two square aluminum plates, each 8 cm × 8 cm, mounted parallel to each other as shown in Figure 2.1. What is the capacitance in picofarads, if the distance between the plates is 1 cm and the relative permittivity is 1.0? If the space between the plates is filled with mica of relative permittivity of 5.0, what will be the capacitance? In the second case,

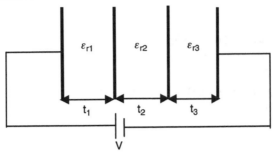

**Figure 2.2** Capacitance with a composite medium.

if a charge of 300 pC is given, what is the potential difference between the plates?

Solution

$$\varepsilon_0 = 8.854 \times 10^{-12}\,\text{F/m}, \quad \varepsilon_r = 1.0$$

$$A = 8 \times 8 = 64\,\text{cm}^2\ (0.0064\,\text{m}^2), \quad t = 1\,\text{cm}\ (0.01\,\text{m})$$

$$C = \frac{\varepsilon_0 \varepsilon_r A}{t} = \frac{(8.854 \times 10^{-12})(1.0)(0.0064)}{0.01} = 5.66 \times 10^{-12}\,\text{F}$$

If the space is filled with mica of $\varepsilon_r$ 5.0, the capacitance is increased five times.

$$C\ (\text{with mica}) = (5)\ (5.66 \times 10^{-12}) = 28.3 \times 10^{-12}$$

$$V = \frac{Q}{C} = \frac{300 \times 10^{-12}}{28.3 \times 10^{-12}} = 10.6\,\text{V}$$

**Example 2.2**

A parallel plate capacitor has a plate area of 1.5 m$^2$ spaced by three plates of different materials. The relative permittivities are 3, 4, and 6, respectively. The plate thicknesses are 0.6, 0.8, and 0.3 mm, respectively. Calculate the combined capacitance and field strength in each material when the applied voltage is 500 V.

Solution

$$\varepsilon_0 = 8.854 \times 10^{-12}\, \text{F/m}$$
$$\varepsilon_{r1} = 3.0, \quad \varepsilon_{r2} = 4.0, \quad \varepsilon_{r3} = 6.0$$
$$A = 1.5\ \text{m}^2$$
$$t_1 = 0.6\ \text{mm},\ t_2 = 0.8\ \text{mm},\ t_3 = 0.3\ \text{mm}$$
$$V = 500\ \text{V}$$

Using Equation (2.8), the combined capacitance can be calculated as:

$$C = \frac{(8.854 \times 10^{-12})(1.5)}{\left(\dfrac{0.6}{3} + \dfrac{0.8}{4} + \dfrac{0.3}{6}\right)10^{-3}} = 0.0295 \times 10^{-6}\, \text{F}$$

Charge on the plates, $Q = C\,V = (0.0295 \times 10^{-6})\ (500\ \text{V})$
$$= 14.75 \times 10^{-6}\, \text{C}$$

$$B = \frac{14.75 \times 10^{-6}}{1.5} = 9.83 \times 10^{-6}\, \text{C/m}^2$$

$$E_1 = \frac{B}{\varepsilon_0 \varepsilon_{r1}} = \frac{9.83 \times 10^{-6}}{(8.854 \times 10^{-12})(3)} = 370\ \text{kV/m}$$

$$E_2 = \frac{B}{\varepsilon_0 \varepsilon_{r2}} = \frac{9.83 \times 10^{-6}}{(8.854 \times 10^{-12})(4)} = 277.6\ \text{kV/m}$$

$$E_3 = \frac{B}{\varepsilon_0 \varepsilon_{r3}} = \frac{9.83 \times 10^{-6}}{(8.854 \times 10^{-12})(6)} = 185\ \text{kV/m}$$

## Spherical Capacitor

Consider a charged sphere of radius $r$ meters having a charge of $Q$ coulombs placed in a medium of relative permittivity $\varepsilon_r$.

The surface potential $V$ of such a sphere with respect to earth is given by:

$$V = \frac{Q}{4\pi\varepsilon_0\varepsilon_r r} \tag{2.9}$$

Rearranging Equation (2.9) and using Equation (2.1):

$$C = 4\pi\varepsilon_0\varepsilon_r r \tag{2.10}$$

Consider a spherical capacitor consisting of two concentric spheres of radii $a$ and $b$ as shown in Figure 2.3. If the inner sphere is given a charge of $+Q$ coulombs, it will induce a charge of $-Q$ coulombs on its outer surface, which will go to earth. The potential difference between the two surfaces is:

$$V = \frac{Q}{4\pi\varepsilon_0\varepsilon_r} \left( \frac{1}{a} - \frac{1}{b} \right) \tag{2.11}$$

The capacitance of the spherical capacitor is given by:

$$C = 4\pi\varepsilon_0\varepsilon_r \, \frac{ab}{b-a} \tag{2.12}$$

Knowing the dimensions $a$ and $b$ along with the relative permittivity, the capacitance value can be calculated.

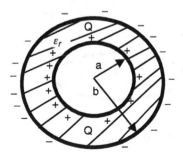

**Figure 2.3** A spherical capacitor.

## 2.2   CYLINDRICAL CAPACITORS

Consider a cylindrical capacitor consisting of two co-axial cylinders of radii $a$ and $b$ as shown in Figure 2.4. A cable conductor can be considered as a simple cylindrical cable, with a metallic conductor inside. Assume the charge per unit length of the cable to be $+Q$ coulombs on the outer side of the inner cylinder and $-Q$ coulombs on the inner side of the outer cylinder. The relative permittivity of the material is taken as $\varepsilon_r$. In order to evaluate the flux density, electric field, and the capacitance of the cylindrical cable, consider an imaginary cylinder at a radius of $x$ as shown in Figure 2.4. The area $A$ of the curved surface at radius $x$ is given by:

$$A = 2\pi x\,(1) = 2\pi x \tag{2.13}$$

The electric flux density $B$ on the surface of the imaginary cylinder is given by:

$$B = \frac{\text{Flux in coulombs}}{\text{Area in square meters}} = \frac{Q}{A}\ \text{C/m}^2 \tag{2.14}$$

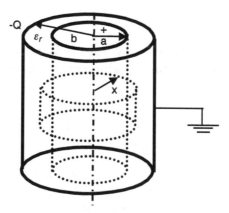

**Figure 2.4**   A cylindrical capacitor.

The electric field $E$ is given by:

$$E = \frac{B}{\varepsilon_0 \varepsilon_r} = \frac{Q}{2\pi\varepsilon_0\varepsilon_r x} \text{V/m} \qquad (2.15)$$

The incremental voltage change is due to the field and incremental distance, given by:

$$dV = -E \cdot dx \qquad (2.16)$$

The voltage $V$ between the inner and outer cylinders is calculated as:

$$V = \int_a^b -E\,dx = \int_a^b \frac{Q}{2\pi\varepsilon_0\varepsilon_r x} = \frac{Q}{2\pi\varepsilon_0\varepsilon_r} \log_e \frac{b}{a} \qquad (2.17)$$

The capacitance $C$ of the cylindrical capacitor is given by:

$$C = \frac{Q}{V} = \frac{2\pi\varepsilon_0\varepsilon_r}{2.3 \log_{10}(b/a)} \qquad (2.18)$$

For a cylinder of length $L$ meters, the capacitance $C$ is given by:

$$C = \frac{2\pi\varepsilon_0\varepsilon_r L}{2.3 \log_{10}(b/a)} \qquad (2.19)$$

**Example 2.3**

A simple co-axial cable consists of a 4 mm diameter copper conductor with an insulation of 3 mm. The relative permittivity of the insulation is 3. Calculate the capacitance of the cable for a length of 100 m.

Solution

$$A = 4 \text{ mm}, \quad b = (4 + 3 + 3) = 10 \text{ mm}, \quad L = 100 \text{ m}$$

$$C = \frac{2\pi\varepsilon_0\varepsilon_r L}{2.3 \log_{10}(b/a)} = \frac{(2)(3.14)(8.854 \times 10^{-12})(3)}{2.3 \log_{10}(10/4)}(100)$$
$$= 18.225 \times 10^{-9} \text{ F}$$

## 2.3  ENERGY STORED IN A CAPACITOR

The energy is stored in a capacitor in an electrostatic field and it produces a field flux and flux density. Charging of such an electric field requires energy. The stored energy in the dielectric medium is important in various calculations. When discharging the capacitor, the field energy is released. In order to calculate the stored energy in a capacitor, consider the potential difference between the capacitor plates as $v$. This potential is equal to the work done in transferring 1 coulomb of charge from one plate to another. If d$q$ is the incremental charge to be transferred, the work done is given by:

$$\mathrm{d}W = v \cdot \mathrm{d}q \tag{2.20}$$

$$\text{where } q = Cv; \ \mathrm{d}q = C \cdot \mathrm{d}v \tag{2.21}$$

Using Equations (2.20) and (2.21), the total work done $W$ due to $V$ units of potential difference is:

$$W = \int_0^V Cv \, \mathrm{d}v = \frac{1}{2}CV^2 \tag{2.22}$$

If capacitance $C$ is in farads and the potential difference $V$ is in volts, then:

$$W = \frac{1}{2}CV^2 = \frac{1}{2}QV = \frac{Q^2}{2C} \tag{2.23}$$

If the charge $Q$ is expressed in coulombs and the capacitance $C$ in farads, then the energy stored is given in joules.

## Example 2.4

A capacitance of $0.01\,\mu F$ is connected to a circuit with a DC voltage of 100 V. Then the capacitor is disconnected and immersed in oil with a relative permittivity of 2. Find the energy stored in the capacitance before and after immersion in the oil.

Solution

$C = 0.01\,\mu F$; $V = 100$ V

$$\text{Energy stored in the capacitor} = \frac{1}{2}CV^2 = \frac{1}{2}(0.01 \times 10^{-6})(100)^2$$

$$= 50 \times 10^{-6}\,\text{J}$$

When immersed in oil, the capacitance is increased by a factor of two. The charge is constant, assuming no losses. The voltage is decreased by a factor of 2.

$C$ (new) $= (2)\,(0.01) = 0.02\,\mu F$
$V$ (new) $= 100/2 = 50$ V

$$\text{Energy stored in the capacitor} = \frac{1}{2}(0.02 \times 10^{-6})(50)^2$$

$$= 50 \times 10^{-6}\,\text{J}$$

The energy in the capacitor remains the same.

## 2.4  CHARGING AND DISCHARGING OF A CAPACITOR

The charging and discharging of a capacitor is an important aspect of electrical circuits. Both functions are controlled by the circuit time constant, which is proportional to the value of

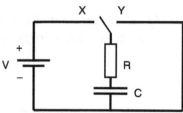

**Figure 2.5** Circuit diagram for charging and discharging a capacitor.

the capacitance. In order to explain the concept of the time constant, a simple RC circuit, shown in Figure 2.5, is used. The capacitor is charged from the battery, when the circuit switch is connected to "X." At any instant during the charging, the Kirchhoff's voltage equation is given by:

$$V = v + iR \tag{2.24}$$

$$i = \frac{dq}{dt} = \frac{d}{dt}(Cv) = C\frac{dv}{dt} \tag{2.25}$$

where

$V = $ Applied voltage

$v = $ Voltage across the capacitor

$q = $ Charge on the capacitor plates

$R = $ Resistance

$$V = v + RC\frac{dv}{dt} \tag{2.26}$$

Rearranging Equation (2.26):

$$-\frac{dv}{V - v} = -\frac{dt}{CR} \tag{2.27}$$

Integrating both sides of Equation (2.27):

$$-\int \frac{dv}{V-v} = -\frac{1}{CR}\int dt \qquad (2.28)$$

$$\log_e(V-v) = -\frac{t}{CR} + K \qquad (2.29)$$

where $K$ is the constant of integration. At the start of charging, $t=0$ and $v=0$. Then $K = \log_e V$. Using this value of $K$ and rearranging Equation (2.29):

$$v = V(1 - e^{-t/RC}) = V(1 - e^{-t/\tau}) \qquad (2.30)$$

where $\tau = RC =$ time constant.

The time constant is defined as the time during which the voltage across the capacitor would have reached its maximum value had it maintained the initial rate of rise. This concept is shown in Figure 2.6. In Equation (2.30), let $t = \tau$, then:

$$v = V(1 - e^{-1}) = 0.632 \text{ V} \qquad (2.31)$$

Hence the time constant can also be defined as the time during which the capacitor voltage reaches 0.632 of the final steady-state voltage.

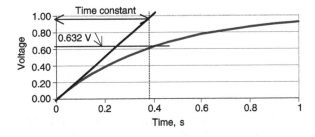

**Figure 2.6**  The time constant; the capacitor voltage reaches 0.632 of the final voltage.

## Charging Current

It is known that $v = q/c$ and $V = Q/C$. Using these relations in Equation (2.30):

$$\frac{q}{C} = \frac{Q}{C}(1 - e^{-t/\tau}) \qquad (2.32)$$

Differentiating both sides:

$$\frac{dq}{dt} = i = Q\frac{d}{dt}(1 - e^{-t/\tau}) = I_m e^{-t/\tau} \qquad (2.33)$$

where $I_m = V/R = $ maximum current.

## Discharging of a Capacitor

In Figure 2.5, if the switch is connected to "Y," then the source voltage is disconnected, the capacitor circuit is shorted, and it will discharge through the resistance. Since the battery is disconnected, $V = 0$ in Equation (2.26) gives the following relation:

$$0 = v + RC\frac{dv}{dt} \qquad (2.34)$$

$$\frac{dv}{v} = -\frac{dt}{RC} \qquad (2.35)$$

Integrating both sides of Equation (2.35):

$$\int\frac{dv}{v} = -\frac{1}{RC}\int dt \qquad (2.36)$$

$$\log_e v = -\frac{t}{RC} + K \qquad (2.37)$$

where $K$ is the constant of integration. At the start of discharge, $t = 0$ and $v = V$. Solving for $K$ and rearranging:

$$v = Ve^{-t/\tau} \qquad (2.38)$$

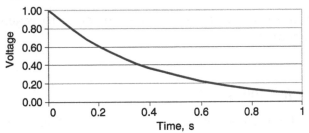

**Figure 2.7**  Voltage across the capacitor during de-energization.

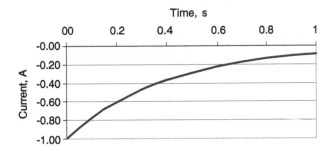

**Figure 2.8**  Decaying current in the capacitor circuit.

The voltage across the capacitor decreases exponentially and the decaying curve is shown in Figure 2.7. It also can be proved that the capacitor current decays according to Equation (2.39):

$$i = -I_m \, \mathrm{e}^{-t/\varepsilon} \tag{2.39}$$

where $I_m$ is the maximum value of the current. The capacitor current decreases from the maximum value of $-1.0$ per unit in an exponential manner and is shown in Figure 2.8.

**Example 2.5**

A $5\,\mu\mathrm{F}$ capacitor is charged through a resistance of $100\,\mathrm{k}\Omega$. (a) Calculate the time constant of the circuit.

(b) Find the time it takes for the capacitor to reach 90% of the final charge.

Solution

$C = 5\,\mu\text{F} \quad R = 100\,\text{k}\Omega$

(a)   Time constant of the circuit $= 100,000 \times 5 \times 10^{-6}$
$$= 0.5 \text{ s}$$

(b)   When $q = 0.9\,Q$

$$0.9 \times Q = Q(1 - e^{-t/0.5})$$

Rearranging and taking log on both sides:

$$0.1 = e^{-t/0.5}$$

$$\log_e(0.1) = \frac{-t}{0.5}$$

$$t = -(0.5)\,\log_e(0.1) = 1.15\,\text{s}$$

## PROBLEMS

2.1.   Consider the parallel plate capacitor shown in Figure 2.1. If the medium consists partly of air and partly of a dielectric slab of thickness $t$ and relative permittivity $\varepsilon_r$, find an expression for the capacitance.

2.2.   Consider a spherical capacitor as shown in Figure 2.3. If the inner sphere is grounded and the outer sphere is given a charge of $+Q$ coulombs, find an expression for the capacitance.

2.3.   The capacitance of a three-plate arrangement is derived based on Figure 2.2. Draw a one-line diagram of the capacitor section for a 5-parallel plate arrangement and derive the equation for the capacitance. State the assumptions.

2.4. A capacitor is made of two plates with a distance of 5 mm between them. The area of each of the plates is 20 cm$^2$. The charge on each plate is $0.7 \times 10^{-6}$ C. Calculate the electric field strength between the plates. What is the potential difference between the plates?

2.5. A power system cable consists of a 6 mm diameter aluminum conductor and a 3 mm insulation. The relative permittivity of the insulation material is 4. Calculate the capacitance of the cable for a length of 1 km.

2.6. A 4 µF capacitor is charged by a 110 V DC supply through a 100 kΩ resistance. How long will it take the capacitor to develop a voltage of 90 V? What is the energy stored in the capacitor at that instant?

2.7. A capacitor is charged to 100 V and then is connected to a voltmeter with a resistance of 10 MΩ. The voltmeter reading has fallen to 20 V at the end of an interval of 1 min. Find the value of the capacitance.

2.8. Construct a parallel plate capacitor with an insulation based on the ideas discussed in this chapter. Prepare two aluminum electrodes using aluminum foil, each 5 in. × 5 in. Prepare two paper insulations of 6 in. × 6 in. Arrange the parallel plate capacitor and put a charge on the plate using a 6 V battery. Remove the battery and measure the voltage. Conduct additional tests and prepare a report.

# REFERENCES

1. Christie, C. V. (1938). *Electrical Engineering*, McGraw-Hill, New York.

2. Coursey, P. R. (1927). *Electrical Condensers*, Sir Isaac Pitman & Sons, London.

# 3

# POWER FACTOR CONCEPTS

## 3.1  THE POWER FACTOR

The current required by motors, lights, and computers is made up of real and reactive components. This concept of a two-component current is helpful in understanding the capacitor current. Loads such as a heater require the supply of only the real component of current. Some loads, such as an induction motor, require both real and reactive currents [1].

The real current is that component that is converted by the equipment into useful work such as production of heat through a heater element. The unit of measurement of this current is ampere (A) and of power (voltage × real current) is watts (W).

The reactive current is that component that is required to produce the flux necessary for the functioning of induction devices. The current is measured in ampere (A) and the reactive power (voltage × reactive current) in VARs.

### Total Current

The total current is the algebraic sum of the real and reactive current, measured in amperes. VA represents the voltage-ampere product. The relation between the real, reactive, and

*35*

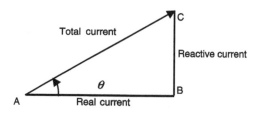

**Figure 3.1** The relation between the real, reactive, and total current.

total current is shown in Figure 3.1. The total current is given by:

$$\text{Total current} = \sqrt{(\text{Real current})^2 + (\text{Reactive current})^2} \quad (3.1)$$

The square root appears in the equation because these currents are resolved using orthogonal relations in a triangle. The other two currents can be calculated using the following relations:

$$\text{Real current} = \sqrt{(\text{Total current})^2 - (\text{Reactive current})^2} \quad (3.2)$$

$$\text{Reactive current} = \sqrt{(\text{Total current})^2 - (\text{Real current})^2} \quad (3.3)$$

These current components also are useful for calculating both the real and reactive power.

## Power Factor

The power factor may be expressed as the ratio of the real current to the total current in a circuit. Alternatively, the power factor is the ratio of kW to total kVA:

$$\text{Power factor} = \frac{\text{Real current}}{\text{Total current}} = \frac{\text{kW}}{\text{kVA}} \quad (3.4)$$

Referring to the triangle in Figure 3.1:

$$\text{Power factor} = \frac{AB}{AC} \qquad (3.5)$$

The angle $\theta$ is called the power factor angle. This is the angle included between the total current and the real current. The cosine of this angle (cos $\theta$) is the power factor.

## Example 3.1

In a 110 V AC circuit, the real and reactive currents are 40 A and 30 A, respectively. Calculate the total current. What are the corresponding total kVA, kW, and kVAR? Verify the results. What is the power factor of this circuit? What is the corresponding power factor angle?

Solution

$$\text{Total current} = \sqrt{40^2 + 30^2} = 50\,\text{A}$$

$$\text{Total kVA} = \frac{(110\,\text{V})(50\,\text{A})}{1000} = 5.5$$

$$\text{kW} = \frac{(110\,\text{V})(40\,\text{A})}{1000} = 4.4$$

$$\text{kVAR} = \frac{(110\,\text{V})(30\,\text{A})}{1000} = 3.3$$

$$\text{kVA} = \sqrt{(4.4)^2 + (3.3)^2} = 5.5$$

$$\text{Power factor} = \frac{40}{50} = 0.8$$

$$\text{Power factor angle} = \cos^{-1}(0.8) = 36.8°$$

### The Lagging Power Factor

Consider an inductive load as shown in Figure 3.2. In this circuit, both watts and VARs are delivered from the source. The corresponding phasor diagram is shown in Figure 3.2. The power factor angle in this case is negative, and therefore the power factor is lagging.

### The Leading Power Factor

Consider a capacitive load as shown in Figure 3.3. In this circuit, the watts are delivered from the source. The reactive power (VARs) is delivered from the load to the source. The corresponding phasor diagram is shown in Figure 3.3. The power factor angle in this case is positive, and therefore the power factor is leading.

In order to understand this concept, consider the following generator load circuits, shown in Figure 3.4.

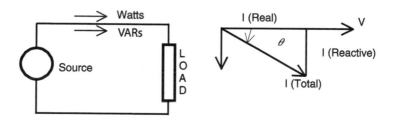

**Figure 3.2**   The concept of lagging power factor.

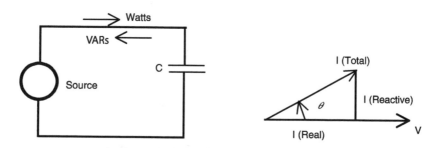

**Figure 3.3**   The concept of leading power factor.

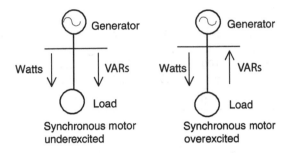

**Figure 3.4**   The power flow in a synchronous motor.

**TABLE 3.1**   The Real and Reactive Power Flow Concepts in a Synchronous Motor

| | At Generator | | | At Load | | |
|---|---|---|---|---|---|---|
| Load | kW | kVAR | PF | kW | kVAR | PF |
| Synch. motor underexcited | Out | Out | Lag | In | In | Lag |
| Synch. motor overexcited | Out | In | Lead | In | Out | Lead |

The power flow at the generator and load can be identified along with the power factor as shown in Table 3.1. The power factor is lagging if the load requires reactive power and leading if the load delivers reactive power. The power factor of group loads needs careful consideration and is explained through an example.

**Example 3.2**

An industrial substation supplies lighting, induction motor, and synchronous motor load as given below:

> Lighting load = 40 kW at unity power factor
> Induction motor load = 100 kW at 0.7 power factor
> Synchronous motor load = 200 kW at 0.9 power factor

Find the kW, kVAR, and power factor of each load. Also find the kW, kVAR, and power factor at the substation level.

Solution

| Load | kW | PF | kVA | kVAR | Angle in degrees |
|------|-----|-----|-----|------|------------------|
| Lighting | 40 | 1.0 | 40 | 0.0 | 0 |
| Induction motor | 100 | 0.7 | 143 | 71.4 | 45.6 |
| Synchronous motor | 200 | 0.9 | 222 | 87.1 | 25.8 |
| Total | 340 | | | 158.5 | |

$$kVA = \sqrt{340^2 + 158.5^2} = 375$$

$$\text{Overall power factor} = \frac{340}{375} = 0.907 \, (\text{lag})$$

**Power Factor Improvement**

Many utilities prefer a power factor of the order of 0.95. Since industrial equipment such as an induction motor operates at a much lower power factor, the overall power factor of the industrial load is low. In order to improve the power factor, synchronous condensers or capacitors are used.

The synchronous machines, when operated at leading power factor, absorb reactive power and are called synchronous condensers. These machines need operator attendance and require periodical maintenance. Power factor capacitors are static equipment without any rotating parts and require less maintenance. Therefore, shunt capacitors are widely used in power factor correction applications. The shunt capacitors provide kVAR at leading power factor and hence the overall power factor is improved.

**Example 3.3**

In Example 3.2, the power factor is to be improved to 0.95 with capacitors. How much shunt capacitor kVAR is needed? Draw a one-line diagram of the circuit and phasor diagram.

Solution

$$kW = 340, \quad kVAR = 158.5, \quad \text{Desired power factor} = 0.95$$

Required kVA $= \dfrac{340}{0.95} = 358$

Power factor angle, $\cos^{-1} (0.95) = 18.2$

kVAR, $[358 \sin (18.2)] = 112$

kVAR supplied by the capacitors $(158.5 - 112) = 46.5$

The one-line diagram of the load circuit and the corresponding power factor improvement phasor relations are shown in Figure 3.5.

From the right angle triangle in Figures 3.1 and 3.5, the equations related to the power factor can be written. The kW component usually remains constant. The kVAR and the kVA components change with the power factor. Knowing the desired and existing power factor, the kVAR to be supplied using shunt capacitors can be calculated as:

$$\cos \theta = \text{Power factor} = \frac{kW}{kVA} \qquad (3.6)$$

$$\tan \theta = \frac{kVAR}{kW} \qquad (3.7)$$

$$\sin \theta = \frac{kVAR}{kVA} \qquad (3.8)$$

$$kVAR \ (\text{capacitor}) = kW \ (\tan \theta 1 - \tan \theta 2) \qquad (3.9)$$

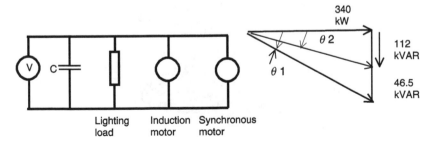

**Figure 3.5** One-line diagram of the circuit and phasor.

## 3.2   THE SYNCHRONOUS CONDENSER

The power factor of a load can be improved by using synchronous machines operating in the leading power factor mode. Such an operation of a synchronous machine is called a synchronous condenser. A photograph of a synchronous condenser is shown in Figure 3.6. In order to demonstrate this concept for power factor correction, consider a synchronous machine operating without any losses. The terminal voltage of the machine is represented as $V$ and the back emf is represented as $E$. If a restraining force is applied to the rotor of the machine, it will tend to slow down until a torque is developed equal to the opposite restraining force. Then the rotor will operate at the same speed as the generator and the back emf will be at an angle to the terminal voltage vector. There is a resultant voltage vector, $E_1$, and a current flow occurs ($I$ per phase) given by:

$$I = \frac{(V - E)}{\sqrt{3}\,Z} \tag{3.10}$$

where $Z$ is the impedance per phase. Since the impedance is dominantly reactive, the current will be lagging the applied voltage. The power input to the synchronous motor under this condition will be:

$$P = VI \cos \phi \tag{3.11}$$

**Figure 3.6**  Photograph of a synchronous condenser. (Courtesy of OSHA, www.osha.gov.)

where $\phi$ is the phase angle between the terminal voltage and the current. This condition is represented in Figure 3.7a. If the excitation is increased on the synchronous motor, the unity power factor condition will occur as shown in Figure 3.7b. The back-emf will be increased from the previous condition. The resultant voltage has caused the change in the position and value. The current $I$ is in phase with the terminal voltage and hence the power factor is unity [2,3].

A further increase in the excitation current will increase the motor back-emf as shown in Figure 3.7c. The resultant voltage is modified in value and position. The current is leading the terminal voltage by an angle $\phi$. The in-phase current can be seen to be equal in all three power factor cases. This must be true since the same speed and torque are assumed in all three operating conditions. Then the real power will remain the same in all three conditions. Since the power factor angle is changed in each condition, the corresponding reactive power at the terminal will be different.

## V-Curves

Since the line voltage in each operating condition is the same, it is apparent from the above that the load, excitation, line

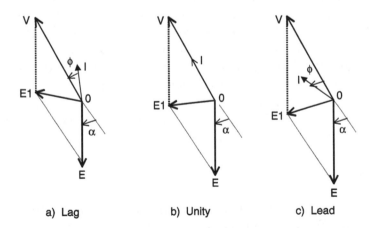

a) Lag  b) Unity  c) Lead

**Figure 3.7** Phasor diagram of the synchronous motor for lag, unity, and lead power factor conditions.

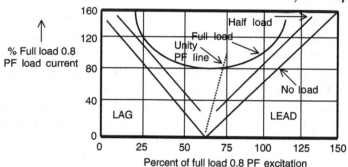

**Figure 3.8** V-curves showing the relation of load current to excitation.

current, and the power factor are closely related. A family of curves known as V-curves expresses the relationship. This is represented in a general format in Figure 3.8. The minimum value of line current for each load condition occurs at unity power factor and at a specific excitation current. As excitation is decreased, the line current will increase and the motor will operate at lagging power factor. If the excitation is increased from the unity power factor condition, the line current will again increase, but will operate under a leading power factor condition.

Therefore, the synchronous condenser can be used to improve the power factor of the system like a shunt capacitor bank; however, there are some differences between the synchronous condenser and the shunt capacitor bank, as listed in Table 3.2. Since the synchronous condenser is expensive to install and maintain, these devices are not widely used in power system applications. Since capacitor banks are easier to install and maintain, these devices are used for power factor correction at all voltage levels in the utility.

## PROBLEMS

3.1. Define power factor using the power triangle.
3.2. What is the advantage of a better power factor? What are the undesirable effects of a poor power factor?

TABLE **3.2** Differences between a Synchronous Condenser and Shunt Capacitor

| Synchronous Condenser | Shunt Capacitor |
| --- | --- |
| Rotating machine | Static equipment |
| Fine control of Q using excitation | In steps |
| More stabilizing effect | Not so |
| For short periods the machine can supply excess kVAR | Overloading is not recommended |
| Losses in synchronous machines are much higher than the capacitors | Lower losses |
| One installation | Capacitor can be installed at several locations in the distribution system |
| kVAR rating is fixed | kVAR can be added/decreased |
| Failure: unit up or down | Can failure is possible |
| Overvoltage performance is good | Overvoltage: moderate or limited |
| Harmonics: none | May produce harmonics or resonance with system inductance |
| Fast enough response | Response is system dependent |

3.3. The real and reactive current components of a 110 V, 60 Hz, single-phase load are 7 A and 4 A, respectively. What is the total line current? What are the real and reactive power supplied to the load? What are the kVA and power factor of the load?

3.4. An induction motor load is operating at 0.7 power factor drawing 600 kW. A synchronous motor is connected to the same bus and can be operated at lagging, unity, or leading power factor. Find the resultant power factor if the synchronous motor is operated at (a) $(200 + j\,0)$ kVA; (b) $(200 + j\,152)$ kVA.

## REFERENCES

1. Porter, G. A., McCall, J. C. (1990). *Application and Protection Considerations in Applying Distribution Capacitors,* Pennsylvania Electric Association, September 25–26, 1–36.

2. Kekela, J., Firestone, L. (1963). Under Excited Operation of Generators, *AIEE Transactions*, 82, 722–727.

3. Dalziel, C. F. (1939). Static Power Limits of Synchronous Machines, *AIEE Transactions*, 58, 93–102.

# 4

## INDUSTRY STANDARDS

### 4.1 STANDARDS ORGANIZATIONS

There are several organizations responsible for providing leadership in developing standards for the capacitor industry. These organizations play a vital role by conducting seminars and conferences, producing publications, and disseminating technical material from various companies or universities. Otherwise, copyrights, patents, and employer discretion protect most technical materials. Such a process helps in developing and maintaining the necessary standards for devices such as the capacitor. These standards provide guidelines for the design, manufacture, testing, installation, and maintenance of power system capacitors. The important organizations responsible for capacitor standards are identified below along with their contributions in the form of standards. The mission, vision, objectives, and services offered by these organizations are provided at their respective web sites.

### 4.1.1 The Institute of Electrical and Electronics Engineers (IEEE)

IEEE is a worldwide technical professional society. The technical objectives of IEEE focus on advancing the theory and practice of electrical, electronics, computer engineering, and computer science. Society membership is open to all electrical and computer professionals, including students. IEEE promotes the engineering process of creating, developing, integrating, sharing, and applying knowledge about electric, information technologies, and sciences for the benefit of humanity and the profession. The IEEE standard documents are developed by the Technical Committees of the different IEEE Societies. The software engineering standards are developed within the Computer Society. Through its members, IEEE is a leading authority in technical areas ranging from computer engineering, biomedical technology, telecommunications, electric power, aerospace, and consumer electronics. The standards activities within IEEE are coordinated through various conferences and committees working on various technical issues. Through its technical publishing, conferences, and consensus-based standards activities, IEEE is responsible for:

- Approximately 30% of the world's published literature in electrical engineering.
- Many conferences and technical exhibitions on various technologies.
- More than 900 active standards with an additional 700 under development.
- A platform for meetings and networking among professionals.
- Educational opportunities to ensure the vitality of engineers.
- Publishing papers on all aspects of electrical and computer engineering.
- Recognition of technical and professional achievements through awards.

### 4.1.2 The International Electrotechnical Commission (IEC)

IEC is the international standards and conformity assessment body for all fields of electrotechnology. It provides a forum for the preparation and implementation of consensus-based voluntary international standards, facilitating international trade in the relevant fields and helping meet expectations for an improved quality of life. The membership currently comprises 52 National Committees, representing all electrotechnical interests in each member country, from manufacturing and service industries, to government, scientific, research and development, academic, and consumer bodies. For information technology standards, ISO and IEC have formed a Joint Technical Committee. IEC is the leading global organization that prepares and publishes international standards for all electrical, electronic, and related technologies. These serve as a basis for national standardization and as references when drafting international tenders and contracts. Through its members, IEC promotes international cooperation on all questions of electrotechnical standardization and related matters, such as the assessment of conformity to standards in the fields of electricity, electronics, and related technologies. The IEC charter embraces all electrotechnologies, including electronics, magnetics, electromagnetics, electro-acoustics, multimedia, telecommunication, energy production, distribution, electromagnetic compatibility, measurement, dependability, design and development, safety, and the environment. The commission's objectives are to:

- Meet the requirements of the global market efficiently.
- Ensure worldwide use of its standards and conformity assessment schemes.
- Assess and improve the quality of products covered by its standards.
- Establish the conditions for the interoperability of complex systems.
- Contribute to the improvement of human health and safety.
- Contribute to the protection of the environment.

IEC's international standards facilitate world trade by removing technical barriers to trade, leading to new markets and economic growth. IEC's standards are vital since they also represent the core of the World Trade Organization's

Agreement on Technical Barriers to Trade, whose 100-plus central government members explicitly recognize that international standards play a critical role in improving industrial efficiency and developing world trade. The IEC standards provide the industry and its users with the framework for economies of design, greater product and service quality, increased interoperability, and better production and delivery efficiency. At the same time, IEC's standards also encourage an improved quality of life by contributing to safety, human health, and the protection of the environment.

IEC's multilateral conformity assessment schemes, based on its international standards, are truly global in concept and practice, reducing trade barriers caused by different certification criteria in various countries and helping industries open up new markets. Removing the significant delays and costs of multiple testing and approval allows the industry to deliver its products to the market more quickly and more cheaply. As technology becomes more complex, users and consumers are becoming more aware of their dependence on products whose design and construction they may not understand. In this situation, reassurance is needed that the product is reliable and will meet expectations in terms of performance, safety, durability and other criteria. IEC's conformity assessment and product certification schemes exist to provide just such a reassurance, and the regulatory nature of some products now also reflects recognition of the schemes among some government regulators.

Using the IEC standards for certification at the national level ensures that a certified product has been manufactured and type tested to well-established international standards. The end user can be sure that the product meets minimum quality standards, and need not be concerned with further testing or evaluation of the product.

### 4.1.3  The National Electrical Manufacturers Association (NEMA)

NEMA was created in the fall of 1926 by the merger of the Electric Power Club and the Associated Manufacturers of Electrical Supplies. It provides a forum for the standardization of electrical equipment, enabling consumers to select from a range of safe, effective, and compatible electrical products. The organization has also made numerous contributions to the electrical industry by shaping public policy and operating as a central confidential agency for gathering, compiling, and analyzing market statistics and economic data. NEMA attempts to promote the competitiveness of its member companies by providing a forum for:

- Development of standards for the industry and the users of its products.
- Establishment and advocacy of industry policies on legislative and regulatory matters that might affect the industry and those it serves.
- Collection, analysis, and dissemination of industry data.

NEMA publishes more than 500 standards along with certain standards originally developed by the American National Standards Institute (ANSI) or the International Electrotechnical Commission (IEC). The association promotes safety in the manufacture and use of electrical products, provides information about NEMA to the media and the public, and represents industry interests in new and developing technologies.

NEMA, with headquarters in Rosslyn, Virginia, has more than 400 member companies, including large, medium, and small businesses that manufacture products used in the areas of generation, transmission, distribution, control, and end-use of electricity. NEMA is the trade association of choice through which the electric industry develops and promotes positions on standards and government regulations. It also helps members acquire information on industry and market economics. NEMA's mission is to promote the competitiveness of its member companies by providing quality services that will impact positively on standards, government regulations, and market economics.

### 4.1.4   The American National Standards Institute (ANSI)

ANSI is a private, nonprofit organization that administers and coordinates the U.S. voluntary standardization and conformity assessment system. The institute's mission is to enhance both the global competitiveness of U.S. business and the quality of life by promoting and facilitating voluntary consensus standards and conformity assessment systems, and safeguarding their integrity. ANSI currently provides a forum for more than 270 ANSI-accredited standards, developers representing approximately 200 distinct organizations in the private and public sectors. These groups work cooperatively to develop voluntary national consensus standards and American National Standards. IEEE is one such organization that develops standards related to electrical and computer engineering. The approved standards are reviewed by volunteer groups and updated on a regular basis.

## 4.2   STANDARDS RELATED TO CAPACITORS

The following four standards address power system capacitor specifications, applications, and protection of shunt [1–3] and series capacitors [4].

IEEE Standard 18 provides definitions, specifications and testing approaches for the power factor correction shunt capacitors. There is also a table from which to select the ratings of capacitors for a given voltage rating [1].

IEEE Standard 1036 outlines the purpose of shunt capacitors, ratings, application approach to distribution lines, shunt capacitors for substations, and related issues. Special applications for harmonic filters, motor applications, and surge capacitor-related application issues are discussed. Guidelines for safety and personnel protection, initial inspection, measurements, and energization issues of shunt capacitors are discussed. Periodic inspection, maintenance, and field-testing issues are outlined in this standard [2].

ANSI/IEEE Standard C37.99 discusses the basics of unit ratings, bank arrangements such as delta or wye configurations, rating requirements, protection issues, effect of unbalance protection, and system considerations such as harmonics [3].

IEEE Standard 824 provides guidelines for the application of series capacitors in power systems. The series capacitor compensation is a very special application of power system capacitors and requires careful consideration [4].

ANSI/IEEE Standard C37.04 establishes the rated short circuit current as the highest value of the symmetrical component of the short circuit current measured from the envelope of the current wave [5].

ANSI/IEEE Standard C37.06 provides the standard ratings for various circuit breakers, including the following:

- Preferred ratings for indoor circuit breakers.
- Preferred capacitance current switching ratings for indoor oilless circuit breakers.
- Preferred rated voltage levels of 72.5 kV and below for outdoor circuit breakers including circuit breakers applied in gas insulated substations.
- Preferred capacitance current switching ratings for outdoor circuit breakers of 72.5 kV and below, including circuit breakers applied in gas insulated substations.
- Preferred ratings for outdoor circuit breakers of 121 kV and above, including circuit breakers for gas insulated substations.
- Preferred capacitance current switching ratings for outdoor circuit breakers of 121 kV and above, including circuit breakers for gas insulated substations.

This standard also contains the schedule of dielectric test values and external insulation for AC high voltage circuit breakers and gas insulated substations. Particular attention should be paid when selecting the circuit breakers for capacitance current switching applications [6].

ANSI/IEEE Standard C37.09 deals with the specifications of the design tests, production tests, tests after delivery, field tests, and conformance tests on various circuit breakers.

This standard specifies the allowable transient recovery voltage (TRV) on various switching conditions including capacitance current switching [7].

ANSI/IEEE Standard C37.010 provides guidelines for the application of high voltage circuit breakers. This includes applicable interrupting time, permissible tripping delay, reclosing time, TRV rate, capacitance current switching, shunt reactor switching, and short-circuit-related considerations [8].

ANSI/IEEE Standard C37.011 examines transient recovery voltage concepts. The calculation of transient recovery voltage for various capacitance-based systems and the TRV calculations in faulted power systems are also discussed. The TRV is an important parameter in the selection and operation of any circuit breaker. Typical capacitance values for various equipment such as generators, buses, transmission lines, and instrument transformers are given in this standard [9].

ANSI/IEEE Standard C37.012 deals mainly with capacitance current switching application issues such as interrupting time, transient overvoltage, and effect of reclosing. The considerations required for capacitance currents and recovery voltages under fault conditions are also discussed in this standard [10]. The GIS substation-related guidelines are presented in Reference [11].

IEEE Standard 519 is the harmonic standard that deals with the various sources of generation, system response characteristics, and effect of harmonics on various power system components such as motors, generators, cables, and capacitors. The effects of reactive power compensation and harmonic control are outlined in detail. The analysis methods, measurement approaches, and recommended practices for individual consumers are provided. The standard also provides recommendations from the utility point of view. Further, the standard deals with the methodology for evaluating new harmonic sources. Finally, some application examples are presented. Because power factor correction and harmonic control are performed simultaneously using shunt capacitors this standard is important with respect to capacitor applications [12].

ANSI Standard 141 provides guidelines for system planning, voltage considerations, surge voltage protection, coordination of protection devices, fault calculations, grounding, power factor related topics, power switching, instrumentation, cables, bus ways, and energy conservation. The use of shunt capacitors for improving power factors of industrial loads is discussed in the chapter on power factor and related considerations. The application of the shunt capacitor in induction motors is also presented from the viewpoint of efficiency and safety [13]. ANSI Standard 399 [14] deals with computer-aided analysis of power systems.

IEEE Standard C62.11 provides voltage ratings, maximum continuous overvoltage values, performance testing evaluation procedures for surge arresters, design tests, construction guidelines, and protective characteristics for the metal oxide varistor type of surge arresters [16].

IEEE Standard C62.22 presents the general considerations of surge arrester applications such as overvoltage, separation effects, and insulation coordination from the systems viewpoint. The protection of transmission systems, surge arrester selection approaches, and location of equipment are discussed in this standard. For individual equipment, the required considerations for the transformer, shunt capacitors, underground cables, GIS substations, rotating machines, power line insulation, series capacitors, and circuit breaker TRV control approaches are also discussed. The required procedures for the selection of surge arresters for the distribution system are outlined in this standard [15].

ANSI Standard C37.40 deals with distribution cutouts and fuses suitable for all electric circuit applications. The definitions and specifications for both the expulsion and current limiting fuses are provided in this standard [17].

ANSI Standard C37.42 discusses the distribution cutouts and fuses suitable for capacitor circuit applications. The specifications of the cutouts suitable for mounting the fuses are outlined along with the testing methods. The rated voltage, continuous current, interrupting current, short time

ratings, and basic impulse insulation ratings of cutouts
are presented for voltage levels of 5.2–18 kV. This includes
both the enclosed and open cutout ratings. The melting
currents of expulsion type K (fast) and type T (slow) fuses
are presented in tables for preferred ratings, intermediate
ratings, and ratings below 6 A. These two tables are
reproduced in Appendix E [18].

ANSI Standard C37.47 deals with the specifications
for distribution fuse disconnecting switchgear for fuse
supports and current limiting fuses. The voltage and current
ratings of the current limiting fuses are given in this
standard [19].

IEC has several standards on capacitors. These standards
are widely used throughout Europe and elsewhere [20–27].

NEMA also has certain standards applicable to capacitors [28,29].

## 4.3  CONCLUSIONS

As outlined above, there are many standards regarding
shunt capacitor design, manufacture, installation, and testing. Other capacitor standards on these aspects are available
from other countries (e.g., Canada and the U.K.). The guidelines provided by these standards may not agree with one
another. The numerical values provided in one version of a
standard may be superseded by a later version. For example,
ANSI Standard 18 contains a table on the preferred kV and
kVAR ratings with modified values in the 1980, 1992, and
2002 versions. Such changes in numerical values imply a
change in the trend in manufacture and application of
shunt capacitors. The user is always encouraged to use
the latest version of any standard; however, the older versions are still needed for calculations and analysis of the
capacitor banks installed in previous years. In addition to
capacitors, there are many standards associated with other
related equipment such as circuit breakers, surge arresters,
and fuses.

# REFERENCES

## Standards on Power System Capacitors

1. ANSI/IEEE Standard 18 (2002), IEEE Standard for Shunt Capacitors.

2. IEEE Standard 1036 (1992), IEEE Guide for Application of Shunt Capacitors.

3. ANSI/IEEE Standard C37.99 (1980), IEEE Guide for Protection of Shunt Banks.

4. IEEE Standard 824 (1994), IEEE Standard for Series Capacitors in Power Systems.

## Standards on Circuit Breakers Related to Power System Capacitors

5. ANSI/IEEE Standard C37.04 (1979, Reaff. 1988), IEEE Standard Rating Structure for AC High Voltage Circuit Breakers Rated on a Symmetrical Current Basis.

6. ANSI Standard C37.06 (2000), Preferred Ratings and Related Required Capabilities for AC High Voltage Breakers.

7. ANSI/IEEE Standard C37.09 (1979), Test Procedure for AC High Voltage Circuit Breakers Rated on a Symmetrical Current Basis.

8. ANSI/IEEE Standard C37.010 (1979), Application Guide for AC High Voltage Circuit Breakers Rated on a Symmetrical Current Basis.

9. ANSI/IEEE Standard C37.011 (1979), Application Guide for Transient Recovery Voltage for AC High Voltage Circuit Breakers Rated on a Symmetrical Current Basis.

10. ANSI/IEEE Standard C37.012 (1979, Reaff. 1988), Application Guide for Capacitor Switching of AC High Voltage Circuit Breakers Rated on a Symmetrical Current Basis.

11. ANSI Standard C37.122 (1993), IEEE Standard for Gas Insulated Substations.

## Standards on Power System Applications and Harmonics

12. ANSI Standard 141 (1993), Recommended Practice for Electric Power Distribution for Industrial Plants (Red Book).

13. IEEE Standard 519 (1991), IEEE Standard Practices and Requirements for Harmonic Control in Electric Power Systems.

14. ANSI/IEEE Standard 399 (1990), IEEE Recommended Practice for Power System Analysis (Brown Book).

## Standards on Surge Arresters

15. IEEE Standard C62.22 (1997), IEEE Guide for the Application of Metal Oxide Surge Arresters for Alternating-Current Systems.

16. IEEE Standard C62.11 (1999), IEEE Standard for Metal Oxide Surge Arresters for AC Power Circuits (> 1 kV).

## Standards on Cutouts and Fuses

17. ANSI Standard 37.40 (1981), IEEE Standard Service Conditions and Definitions for High Voltage Fuses, Distribution Enclosed Single-Pole Air Switches, Fuse Disconnecting Switches and Accessories.

18. ANSI Standard 37.42 (1981), Specifications for Distribution Cutouts and Fuse Links.

19. ANSI Standard 37.47 (1981), Specifications for Distribution Fuse Disconnecting Switchgear, Fuse Supports and Current Limiting Fuses.

## IEC Standards

20. IEC Standard 60143-1 (1992), Series Capacitors for Power Systems. General Performance, Testing, and Rating. Safety Requirements. Guide for Installation.

21. IEC Standard 60143-2 (1994), Series Capacitors for Power Systems. Protective Equipment for Series Capacitor Banks.

22. IEC Standard 60831-1 (1996), Shunt Power Capacitors of the Self-Healing Type for AC Systems Having Rated Voltage Up To and Including 1000 V. General Performance, Testing, and Rating. Safety Requirements. Guide for Installation and Operation.

23. IEC Standard 60831-2 (1995), Shunt Power Capacitors of the Self-Healing Type for AC Systems Having Rated Voltage Up To and Including 1000 V. Ageing Test, Self-Healing Test, and Destruction Test.

24. IEC Standard 60871-1 (1997), Shunt Capacitors for AC Power System Having Rated Voltage above 1000 V. General Performance, Testing, and Rating—Special Requirements—Guide for Installation and Operation.

25. IEC Standard 60871-2 (1987), Shunt Capacitors for AC Power System Rated Voltage above 1000 V. Endurance Testing.

26. IEC Standard 60931-1 (1996), Shunt Power Capacitors of the Non-Self-Healing Type Up To and Including 1000 V. General Performance, Testing, and Rating. Safety Requirements. Guide for Installation and Operation.

27. IEC Standard 60931-2 (1995), Shunt Power Capacitors of Non-Self-Healing Type Up To and Including 1000 V. Ageing Test and Destruction Test.

## NEMA Standards

28. NEMA CP-1 (1992), Shunt Capacitors.

29. NEMA CP-9 (1992), External Fuses for Shunt Capacitors.

# 5

## CAPACITOR SPECIFICATIONS

### 5.1 TYPES OF CAPACITOR BANKS

Capacitor units are manufactured as single-phase units. They are connected in wye or delta for three-phase applications. Three-phase capacitor units are also available from certain manufacturers. Based on fusing arrangement, capacitor banks are divided into three types:

- Externally fused capacitor banks
- Internally fused capacitor banks
- Fuseless capacitor banks

The features of these capacitor banks are described below.

### 5.1.1 Externally Fused Capacitor Banks

These are units with fuses connected externally to the capacitor units. The fuses are intended to disconnect the faulted capacitor unit in a bank, preventing rupture of the capacitor case. The remaining units can continue in service without interruption. This type of capacitor bank design consists of several capacitor units in parallel. A typical externally fused capacitor bank design is shown in Figure 5.1 [1]. When a capacitor unit is isolated from a bank in one phase,

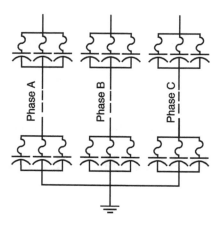

**Figure 5.1**   Externally fused capacitor bank [1].

the capacitance in other phases is higher, resulting in higher phase voltages. To limit the incremental overvoltage due to the removal of a unit in the case of a fuse operation, the unit kVAR size is limited in this type of design. The faulted unit can be identified by visual inspection of the failed fuse unit. However, a failed fuse unit is not an indication of the state of the capacitor unit (failed or not failed). Further, the external fuse unit can deteriorate due to pollution or environmental factors. Such capacitor banks are used in distribution systems on pole-mounted capacitor banks. Typical specifications of the externally fused capacitor bank units are:

- Capacitor unit size varies from 50–400 kVAR.
- The bank consists of series–parallel combinations.
- When the fuse of a capacitance unit fails, the voltage in the other phases increases.
- Warning of a defective unit is received through fuse activation or neutral unbalance detection.

### 5.1.2   Internally Fused Capacitor Banks

In this design, the entire capacitor is constructed with several series–parallel combinations. A typical internally fused capacitor bank design is shown in Figure 5.2 [1]. Usually, the current

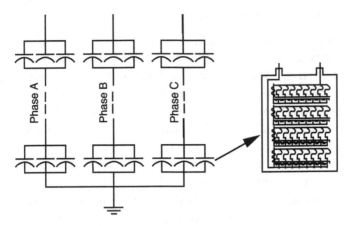

**Figure 5.2** Internally fused capacitor bank and capacitor unit [1].

limiting fuses are used inside the capacitor bank. These fuses are designed to isolate the failed capacitor unit, allowing the continued service of the remaining units. The effect of the single element on the voltage rise will be small and hence the capacitor bank can operate in normal service. It is claimed that the capacitor bank can operate satisfactorily even with several failed capacitor elements. The following advantages are claimed:

- No sustained arcing and hence no risk of capacitor case rupture.
- Easy installation and maintenance.
- Fuse elements are not exposed to external environmental factors.
- The capacitor and fuse assemblies are tested together in the factory.
- They are suitable for substation, HVDC, SVC, and filter applications.
- The failure of the fuse can be identified through unbalance in the phase currents or neutral current.

The limitation of an internally fused capacitor unit is the inability to identify the failed capacitor or fuse unit. If a significant number of capacitor or fuse elements fail, then the capacitor bank should be replaced.

### 5.1.3 Fuseless Capacitor Banks

In a fuseless capacitor bank, a number of individual capacitor units are connected in series, and the group of units is referred to as a string. One or more strings are then connected in parallel per phase as shown in Figure 5.3, which is a typical schematic diagram of one phase of a two-string fuseless bank [2]. Since individual capacitor units are not connected in parallel within a string, the capacitor unit's internal series groups can be viewed as being in series with those of the other capacitors in that string. Therefore, if one of the internal series groups of a capacitor unit fails due to a short circuit, the resulting increase in current through the capacitor unit is small. Thus the increase in voltage applied to the internal series groups in this and other units in the string is also correspondingly small. Because the fault causing this short-circuited series group within the capacitor unit is stable, the affected unit may remain in service for an indefinite period of time. This is the underlying principle for the serviceability of fuseless banks.

Fuseless capacitor banks have a number of advantages over fused banks, but they also have their own unique set of limitations and concerns. Although both fused and fuseless banks will provide good service when properly designed, before purchasing a fuseless bank, the user should be aware

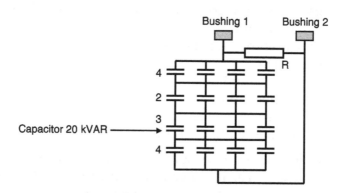

**Figure 5.3**   Fuseless capacitor bank assembly of a 320 kVAR/phase.

of the differences. Some of the advantages of fuseless banks compared to externally fused banks are:

- Typically require one half the substation space of externally fused banks.
- Usually less expensive than fused banks.
- Exhibit greater resistance to animal-related outages as interconnecting material can be insulated.

Some of the problems associated with fuseless capacitors compared to externally fused banks are:

- Bushing faults and internal failures to the tank require the bank to be immediately tripped offline because no fuse is present to clear the faulted unit.
- Replacement capacitor units must be of internal electrical construction similar to those originally supplied. This increases the spares stocking requirements because commonly available standard units cannot be used.
- The bank should be tripped in the event of loss of power to the relay control. The loss of a capacitor unit can be detected through the neutral current flow.
- Transient current duties are higher for the individual capacitor units, possibly requiring the addition of a current-limiting reactor.

The maintenance time required to locate faulty units is longer than for externally fused banks. Although the concept is very attractive and simple to assemble, in the event of a large number of capacitor units, the identification and replacement of the failed units require significant effort.

## 5.2 CAPACITOR SPECIFICATIONS

The specifications for capacitors are identified by the industry standards. They include tolerances and acceptable operating ranges. The capacitors shall be capable of continuous operation without exceeding the following limits [3]:

- 110% of peak voltage
- 120% of rms voltage

**TABLE 5.1**   Maximum Permissible Operating Voltages for Shunt Capacitors

| Capacitor Voltage, V | Maximum Permissible Voltage, V |
|---|---|
| 230 | 253 |
| 460 | 506 |
| 575 | 633 |
| 2,400 | 2,640 |
| 4,160 | 4,575 |
| 6,900 | 7,590 |
| 7,200 | 7,920 |
| 7,970 | 8,767 |
| 12,000 | 13,200 |
| 13,800 | 15,180 |

- 135% of nameplate kVAR
- 180% of nominal rms current based on rated kVAR and rated voltage

Other important specifications are identified below.

### 5.2.1   Voltage

Nominal system voltages are specified line to line. The capacitor units are single phase and appropriate phase voltage is to be used. Capacitors are capable of operation at 110% of rated rms voltage and the crest should not exceed $1.2\sqrt{2}$ rated rms voltage. The maximum operating voltages for shunt capacitors are listed in Table 5.1.

### 5.2.2   kVAR Ratings

Capacitor unit ratings are specified in kVAR. Normally available capacitor ratings are 50, 100, 150, 200, 300, and 400 kVAR per unit. The capacitor units are capable of continuous operation but not exceeding 135% of nameplate kVAR. The typical kVAR ratings are listed in Table 5.2. If the operating voltage increases or decreases from the

**TABLE 5.2** Voltage and kVAR Ratings of the Capacitor Units

| Terminal-to-Terminal Voltage | kVAR | No. of Phases | BIL, kV |
|---|---|---|---|
| 216 | 5, 7.5, 13.3, 20, and 25 | 1 and 3 | 30 |
| 240 | 2.5, 5, 7.5, 10, 15, 20, 25, and 50 | 1 and 3 | 30 |
| 480 | 5, 10, 15, 20, 25, 35, 50, 60, and 100 | 1 and 3 | 30 |
| 600 | 5, 10, 15, 20, 25, 35, 50, 60, and 100 | 1 and 3 | 30 |
| 2,400 | 50, 100, 150, and 200 | 1 | 75 |
| 2,770 | 50, 100, 150, and 200 | 1 | 75 |
| 4,160 | 50, 100, 150, and 200 | 1 | 75 |
| 4,800 | 50, 100, 150, and 200 | 1 | 75 |
| 6,640 | 50, 100, 150, 200, 300, and 400 | 1 | 95 |
| 7,200 | 50, 100, 150, 200, 300, and 400 | 1 | 95 |
| 7,620 | 50, 100, 150, 200, 300, and 400 | 1 | 95 |
| 7,960 | 50, 100, 150, 200, 300, and 400 | 1 | 95 |
| 8,320 | 50, 100, 150, 200, 300, and 400 | 1 | 95 |
| 9,540 | 50, 100, 150, 200, 300, and 400 | 1 | 95 |
| 9,960 | 50, 100, 150, 200, 300, and 400 | 1 | 95 |
| 11,400 | 50, 100, 150, 200, 300, and 400 | 1 | 95 |
| 12,470 | 50, 100, 150, 200, 300, and 400 | 1 | 95 |
| 13,280 | 50, 100, 150, 200, 300, and 400 | 1 | 95 and 125 |
| 13,800 | 50, 100, 150, 200, 300, and 400 | 1 | 95 and 125 |
| 14,400 | 50, 100, 150, 200, 300, and 400 | 1 | 95 and 125 |
| 15,125 | 50, 100, 150, 200, 300, and 400 | 1 | 125 |
| 19,920 | 100, 150, 200, 300, and 400 | 1 | 125 |
| 20,800 | 100, 150, 200, 300, and 400 | 1 | 150 and 200 |
| 21,600 | 100, 150, 200, 300, and 400 | 1 | 150 and 200 |
| 22,800 | 100, 150, 200, 300, and 400 | 1 | 150 and 200 |
| 23,800 | 100, 150, 200, 300, and 400 | 1 | 150 and 200 |
| 23,940 | 100, 150, 200, 300, and 400 | 1 | 150 and 200 |
| 4,160 GrdY/2400 | 300 and 400 | 3 | 75 |
| 4,800 GrdY/2770 | 300 and 400 | 3 | 75 |
| 7,200 GrdY/4160 | 300 and 400 | 3 | 75 |
| 8,320 GrdY/4800 | 300 and 400 | 3 | 75 |
| 12,470 GrdY/7200 | 300 and 400 | 3 | 95 |
| 13,200 GrdY/7620 | 300 and 400 | 3 | 95 |
| 13,800 GrdY/7960 | 300 and 400 | 3 | 95 |
| 14,400 GrdY/8320 | 300 and 400 | 3 | 95 |

nominal operating voltage, then the kVAR delivered changes accordingly.

$$\text{kVAR delivered} = \text{Rated kVAR} \left(\frac{\text{Operating Voltage}}{\text{Rated Voltage}}\right)^2 \quad (5.1)$$

**Example 5.1**

A capacitor unit rated for 200 kVAR, 2.4 kV, 60 Hz is operated at 2.52 kV (1.05 P.U.). What is the kVAR delivered at this voltage? If the operating voltage at the plant during peak load is 0.95 P.U., what is the kVAR delivered at that voltage?

Solution

$$\text{kVAR} = 200, \quad \text{Voltage} = 2.4 \text{ kV}$$

$$\text{kVAR delivered at } 2.52 \text{ V} = 200\left(\frac{2.52}{2.40}\right)^2 = 220.5 \text{ kVAR}$$

The voltage increases by 5% and the kVAR increases by 10.25%.

$$\text{kVAR delivered at } 0.95 \text{ P.U. voltage} = 200\left(\frac{0.95}{1.0}\right)^2 = 180.5 \text{ kVAR}$$

In this case, the voltage decreases by 5% and the kVAR delivered decreases by 9.8%.

The allowable basic insulation levels (BIL) of standard capacitors are defined by the standards. Typical values are shown in Table 5.2.

**5.2.3  Frequency**

Power factor capacitors are designed for operation at 50 or 60 Hz. The capacitor kVAR output is directly proportional to the system operating frequency. If the capacitor operates at a different frequency than the rated frequency, then the kVAR delivered is:

$$\text{kVAR delivered} = \text{Rated kVAR}\left(\frac{\text{Operating Frequency}}{\text{Rated Frequency}}\right) \quad (5.2)$$

## Example 5.2

A capacitor unit rated for 200 kVAR, 2.4 kV, 60 Hz is operated at 50 Hz supply. What is the kVAR delivered at this frequency?

Solution

$$kVAR = 200, \qquad Voltage = 2.40\,V, \qquad f = 60\,Hz$$

$$kVAR \text{ delivered at } 50\,Hz = 200\left(\frac{50}{60}\right) = 166.6 \text{ kVAR}$$

The frequency decreases by 16.7% and the kVAR decreases by 16.7%.

### 5.2.4 Ambient Temperature

Capacitors are designed for switched and continuous operation in outdoor locations with unrestricted ventilation and direct sunlight, at ambient temperatures, for various mounting arrangements. The acceptable ambient temperatures for various capacitor installations are presented in Table 5.3. The ambient and operating temperature of the capacitor banks should take the following additional factors into consideration:

1. Radiation from the sun and other sources such as arc furnaces.
2. If the kVAR delivered is increased, the losses are increased.

**Table 5.3** Maximum Ambient Temperatures

| Mounting Method | 24-h Average | Normal Annual |
|---|---|---|
| Isolated capacitor | 46°C | 35°C |
| Single row | 46°C | 35°C |
| Multiple rows and tiers | 40°C | 25°C |
| Metal enclosed | 40°C | 25°C |

3. During installations, if the capacitors are mounted close to each other, then the effective cooling is reduced. Proper mounting and ventilation are required for long capacitor life.

Some of the approaches used to improve the cooling are forced air cooling, increased space between capacitor units, and locating capacitor banks at lower ambient temperatures.

### 5.2.5 Service Conditions

The capacitor units can be operated at their specified ratings and the following service conditions are to be observed:

1. The ambient temperature is to be within allowed limits.
2. The altitude does not exceed 6,000 ft (1,800 m) above mean sea level.
3. The voltage applied between terminals and case is not to exceed the insulation class.
4. The applied voltage does not contain excess harmonics.
5. The nominal operating frequency is equal to the rated frequency.

Any abnormal operating service condition such as exposure to fumes, explosive dust, mechanical shock or vibrations, radiated heat from nearby sources, restricted mounting, outside ambient temperature limits, higher altitudes, and excessive momentary duties may pose stress to the capacitor units.

### 5.2.6 Types of Bushing

Capacitors are manufactured as single-phase or three-phase units. Single bushing or double bushing capacitor units are available for single-phase capacitors. The single bushing outdoor units are terminated with only one bushing. The other electrode is terminated through the case and has a suitable connection point at the other terminal. In two-bushing capacitor units, the two electrodes are terminated through two bushings. The case is insulated from the electrodes. The cost of the two-bushing unit is higher than the cost of the one-bushing unit. Typical one- and two-bushing capacitor units are shown in Figures 5.4 and 5.5 [5], respectively. Low voltage capacitor units and distribution capacitor units are generally single-phase units. The ratings can be from 20 to 1000 kVAR per unit. These units may be:

1. Two-bushing units with open terminal and dead casing
2. Two-bushing units with terminal enclosure and dead casing
3. One-bushing unit with open terminal and live casing

These units are suitable for indoor or outdoor applications.

Three-phase capacitor units are available from 1 through 24 kV. The ratings may be from 20 through 800 kVAR/can. These units may be:

1. Three-bushing, open terminal, and dead casing for indoor or outdoor application

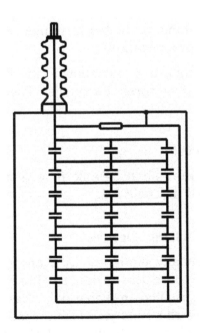

**Figure 5.4** One-bushing type of capacitor [5].

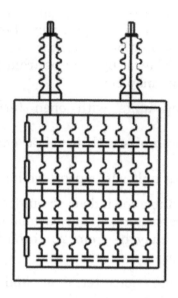

**Figure 5.5** Two-bushing type of capacitor [5].

2. Three-bushing, protected terminals suitable for indoor or outdoor applications

A three-phase capacitor is shown in Figure 5.6. The three-phase unit has three terminals. There is no neutral terminal.

### 5.2.7 Impulse Level

The power capacitors shall be able to withstand the impulse levels as identified in Table 5.4.

### 5.2.8 Internal Discharge Devices

Capacitors are usually equipped with an internal discharge device that will reduce the residual voltage to 50 V or less within the time specified in Table 5.5, after the capacitor is disconnected from the supply. The internal discharge device may not ensure adequate discharge of the capacitors in the

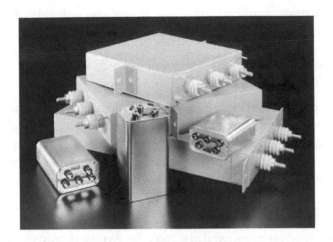

**Figure 5.6** Three-phase capacitor unit. (Courtesy of Aerovox Division PPC, New Bedford, MA.)

TABLE 5.4 Impulse Levels for Capacitors per IEEE Standard 18 [3]

| Capacitor Voltage Rating, Terminal-to-Terminal, V, rms | Impulse Level, Terminal-to-Terminal, Peak kV |
|---|---|
| 216–1,199 | 30 |
| 1,200–5,000 | 75 |
| 5,001–15,000 | 95 |
| 13,200–25,000 | 125 |

TABLE 5.5 Time Limits for Discharging the Capacitors after Disconnection [4]

| Capacitor Voltage Rating, Terminal-to-Terminal, V, rms | Maximum Time Limit |
|---|---|
| 600 V or less | 1 min |
| Over 600 V | 5 min |

power system applications, and therefore suitable discharge resistances should be considered.

### 5.2.9 Momentary Power Frequency Overvoltage

During normal service life, a capacitor may be expected to withstand a combined total of 300 applications of power frequency terminal-to-terminal overvoltages without superimposed transients or harmonic content. The acceptable magnitudes and duration are presented in Table 5.6.

### 5.2.10 Transient Overcurrent

The power system capacitors may be exposed to overcurrents due to energization, de-energization, and similar switching operations. The allowable overcurrent magnitudes and duration are specified in Table 5.7. The total rms voltage and rms

TABLE 5.6  Maximum Permissible Power Frequency Overvoltages and Durations [4]

| Duration | Maximum Permissible Overvoltage, P.U. |
|---|---|
| 6 cycles | 2.20 |
| 15 cycles | 2.00 |
| 1 s | 1.70 |
| 15 s | 1.40 |
| 1 min | 1.30 |

TABLE 5.7  Allowable Overcurrent Magnitudes and Durations [3]

| Probable Number of Transients per Year | Permissible Overcurrent, P.U. |
|---|---|
| 4 | 1,500 |
| 40 | 1,150 |
| 400 | 800 |
| 4000 | 400 |

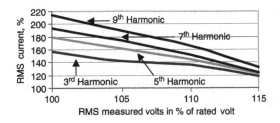

**Figure 5.7** Thermal operating limits of standard capacitors.

current may be determined on a particular circuit from a conventional voltmeter and ammeter. It is possible to determine the percent of rated rms current allowable based on 135% permissible working kVA from the curves shown in Figure 5.7.

**Example 5.3**

What is the maximum current in a 500 kVAR, 4.8 kV capacitor bank? If the bank is subjected to 105% of the rated voltage, what is the permissible capacitor current? Is there any possibility of exceeding the thermal rating?

Solution

$$I\,(\text{rated}) = \frac{500\,\text{kVAR}}{(\sqrt{3})\,(4.8\,\text{kV})} = 60.142\,\text{A}$$

Assuming that only the third harmonic voltages are present in the system to cause overcurrent, 105% voltage will cause 146% of rated current.

$$I\,(\text{permissible}) = (60.142\,\text{A})\,(1.46) = 87.807\,\text{A}$$

If the capacitor current (measured) is below the permissible current, then there is no overload.

**5.2.11 Connections for Three-Phase Configurations**

The single-phase capacitor units are applied in a three-phase power system using ungrounded wye, grounded wye, delta, ungrounded split-wye, and H-configuration. The configurations are to be verified carefully for proper ground connections in order to avoid ferroresonance in the system.

## 5.2.12   Leakage Currents

This is the current flowing from the capacitor roll to the capacitor case. Leakage current can be tested by applying a voltage between the terminals tied together to the case. Alternatively, the voltage can be applied to one terminal to the case and the leakage current can be measured.

## 5.2.13   Dissipation Factor

All capacitors have some resistance associated with them due to the electrodes and the terminal connections. There are also dielectric losses. The total losses can be represented by an equivalent resistance loss due to $R_e$. The dissipation factor is defined as the tangent of the angle $\delta$ between the reactance $X_e$ and the impedance $Z_e$ of the capacitor. The phasor diagram for the dissipation factor is shown in Figure 5.8. The dissipation factor varies with the temperature and system frequency.

*Capacitor Specifications*

| | |
|---|---|
| System line-to-line voltage, kV rms | System phase voltage, kV rms |
| Supply frequency, Hz | Capacitor nominal voltage, kV rms |
| Capacitor peak voltage, kV peak | Nominal capacitance, μF |
| Harmonic current, A | Energization transient, kV peak |
| Rating of single-phase unit, kVAR | Total rating per phase, kVAR |
| Total rating of three-phase unit, kVAR | |
| Capacitor bank connection | (a) Grounded wye, (b) Ungrounded wye, (c) Delta, (d) Split wye |
| Type of connection | (a) Fixed, (b) Switched |
| Type of circuit breaker | (a) Vacuum, (b) Oil, (c) Other, specify |
| Type of bushing | (a) Single, (b) Double |

| Type of mounting | (a) Pole mounted, |
| | (b) Substation, |
| | (c) Metal enclosed, |
| | (d) Rack mounted |
| Type of fusing | (a) Individual, (b) Group, |
| | (c) Fuseless |

*Other Options*

| Current transformer | Yes or No |
| Potential transformer | Yes or No |
| Ground switch for phase | Yes or No |
| Ground switch for neutral | Yes or No |
| Surge arrester | (a) Distribution, |
| | (b) Intermediate, |
| | (c) Station |
| Reactor | (a) Current limiting, |
| | (b) Harmonic notch filter |

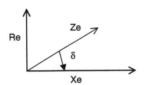

**Figure 5.8**  Phasor diagram related to the dissipation factor.

**Example 5.4**

Power factor correction capacitors are required for a 4.16 kV, three-phase, 60 Hz industrial load. The design requirements are 150 kVAR single-phase units. Select and verify the rating of the capacitor from Table 5.2.

Solution

Voltage (line-to-line) = 4.16 kV
Phase voltage (4.16 kV/1.732) = 2.402 kV
Peak voltage/phase 2.402 × 1.414) = 3.396 kV

Select capacitor unit voltage, kVAR, and current levels of 150 kVAR from Table 5.2.

> Capacitor rms voltage $= 2.77$ kV
> Peak voltage $(2.77\,\text{kV} \times 1.414) = 3.917$ kV
> Current rating $(150/2.77) = 54.2$ A

$$\text{kVAR delivered at } 2.402\,\text{kV} = 150\left(\frac{2.402\ \text{kV}}{2.77\ \text{kV}}\right)^2 = 112.8\ \text{kVAR}$$

Actual current at rated voltage $(112.8\ \text{kVAR}/2.402\ \text{kV}) = 46.96$ A.

Verify the actual ratios versus the acceptable limit.

| Item | Limit | Actual |
|------|-------|--------|
| rms Voltage | 110% | $\dfrac{2.402\,\text{kV}}{2.77\,\text{kV}} \times 100 = 86.7\%$ |
| Peak voltage | 120% | $\dfrac{3.396\,\text{kV}}{3.917\,\text{kV}} \times 100 = 86.7\%$ |
| Current | 180% | $\dfrac{46.96\,\text{A}}{54.2\,\text{A}} \times 100 = 86.7\%$ |
| kVAR | 135% | $\dfrac{112.8\,\text{kVAR}}{150\,\text{kVAR}} \times 100 = 75.2\%$ |

The voltage, current, and kVAR ratios are within acceptable limits.

**Example 5.5**

Select suitable capacitor units for a 75 MVAR, three-phase, 138 kV, 60 Hz power factor correction bank. Specify the assumptions made.

Solution

> Three-phase MVAR $= 75$
> MVAR per phase $= 25$
> Line-to-line voltage $= 138$ kV
> Phase voltage $(138\,\text{kV}/1.732) = 79.7$ kV

Select 400 kVAR per unit, 20.8 kV per unit, 200 kV BIL from Table 5.2. Select four units per series group.

Voltage across the series group $(4 \times 20.8) = 83.2$ kV
No. of parallel groups/phase $(25{,}000/(4 \times 400)) = 15.5$

Select 16 parallel groups per phase.

KVAR per phase $(4 \times 16 \times 400) = 25{,}600$
Three-phase kVAR $(3 \times 25.6 \text{ MVAR}) = 76.8$ MVAR
Delivered MVAR $[(134/144)^2 \times 76.8] = 74.9$ MVAR

## 5.3 OTHER TYPES OF CAPACITORS

Other types of capacitors used in power system applications include electrolytic capacitors, metalized-paper capacitors, dry metalized-film capacitors [6], ultra capacitors, and high frequency capacitors. Some features of these capacitors are discussed below.

### 5.3.1 Electrolytic Capacitors

Electrolytic capacitors use aluminum foil that has been treated anodically to produce an oxide film. One electrode is the aluminum itself, and for unidirectional application, a plain aluminum foil is used. For alternating current operation, a second anodized foil is used as the other electrode. The foils are separated by paper insulation that is soaked in electrolyte to provide contact between the oxide layers. The aluminum can be treated by etching to increase the surface area. This is known as etched foil. In some other capacitors, the aluminum electrode can be sprayed on the fabric before anodizing to increase the surface area. This is known as sprayed foil. In this way, the capacitance per unit is increased significantly. The electrolytic capacitors are for short time applications only. For example, the electrolytic capacitors for motor start applications in the voltage range of 125–350 V rms are suitable for 30-second operation. The starting capacitors are used in single-phase motors. Electrolytic capacitors are not suitable for continuous duty. A photograph of a typical electrolytic

**Figure 5.9** Electrolytic capacitors used for motor starting. (Courtesy of Capacitor Industries Inc., Chicago.)

capacitor is shown in Figure 5.9. Usually the electrolytic capacitors are housed in cylindrical aluminum cans.

### 5.3.2 Metalized-Paper Capacitors

One way to reduce the size of the capacitor is to use thin electrodes and reduced insulation thickness. Using metal deposition techniques, it is possible to reduce the thickness of the electrodes. In this type of capacitor, one layer of insulation is used instead of the two layers used in conventional capacitors. The paper insulation has to be impregnated in order to withstand excessive voltage stress. These capacitors are also used for motor start and run applications and are short time rated. They are much heavier than the electrolytic capacitors but tend to provide longer service life with lower losses. These capacitors can be operated continuously and are therefore suitable for motor start and run applications. The capacitor can be fixed on the motor frames.

### 5.3.3 Dry Metallized-Film Capacitors

The power factor of the fluorescent and discharge lighting circuits is around 0.4 and needs to be compensated. These capacitor units should be low cost and highly reliable. Since these capacitors used in lighting fittings are mounted in ballast assembly, the operating temperatures can be high. The lighting capacitor consists of a thin oriented polypropylene film, metalized on one side, and two films wound together as a cylindrical element. The dry metalized-film type capacitors are available up to 440 V.

### 5.3.4 Impregnated Paper/Foil Capacitors

With the use of impregnated paper instead of paper as the insulation, the size of the capacitor can be reduced. This method has been used to make capacitors for motor start and run applications. These capacitors have been replaced by other designs such as metalized paper capacitors.

### 5.3.5 Ultra Capacitors

An ultra capacitor is an electrochemical device consisting of two porous electrodes in an electrolyte solution that stores charge electrostatically. A typical view of an ultra capacitor is shown in Figure 5.10. The device consists of two porous electrodes, usually made up of activated carbon immersed in an electrolyte solution that flows into and around the electrode plates. The electrolytic solution is typically potassium hydroxide or sulfuric acid. This structure effectively creates two capacitors, one at each carbon electrode, connected in series. The ultra capacitors share the same battery chemistry, but the approach is to operate them at a cell voltage range that leads to electrostatic storage of charge. The resistance between the cells due to the electrolyte is much smaller than that of a battery. Therefore, ultra capacitors can achieve much higher power density than batteries, by a factor of 10 or more. Other

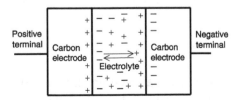

**Figure 5.10**  An ultra capacitor.

designs of ultra capacitor with different electrodes [7] are used in energy storage applications of very short duration.

### 5.3.6 Capacitor Use for Commutation

One application for capacitors is in thyristor controlled chopper and inverter circuits for commutation. Once the thyristor begins to conduct, it continues to conduct until turned off (commutated) by interruption of the supply or by the application of an opposing current supplied by a commutating capacitor. The capacitors available for this purpose are those with impregnated paper dielectric, metalized paper, and plastic film dielectric/mixed dielectric. The operating voltages can be 200–1,200 V DC. These capacitors are subject to charging and discharging at supply frequency or at higher levels. Therefore, proper care should be taken in choosing the capacitors for the application.

### 5.3.7 High Frequency Capacitors

In high frequency applications, the amount of heat generated in a capacitor is greater and requires special considerations. The high frequency capacitors are used in induction heating, pulse, commutation, broadcast transmission, drives, bypass equipment, frequency converters, filters, high voltage power supplies, snubbers, couplers, voltage dividers, spark generators, and harmonic filters. The voltage range may be 1–300 kV and capacitance size may be 100 pF–5000 μF. Typical high frequency capacitors are shown in Figure 5.11. These capacitors are oil filled. The oil serves as insulation and acts as a cooling agent [8]. An example rating of a high frequency capacitor is 800 kVAR, 800 V, 10,000 Hz, 1000 A, and 19.19 MFD.

### PROBLEMS

5.1.  What are the allowable limits for rms voltage, peak voltage, current, and kVAR in applying power capacitors? Why are these factors important?

**Figure 5.11** High frequency capacitors with several taps. (Courtesy of High Energy Corporation, Parkesburg, PA.)

5.2. Power factor correction capacitors are required for a 13.8 kV, three-phase, 60 Hz industrial load. The design requirements are 200 kVAR single-phase units. Select and verify the rating of the capacitor from Table 5.2.

5.3. What considerations are necessary when selecting single bushing capacitors versus double bushing capacitors?

5.4. Is there a possibility of current overloading in a capacitor circuit when the voltages are kept within nominal ranges? If there is any, describe how it can happen and how to mitigate it.

5.5. In Example 5.1, if the supply voltage is raised to 1.10 P.U. for a duration of 6 h, what will happen to the capacitors?

5.6. A capacitor unit is rated at 200 kVAR, 4.16 kV, 60 Hz, and single-phase. Calculate the kVAR delivered if it is operated in a 2.4 kV circuit. Calculate the kVAR delivered.

5.7. A capacitor unit is rated at 200 kVAR, 4.16 kV, 60 Hz, and single-phase. Determine the kVAR delivered if the capacitor is used in a 16.67 Hz traction power system application.

5.8.  Describe an ultra capacitor. How is it similar to and different from a battery?

## REFERENCES

1. Sévigny, R., Ménard, S., Rajotte, C., McVey, M. (2000). Capacitor Measurement in the Substation Environment: A New Approach, *Proceedings of the IEEE 9th International Conference on Transmission and Distribution*, 299–305.

2. Andrei, R. G. (1999). A Novel Fuseless Capacitor Bank Design Using Conventional Single Bushing Capacitors, *IEEE Transactions on Power Delivery*, 14(3), 1124–1133.

3. ANSI/IEEE Standard 18 (1992), IEEE Standard for Shunt Capacitors.

4. IEEE Standard 1036 (1992), IEEE Guide for Application of Shunt Capacitors.

5. www.captech.com.au, Capacitor Technologies, Victoria, Australia.

6. Longland, T., Hunt, T. W., Brecknell, W. A. (1984). *Power Capacitor Handbook,* Butterworth-Heinemann, London.

7. Baker, P. P. (2002). Ultra Capacitors for Use in the Power Quality and Distributed Resource Applications, *IEEE Power Engineering Society Summer Meeting,* 1, 316–320.

8. www.highenergycorp.com, website of High Energy Corporation, Parkesburg, PA.

# 6

## TESTING OF CAPACITORS

Power capacitors are manufactured and tested according to industry standards. ANSI, IEEE, NEMA, or IEC provides the standard guidelines for ratings, testing, application, and operating service conditions. The membership of these organizations consists of manufacturers, industry, utilities, and system engineers. The tests prescribed for the capacitor equipment take into account all the necessary application factors. These tests are classified as design tests, production tests, and field tests [1,2].

### 6.1 DESIGN TESTS

The manufacturer performs design tests on a sufficient number of selected units in order to show compliance with the standards. These tests are not to be repeated unless there is a specific design change in the capacitor unit that would change the characteristics of the unit. The following design tests are performed.

#### 6.1.1 Impulse Withstand Test

Impulse withstand tests are performed between terminals and case, with the terminals connected together. For capacitors with bushings of two different BIL ratings, this test is based

TABLE **6.1**   Voltages for Impulse Withstand Test from Standard 18

| | | Withstand Test Voltage | | |
| --- | --- | --- | --- | --- |
| BIL, kV | Minimum Insulation Creepage Distance, mm | 60 Hz Dry 1 min, kV, rms | 60 Hz Wet 10 s, kV, rms | Impulse 1.2/50 µs Full Wave kV Crest |
| 30 | 51 | 10 | 6 | 30 |
| 75 | 140 | 27 | 24 | 75 |
| 95 | 250 | 35 | 30 | 95 |
| 125 | 410 | 42 | 36 | 125 |
| 150 | 430 | 60 | 50 | 150 |
| 200 | 660 | 80 | 75 | 200 |

on the bushing of the lower BIL. Single bushing capacitors having one electrode shall not be subjected to the impulse withstand test. The impulse voltage recommended for this test is 1.2/50 µs full wave with a tolerance on the crest value of ±3%. The time to crest of a 1.2/50 µs impulse wave is measured as 1.67 times the time required for the voltage to rise from 30 to 90% of the crest value. The capacitor shall successfully withstand three consecutive positive impulses. The impulse test voltages for various capacitor designs are listed in Table 6.1.

### 6.1.2   Bushing Test

This is a supply frequency test. The bushing successfully passes the impulse test if no flashover occurs with the first three applications of the test voltage. If there is a flashover, then three additional impulses shall be applied. If no additional flashover occurs, the bushings shall be considered as having passed the test successfully.

### 6.1.3   Thermal Stability Test

The test capacitor is considered thermally stable if the hotspot case temperature reaches and maintains a constant value within a variation of 3°C for 24 hours. One sample shall be selected as the test capacitor. Two other capacitors having the same ratings and approximately the same power factor and capacitance are selected for barrier capacitors. The barrier capacitors are units mounted adjacent to the test capacitor during the thermal stability test.

Resistor models with the same power loss, thermal characteristics, and physical dimensions as the test capacitor may be substituted for the barrier capacitors. The test capacitor is mounted in an enclosure between the two barrier capacitors at the manufacturer's recommended center-to-center spacing. The mounting position selected shall be the recommended operating position that produces the highest internal temperatures. The air inside the test enclosure is maintained at an average temperature of 46°C and must not be force circulated. The inside wall temperature of the enclosure shall be within ±5°C of the ambient temperature. The ambient temperature shall be measured by means of a thermocouple on the case, supported and positioned in such a way that it is subjected to the minimum possible thermal radiation from the three energized samples. All three sample capacitors shall be energized at a test voltage given by:

$$V_\mathrm{T} = 1.1 V_\mathrm{R} \sqrt{\frac{W_\mathrm{M}}{W_\mathrm{A}}} \qquad (6.1)$$

where $V_\mathrm{T}$ = Test voltage
$\quad\quad V_\mathrm{R}$ = Capacitor rated rms voltage
$\quad\quad W_\mathrm{M}$ = Maximum allowable power loss
$\quad\quad W_\mathrm{A}$ = Actual power loss of the test capacitor

The test voltage calculated for this test is limited to a value that will result in the operation of the test capacitor at a maximum of 144% of its kVAR rating. This voltage is maintained constant, within ±2% throughout the 24 hours of the test period.

### 6.1.4 Radio Influence Voltage (RIV) Test

The RIV test is performed at rated frequency and 115% of the rated rms voltage of the capacitor. Capacitors with two bushings fully insulated from the case shall be tested with the case grounded. Capacitors having only one bushing per phase and the case as the other terminal should not be tested. The following precautions should be observed when measuring the RIV of the capacitors:

- The capacitor should be at room temperature.
- The capacitor bushings shall be dry and clean.
- The capacitor shall be mounted in its recommended position.

When measured in accordance with the above conditions at a frequency of 1 MHz, the RIV shall not exceed 250 μV.

### 6.1.5 Voltage Decay Test

The capacitor shall be energized at a direct current voltage equal to the peak of rated AC voltage. The decay of the voltage, when de-energized, shall be measured by suitable means. The time for decay of residual voltage to 50 V or less shall not exceed 5 min for capacitors rated higher than 600 V or 1 min for capacitors rated 600 V or less.

### 6.1.6 Short-Circuit Discharge Test

The purpose of the short-circuit discharge test is to verify the integrity of the internal connections and conductors of the capacitor operating under normal service conditions. The test shall be carried out by the manufacturer for a particular design or on a similar design that has equal or smaller size conductors. As such, the testing of a particular rating will be applicable to a wide range of capacitor ratings. One unit shall be charged to a DC voltage 2.5 times the rated rms voltage and then discharged. It shall be subjected to five such discharges. Before and after the five discharges, the terminal-to-terminal capacitance shall be measured at low voltage. The discharge circuit shall have no inductive or resistive devices included. The discharge device may be a switch or spark gap and may be situated up to 1 m from the capacitor such that the total perimeter of the external discharge loop is less than 3 m. The conductors used to connect the capacitor to the discharge device shall be of copper and shall have a cross section of at least 10 mm$^2$. The difference in capacitance between the initial and final measurements shall be less than an amount corresponding to either the shorting of an element or operation of an internal fuse.

## 6.2 ROUTINE TESTS

Routine tests are production tests performed by the manufacturer on each capacitor. During the test, new and clean capacitors are used. The ambient temperature is maintained at $25 \pm 5°C$. The AC voltage used for the test is 60 Hz. The following tests are applicable.

### 6.2.1 Short Time Overvoltage Tests

Each capacitor shall withstand the following test voltages for at least 10 seconds. Each capacitor shall, with its case and internal temperature at $25 \pm 5°C$, withstand for at least 10 seconds a terminal-to-terminal insulation test at a standard test voltage of either of the following:

- A direct current test voltage of 4.3 times the rated rms voltage, or
- An alternating sinusoidal voltage of two times the rated rms voltage.

For three-phase, wye connected units, either with a neutral bushing or with the neutral connected to the case, the above testing for terminal to neutral shall be followed by a test at $\sqrt{3}$ times the above standard test voltage between each pair of bushings. This test is performed to evaluate the phase-to-phase insulation. For three-phase, wye connected units, where there is no neutral bushing and the neutral is not connected to the case, the rated voltage is the phase-to-phase voltage of the capacitor unit. In order to test both the phase-to-phase insulation and each leg of the wye at the appropriate voltage, the test voltage shall be 1.16 times the above standard test voltage between each pair of bushings $(2/\sqrt{3} \approx 1.16)$. For three-phase, delta connected units, the rated voltage is the phase-to-phase voltage of the capacitor unit. The test voltage shall be the above standard test voltage between each pair of bushings.

The capacitance shall be measured on each unit both before and after the application of the test voltage. The initial capacitance measurement shall be at low voltage. The change in capacitance, as a result of the test voltage, shall be less than either a value of 2% of the originally measured capacitance or that caused by failure of a single element of the particular design, whichever is smaller.

### 6.2.2 Terminal-to-Case Test

This test is not applicable to capacitors having one terminal common to the case. Terminal-to-case tests shall be made on capacitors having terminals that are insulated from the case. The appropriate test voltage from Table 6.2 shall be applied for at least 10 seconds between an insulated terminal connected together and the case. For capacitors with bushings of two different BIL ratings, this test shall be based on the bushing with the lower BIL.

TABLE 6.2  Test Voltages for Short Time Overvoltage Test, Terminal to Case, from Standard 18

| Range of Capacitor RMS Voltage Ratings Terminal to Terminal, V | BIL kV, Crest | Terminal-to-Case Test Voltage, rms V | |
|---|---|---|---|
| | | Indoor or Housed Equipment | Outdoor |
| 216–300 | 30* | 3,000 | 10,000 |
| 301–1,199 | 30* | 5,000 | 10,000 |
| 1,200–5,000 | 75* | 11,000 | 26,000 |
| 1,200–15,000 | 95 | – | 34,000 |
| 1,200–20,000 | 125 | – | 40,000 |
| 1,200–25,000 | 150 | – | 50,000 |
| 1,200–25,000 | 200 | – | 60,000 |

*Outdoor

### 6.2.3 Capacitance Test

Capacitance tests shall be made on each capacitor to demonstrate that it will deliver the rated reactive power and not more than 110% of rated reactive power at rated voltage and frequency, corrected to an internal temperature of 25°C. Measurements made at other than 25°C are corrected by adjusting for temperature differences according to the established temperature relationship for the capacitor tested.

### 6.2.4 Leak Test

A suitable test shall be performed on each capacitor to ensure that it is free from leaks. The completed and sealed capacitor units are heated in an oven to increase the internal hydrostatic pressure and force liquid out if there are leaks. This test is conducted to ensure that all seams, bushes, and filled holes are sealed properly.

### 6.2.5 Discharge Resistor Tests

A suitable test shall be performed on each capacitor to ensure that the internal discharge device will reduce an initial residual voltage equal to $\sqrt{2}$ times the rated voltage rms to 50 V or less in the time limits.

### 6.2.6 Loss Determination Test

Loss measurement shall be made by the manufacturer on each capacitor to demonstrate that the capacitor losses are equal to or less than the maximum allowed power loss for the given unit. A loss determination test is performed to confirm that the losses are within acceptable limits.

### 6.2.7 Fuse Capability Tests for Internally Fused Capacitors

Internally fused capacitors shall be subjected to one short-circuit discharge test, from a DC voltage of 1.7 times the rated voltage through a gap situated as closely as possible to the capacitor, without any additional impedance in the circuit. It is permitted that the DC charging voltage be generated by initially energizing with an AC voltage having a peak value of 1.7 times the rated voltage and disconnecting at a current of zero. The capacitor is then immediately discharged from this peak value. Alternatively, if the capacitor is disconnected at a slightly higher voltage, the discharge may be delayed until the discharge resistor reduces the voltage to 1.7 times the rated voltage.

The capacitance shall be measured before and after the discharge test. The difference between the two measured values shall be less than an amount corresponding to one internal fuse operation. The discharge test may be made before or after the voltage test between terminals. However, if it is made after the voltage test between terminals, a capacitance measurement at rated voltage shall be made after the discharge test to detect fuse operation. If the capacitors are accepted with operated fuses, the voltage test between terminals shall be made after the discharge test.

## 6.3 INSTALLATION TESTS OR FIELD TESTS

When the capacitor banks are installed in the desired location, certain tests are performed to ensure that the units are connected to the specifications. The following tests are performed before energization of the capacitor banks.

### 6.3.1 Capacitance Measurement

A capacitance meter is used to measure the effective capacitance of the bank to determine the accuracy of the connection. The voltage used for this type of test is only a fraction of the rated voltage. Any failure in the capacitor bank may have low impedance for these low applied voltages and partial failures can be identified. A 10% increase in the capacitance indicates a partially failed capacitor unit. The same method can be used to identify partially failed paper or film capacitors with reasonable accuracy. With several swollen capacitors, the capacitance measurement can show normal reading even though the capacitor is failed. This is because the interpack connections are burned open. All bulged capacitor units are considered to be failed units.

### 6.3.2 Low Voltage Energization Test

In this test method, the capacitive reactance is measured by applying a low voltage to the capacitor bank. A circuit diagram suitable for such a test is shown in Figure 6.1. The applied voltage for this type of test will be of the order of 120 V. The low voltage is supplied to the capacitor bank through fuses. If there is a short circuit or a ground fault, the fuses will act and the power supply will be protected. The capacitor voltage and the current are measured as follows:

$$\text{The capacitance, } C = \frac{I}{\omega V} \qquad (6.2)$$

where  $C =$ Capacitance, F
$\quad\quad\; I =$ Current, A
$\quad\quad\; V =$ Voltage, V
$\quad\quad\; \omega =$ Constant, 377.7

### 6.3.3 High Voltage Insulation Strength Tests

High voltage insulation tests can be performed in accordance with NEMA CP-1. Test voltages for this test are presented in Table 6.3. Extreme caution needs to be taken in performing these tests because these are high voltage tests and the capacitor is an energy-storing device. The capacitors should

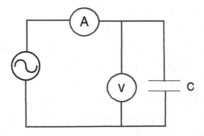

**Figure 6.1**   Circuit diagram to measure the capacitive reactance.

**TABLE 6.3**  Capacitor Field Test Voltage per NEMA Standard CP-1

| Rated Capacitor Voltage, V | Terminal-to-Terminal Voltage, kV | | Rated Capacitor BIL, kV | Terminal-to-Terminal Test Voltage, kV | |
|---|---|---|---|---|---|
| | AC | DC | | AC | DC |
| 2,400 | 3.6 | 7.74 | 75 | 19.5 | 28.5 |
| 2,700 | 4.15 | 8.93 | | | |
| 4,160 | 6.24 | 13.4 | | | |
| 4,800 | 7.20 | 15.4 | | | |
| 6,640 | 9.96 | 21.4 | 95 | 25.5 | 39.0 |
| 7,200 | 10.8 | 23.2 | | | |
| 7,620 | 11.4 | 24.5 | | | |
| 7,960 | 11.9 | 25.6 | | | |
| 9,960 | 14.9 | 32.1 | | | |
| 12,470 | 18.7 | 40.2 | | | |
| 13,280 | 19.9 | 42.8 | 95/125 | 25.5/ 30.0 | 39.0/ 45.0 |
| 13,800 | 20.7 | 44.5 | | | |
| 14,400 | 21.6 | 46.4 | | | |
| 19,920 | 29.8 | 64.2 | 125 | 30 | 45.0 |
| 21,600 | 32.4 | 69.6 | | | |

be protected from tank rupture by an appropriate fuse. The fault current during the test should be limited. It is possible that the stored energy in the capacitor during the direct current test would be enough to rupture a capacitor if the capacitor fails during the test.

When performing the alternating current test, the capacitor should be switched online at or below the rated voltage. The voltage is slowly increased to a level shown in Table 6.3. At the end of the test, the voltage should be decreased to rated voltage or less before the test circuit is opened. The length of time during which a capacitor operates above the rate voltage should not exceed 20 seconds for the test.

When conducting the direct current test, the charging and discharging test should be limited to a maximum of one ampere. This can be achieved by connecting a resistance in series with the test capacitor. The capacitor should be discharged through a suitable resistance after performing the test. Any attempt to short-circuit the capacitor will damage the unit.

## PROBLEMS

6.1. What are the different tests performed on capacitor units? Why are these tests conducted on capacitor units?

6.2. How can you measure the capacitance of a capacitor bank in service? Assume that the bank is a three-phase unit.

## REFERENCES

1. ANSI/IEEE Standard 18 (2002), IEEE Standard for Shunt Capacitors.

2. IEEE Standard 1036 (1992), IEEE Guide for Application of Shunt Capacitors.

3. NEMA Standard CP-1 (1992), Shunt Capacitors.

# 7

## LOCATION OF SHUNT
## CAPACITORS

### 7.1  INTRODUCTION

Power factor correction capacitors can be installed at high voltage bus, distribution, or at the load [1–3]. The following power factor correction approaches are commonly used.

#### 7.1.1  Group Capacitor Bank

A group capacitor bank installation is shown in Figure 7.1. In this approach, the power factor correction is applied to a group of loads at one location. This technique is suitable for utility or industrial customers with distributed load. If the entire load comes on or off together, then it is reasonable to switch the capacitor bank in this manner. If part of the load is switched on and off on a regular basis, then this type of reactive compensation is not appropriate. It is economical to have a large capacitor bank for reactive compensation rather than several smaller banks.

#### 7.1.2  Branch Capacitor Bank

In certain industrial applications, the load is switched on and off based on shifts. Such a load group can be related to

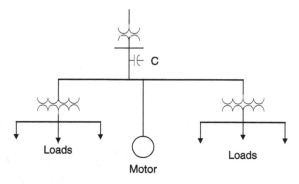

**Figure 7.1**  Group capacitor bank.

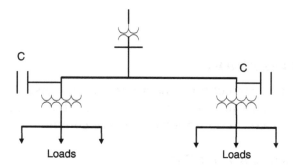

**Figure 7.2**  Branch capacitor bank.

individual feeders or branch circuits. Therefore, it is advantageous to switch the capacitor banks along with the specific branches. An example branch capacitor bank scheme is shown in Figure 7.2. This type of capacitor bank will not help reduce the losses in the primary circuit.

### 7.1.3  Local Capacitor Bank

An example of local capacitor bank application for the power factor correction is shown in Figure 7.3. In this scheme, the individual loads are provided with separate capacitor banks. This type of reactive compensation is mainly suitable for industrial loads. The localized power factor correction can be expensive.

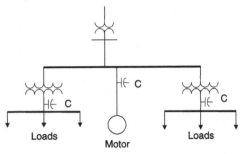

**Figure 7.3**   Local capacitor bank.

## 7.2   CONSIDERATIONS IN LOCATING CAPACITORS

Shunt capacitors provide reactive power locally, resulting in reduced maximum kVA demand, improved voltage profile, reduced line/feeder losses, and decreased payments for the energy. Maximum benefit can be obtained by installing the shunt capacitors at the load. This is not always practical due to the size of the load, distribution of the load, and voltage level. Depending on the need, the capacitor banks are installed at extra-high voltage (above 230 kV), high voltage (66–145 kV), and feeders at 13.8 and 33 kV. In industrial and distribution systems, capacitor banks are installed at 4.16 kV.

### 7.2.1   Pole-Mounted Capacitor Banks

In the distribution systems, the power factor correction capacitors are installed on the poles. These installations are similar to the pole-mounted distribution transformers. In the case of capacitor banks, the following components are installed on a stable platform.

| | |
|---|---|
| Capacitor banks | Fuse units along with mounting |
| Vacuum or oil switches | Distribution class surge arrester |
| Controller to switch the capacitor units | Junction box |
| Control transformer | Current limiter or harmonic filter reactor |

**Figure 7.4** A pole-mounted harmonic filter bank. (Courtesy of Gilbert Electrical Systems and Products, Beckley, WV.)

The interconnections are made using insulated power cables. Pole-mounted capacitor banks can be fixed units or switched units to meet the varying load conditions. The voltage rating can be 460 V–33 kV. The size of the capacitor units can be 300–3,000 kVAR. A typical pole-mounted installation of a capacitor bank is shown in Figure 7.4.

### 7.2.2 Shunt Capacitor Banks at EHV Levels

Usually extra-high voltage (EHV) lines are used to transmit bulk power from remote generations to load centers. These long lines tend to produce significant voltage drops during peak loads. Therefore, shunt capacitors are used at the EHV substations to provide reactive power. Sometimes these capacitor banks are switched as and when required. A typical high voltage harmonic filter bank is shown in Figure 7.5.

**Figure 7.5** A high voltage filter bank. (Courtesy of Gilbert Electrical Systems and Products, Beckley, WV.)

### 7.2.3 Substation Capacitor Banks

When large reactive power is to be delivered at medium or high voltages, then shunt capacitor banks are installed in substation locations. These open stack shunt capacitor units are installed for operating voltages 2.4–765 kV. The open rack construction and exposed connection need significant protection in the sub-station. Such installations contain capacitor banks, cutout units with fuses, circuit breakers, surge arresters, controllers, insulator units at high voltage, and interconnections. A typical substation type capacitor bank installation is shown in Figure 7.6 [7]. At high voltage levels, the shunt capacitor banks are used for reactive power support, voltage profile improvement, reduction in line, and transformer losses. These shunt

**Figure 7.6** Substation capacitor bank. (Courtesy of U.S. Department of Labor, OSHA website.)

capacitor banks are also installed in select substations after careful load flow and stability analysis.

### 7.2.4 Metal-Enclosed Capacitor Banks

When the capacitor banks are installed in industrial or small substations in indoor settings, then metal-enclosed cabinet type construction is employed. Such units are compact and require less maintenance. A typical metal-enclosed capacitor bank is shown in Figure 7.7. The life expectancy of these type of units is longer because they are not exposed to external environmental factors such as severe heat, cold, humidity, and dust.

**Figure 7.7** A metal-enclosed harmonic filter bank. (Courtesy of Gilbert Electrical Systems and Products, Beckley, WV.)

## 7.2.5 Distribution Capacitor Banks

Distribution capacitors are installed close to the load, on the poles, or at the substations. Although these capacitor units provide reactive power support to local load, they may not help reduce the feeder and transformer losses. Low voltage capacitor units are cheaper than high voltage capacitor banks. Protecting distribution capacitor banks from all types of fault conditions is difficult. Sometimes pad-mounted installations are used for low or medium voltage distribution capacitors. A typical pad-mounted capacitor bank is shown in Figure 7.8 [8]. Although the pad-mounted capacitors are outdoor installations, they are protected by metal enclosures from outdoor environment and are similar to pad-mounted transformer installations.

## 7.2.6 Fixed Capacitor Banks

In distribution and certain industrial loads, the reactive power requirement to meet the required power factor is constant.

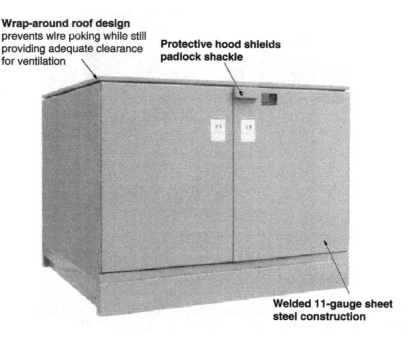

**Figure 7.8** A pad-mounted 15 kV, 1,200 kVAR, three-phase capacitor bank. (Courtesy of S&C Electric Company, Chicago.)

In such applications, fixed capacitor banks are used. Sometimes such fixed capacitor banks can be switched along with the load. If the load is constant for the 24-hour period, the capacitor banks can be on without the need for switching on and off.

### 7.2.7 Switched Capacitor Banks

In high voltage and feeder applications, the reactive power support is required during peak load conditions. Therefore the capacitor banks are switched on during the peak load and switched off during off-peak load. The switching schemes keep the reactive power levels more or less constant, maintain the desired power factor, reduce overvoltage during light load conditions, and reduce losses at the transformers and feeders.

The switching controls are operated using one of the following signals:

- Voltage: since the voltage varies with load.
- Current: as the load is switched on.
- kVAR: as the kVAR demand increases, the capacitor banks can be switched on and vice versa.
- Power factor: as the power factor falls below a predetermined value, the capacitor banks can be switched on.
- Time: sometimes the capacitor banks can be switched on using a timer and switched off at the end of a factory shift.

The general practice is to switch the capacitor in steps in order to accommodate large voltage changes. Several layouts for switching capacitor banks are shown in Figure 7.9. In Figure 7.9a, one capacitor bank is switched by a circuit breaker. Figure 7.9b shows one fixed capacitor and two automatically switched capacitor banks. The circuit breakers must have suitable short-circuit ratings to handle the energization and back-to-back switching requirements.

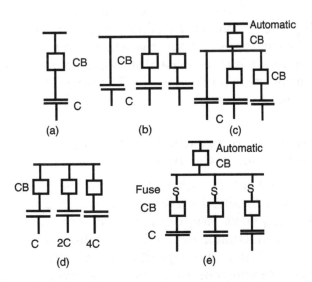

**Figure 7.9** Various configurations of switched capacitor schemes.

**TABLE 7.1** Selection of Capacitors in Binary Order for Power Factor Control

| Item | Bit 0 | Bit 1 | Bit 2 | Remarks |
|------|-------|-------|-------|---------|
| 1 | 0 | 0 | 0 | All switches are open. |
| 2 | 1 | 0 | 0 | Switch 1 is closed. |
| 3 | 0 | 1 | 0 | Switch 2 is closed. |
| 4 | 1 | 1 | 0 | Switches 1 and 2 are closed. |
| 5 | 0 | 0 | 1 | Switch 3 is closed. |
| 6 | 1 | 0 | 1 | Switches 1 and 3 are closed. |
| 7 | 0 | 1 | 1 | Switches 2 and 3 are closed. |
| 8 | 1 | 1 | 1 | All three switches are closed. |

Figure 7.9c shows the capacitor bank switching arrangement with one automatic and two nonautomatic circuit breakers. In certain applications with random variations in the reactive power requirements, the capacitors are to be switched in and out using a binary arrangement. Such a scheme is shown in Figure 7.9d. The corresponding choice of capacitors is listed in Table 7.1. This arrangement can be used to switch seven steps of capacitor banks using three capacitor banks and three circuit breakers. The selection requires careful programming and is achievable using programmable controllers. Figure 7.9e shows another scheme where one automatic circuit breaker can switch three capacitor banks equipped with fuses and nonautomatic circuit breakers. The capacitor banks can be of equal size.

## 7.2.8 Installation of Capacitors on the Low Voltage Side of the Transformer

The capacitor bank is installed close to the load to provide reactive power locally. In a system in which a large number of small equipment are compensated, the reactive power demand may fluctuate, depending on the load. During off-peak load condition, the capacitor bank voltage may go up and hence overcompensation should be avoided. This may result in unwanted fuse operation and failure of capacitor

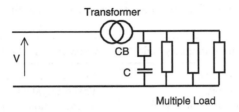

**Figure 7.10** Per phase representation of the low voltage capacitor installation.

units. Therefore, a switched capacitor bank on the low voltage side of the transformer may be a good choice. Harmonics in the system should be checked to determine if the capacitor and the reactance of the power transformer are in series and create resonance. A typical scheme is shown in Figure 7.10.

## 7.2.9 Installation of Capacitors on the High Voltage Side of the Transformer

This type of installation provides the same kind of reactive power compensation as a low voltage capacitor bank. The installation can be safe from overvoltage if it is switched on and off, depending on the reactive power requirement. One of the main advantages of high voltage capacitor installation is that the losses in the stepdown transformer are reduced. The cost of a high voltage capacitor scheme will be higher. As in the low voltage capacitor scheme, the possibility of overcompensation and resonance issues should be checked. A typical scheme is shown in Figure 7.11. Sometimes it may

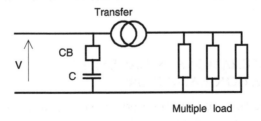

**Figure 7.11** Per phase representation of the high voltage capacitor.

TABLE **7.2**   Power Factor Correction on the HV Side versus at the Load Location

| On the Transformer Primary Side | Capacitor at the Load Location |
|---|---|
| Need one capacitor bank. | Need three capacitor banks. |
| One physical location. | Three physical locations. |
| Rack mounted outdoor or metal enclosed indoor. | Metal enclosed indoor or pole mounted. |
| Easy to maintain. | Multiple locations require more maintenance. |
| Can be designed as a tuned filter. | Filtering with transformer. |
| Controlled resonant point. | Multiple resonant points. |
| Stable system impedance from filter location. | System impedance sees resonance points in the impedance mode. |
| Relatively low cost due to one location. | Higher cost due to multiple locations. |
| May not be able to switch based on load changes. | Load changes can be handled. |
| Need to have one circuit breaker to handle capacitor switching. | Need to have circuit switches to handle capacitor switching. |

be possible to correct the power factor at the individual load location. The relative advantages and disadvantages are presented in Table 7.2.

### 7.2.10   Mobile Capacitor Banks

When there is need to apply shunt capacitors in distribution systems to relieve overloaded facilities until permanent changes are made, portable capacitor banks can be used. These banks are available in three-phase and single-phase units. A typical mobile capacitor bank mounted on a truck is shown in Figure 7.12.

### 7.3   CONSIDERATIONS IN HANDLING SUBSTATION CAPACITORS

Shunt capacitors are connected to the power system at the substation locations for power factor improvement both at distribution and transmission levels. The distribution capacitors

**Figure 7.12** A mobile capacitor bank. (Courtesy of Gilbert Electrical Systems and Products, Beckley, WV.)

are designed for VAR supply at the high voltage side in the transformers. This improves the power factor of the transformer at the high voltage side. Transmission level capacitors are selected based on the power factor and the stability needs of the network. The shunt capacitor banks are used to improve the power factor, minimize losses, increase system voltage, and increase stability margins [4,5].

### 7.3.1 Maximum Capacitor Bank Size

The maximum capacitor bank size is determined by the change in the steady-state voltage, switching transient limits, and circuit breaker capabilities. When a shunt capacitor bank is energized or de-energized, the steady-state voltage increases or decreases. In order to have a minimal effect on the customer load and on the system, the voltage change is often limited to a value in the range of 2–3%. This voltage change can be estimated using the following equation:

$$\Delta V = \left( \frac{\text{MVA}_C}{\text{MVA}_{SC}} \right) \times 100\% \qquad (7.1)$$

where $MVA_C = MVAR$ size of the capacitor bank
$MVA_{SC} =$ Short-circuit MVA at the substation location.

The continuous current rating of the circuit breaker for the use of shunt breaker switching is an important factor in determining the maximum size of the capacitor bank. This rating is determined by the nominal capacitor current, 1.25 times for ungrounded operation and 1.35 times for grounded operation.

## Example 7.1

A capacitor bank installation is considered at a 115 kV, three-phase, 60 Hz substation. The short-circuit rating at the substation is 40 kA. The capacitor bank under consideration consists of two 30 MVAR, wye connected banks. One suggestion is to switch both the banks using the same circuit breaker. The other alternative is to use each bank with a time delay using individual breakers. Discuss the technical issues related to this project.

Solution

One of the important items is the voltage change during the energization process due to the size of the capacitor bank. Therefore, the voltage change for 30 MVAR and 60 MVAR capacitor banks is calculated.

Short circuit MVA at the substation $= 1.732 \times 115\,\text{kV} \times 40\,\text{kA}$
$$= 7,967\,\text{MVA}$$

$$\Delta V(\text{due to 30 MVAR bank}) = \left(\frac{30\,\text{MVAR}}{7,967\,\text{MVA}}\right) \times 100\% = 0.38\%$$

$$\Delta V(\text{due to 60 MVAR bank}) = \left(\frac{60\,\text{MVAR}}{7,967\,\text{MVA}}\right) \times 100\% = 0.75\%$$

The voltage change in both cases is not significant. Therefore, both options are acceptable. By having individual circuit breakers, it is possible to energize only one bank. The two circuit breaker option provides greater flexibility and reliability.

## 7.3.2 Minimum Capacitor Bank Size

Minimum shunt capacitor size is influenced by the capacitor bank unbalance detection schemes and fuse coordination. When a capacitor fuse operates to indicate a failed capacitor, an unbalanced condition can occur that subjects units in the same series group to a 60 Hz overvoltage. A common criterion is to limit this overvoltage to 110% of the rated voltage with one unit out. This requires a minimum number of units in parallel as given in Table 7.3. This condition has to be satisfied for safe and efficient operation of a capacitor bank. When a capacitor is completely shorted, other series groups within the capacitor bank are subject to a 60 Hz overvoltage until the fuse clears. The fuse should clear the fault fast enough so that damage to good capacitors is avoided. A capacitor is expected to withstand power frequency phase-to-phase voltages without superimposed transients or harmonics as listed in Table 5.6, Chapter 5. It is noted that the durations of the allowable overvoltages are significantly small. Within the given duration, the capacitor bank has to be isolated from the supply to avoid further failures. Table 7.4 lists the 60 Hz voltages on the other series groups in the bank when a capacitor

TABLE 7.3  Minimum Numbers of Units in Parallel per Series Group to Limit Voltage

| Number of Series Groups | Grounded Wye or Delta | Ungrounded Wye | Ungrounded Split Wye |
|---|---|---|---|
| 1 | — | 4 | 2 |
| 2 | 6 | 8 | 2 |
| 3 | 8 | 9 | 8 |
| 4 | 9 | 10 | 9 |
| 5 | 9 | 10 | 10 |
| 6 | 10 | 10 | 10 |
| 7 | 10 | 10 | 10 |
| 8 | 10 | 11 | 10 |
| 9 | 10 | 11 | 10 |
| 10 | 10 | 11 | 11 |
| 11 | 10 | 11 | 11 |
| 12 and over | 11 | 11 | 11 |

**TABLE 7.4**  Per Unit Voltage on Good Capacitors

| No. of Series Groups | Grounded Wye or Delta | | | Ungrounded Wye | | | Ungrounded Split Wye | | |
|---|---|---|---|---|---|---|---|---|---|
| | $V_a$ | $V_b$ | $V_c$ | $V_a$ | $V_b$ | $V_c$ | $V_a$ | $V_b$ | $V_c$ |
| 1 | — | 1.00 | 1.00 | — | 1.73 | 1.73 | 0.00 | 1.73 | 1.73 |
| 2 | 2.00 | 1.00 | 1.00 | 1.50 | 1.15 | 1.15 | 1.71 | 1.08 | 1.08 |
| 3 | 1.50 | 1.00 | 1.00 | 1.29 | 1.08 | 1.08 | 1.38 | 1.04 | 1.04 |
| 4 | 1.33 | 1.00 | 1.00 | 1.20 | 1.05 | 1.05 | 1.26 | 1.03 | 1.03 |
| 5 | 1.25 | 1.00 | 1.00 | 1.15 | 1.04 | 1.04 | 1.20 | 1.02 | 1.02 |

is shorted on phase A. The values in Tables 5.6 and 7.4, coupled with the fuse size being used, will indicate the minimum number of capacitor units to be used. The capacitor bank should be designed such that the durations of the overvoltages defined in Table 7.4 are not exceeded. The factors that influence this design include the bank connection, the number of series groups, the number of parallel units, and the fuse characteristics.

### 7.3.3  Bank Size and the Loss Reduction

The loss reduction in the line and transformer is one of the benefits to consider when selecting the shunt capacitor bank. A larger capacitor bank may be preferable from a cost stand-point. Several banks at a given location provide flexibility in switching the needed sizes. For a uniformly distributed load, the impact of the number of capacitor banks on the loss reduction is shown in Figure 7.13. For a given load level, the loss reduction is the greatest when one or two capacitor banks are optimally located. Beyond that point, the incremental gain in the loss reduction becomes insignificant and is soon offset by increased capital costs [5].

### 7.3.4  Effect of Voltage Rise Limit on the Switched Capacitor

The voltage rise limit restricts the amount of kVAR that can be switched. Based on the flicker problems, the voltage rise limit

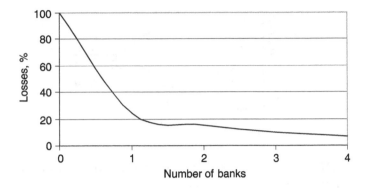

**Figure 7.13** The loss reduction benefit for a given load level.

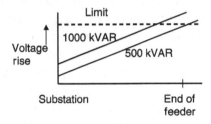

**Figure 7.14** The voltage rise limit restricts the maximum size or its location on the feeder.

either restricts the maximum capacitor bank size for a given location or requires that a given bank be located closer to the substation as illustrated in Figure 7.14 [5].

## 7.4 SHUNT CAPACITOR BANK CONFIGURATIONS

### 7.4.1 Grounded Wye

A grounded wye capacitor configuration is shown in Figure 7.15(a). The advantages of the grounded wye connection compared to the ungrounded wye connection are that the initial cost of the bank may be lower because neutral is not insulated to the system BIL level; the circuit breaker transient recovery voltages are reduced; the mechanical duties

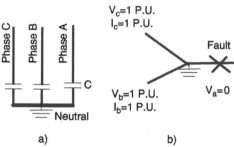

**Figure 7.15** (a) Grounded wye connection; (b) phasor diagram with a single line-to-ground fault.

may be less severe for the structure or racks; and the disadvantages of the grounded wye connection compared to the ungrounded wye connection are high inrush and ground currents. The ground currents also may cause telephone interference. Further, the grounded schemes offer a low impedance path for fault currents and hence require a neutral relay. For grounded wye banks, the current limiting fuses are required because of the line-to-ground fault currents. In the event of a fault on one phase, before the fuse operation, the voltage on the faulted phase can go to zero. Such a condition is shown in Figure 17.15(b). Since the neutral is grounded, the phase voltages on the remaining phases will be 1.0 P.U. and the current through the capacitor units in the unfaulted phases will be 1.0 P.U. This means the capacitor units in the unfaulted phases are safe.

### 7.4.2 Ungrounded Wye

An ungrounded wye capacitor configuration is shown in Figure 7.16(a). In this scheme, the phase voltages and currents are symmetrical during normal operation. If the capacitor in one phase fails before the fuse clears the fault, the neutral is shifted as shown in Figure 7.16(b). Then the voltage across the capacitor units in phases B and C (the unfaulted phases) are equal to the line-to-line voltage (1.732 P.U.). The current through the capacitors in the unfaulted phases reaches 1.732 P.U. of the nominal value. At the fault point, the expected maximum

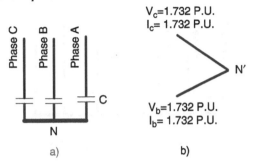

**Figure 7.16** (a) Ungrounded wye connection; (b) phasor diagram with a single line-to-ground fault.

current will be 3.0 P.U. Such an increase in the voltage and current in the unfaulted phases may result in additional failures. Ungrounded wye banks do not permit the flow of zero sequence currents, third harmonic currents, or large capacitor discharge currents during system ground faults. (Phase-to-phase faults may still occur and will result in large discharge currents.) The other advantage is that overvoltages appearing at the CT secondaries are not as high as in grounded banks. However, the neutral should be insulated for full line voltage because it is momentarily at phase potential when the bank is switched or when one capacitor unit fails in a bank configured with a single group of units. For banks above 15 kV this may be expensive.

### 7.4.3 Ungrounded Split-Wye

A typical ungrounded split-wye capacitor bank is shown in Figure 7.17. This scheme is equivalent to the ungrounded wye scheme. The split-wye connection scheme is popular because it is easy to detect unbalance at the neutral. The grounded wye, ungrounded wye, and delta connected capacitors may be subject to ferroresonant overvoltages if they are switched together with transformer units using single-pole switching devices. For the ungrounded capacitor banks, if the transformer has a grounded neutral or even if it consists of many single-phase transformers applied equally along the feeder, a ferroresonant circuit exists only if a single-phase

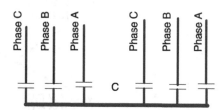

**Figure 7.17** Ungrounded split-wye connection.

switching device is operated upstream. Both transformers and surge arresters will fail under such conditions. If the transformer is three-phase ungrounded, then the grounded capacitor bank should be avoided for the same reason. Although ferroresonance can occur on these circuits, significant resistive load on the transformers can prevent it.

### 7.4.4 Grounded Split-Wye

When a capacitor bank becomes too large, making the parallel energy of a series group too great (above 4650 kVAR) for the capacitor units or fuses, the bank may be split into two wye sections. The characteristics of the grounded double wye are similar to a grounded single-wye bank. The two neutrals should be directly connected with a single connection to ground. The double-wye design allows a secure and faster unbalance protection with a simple uncompensated relay because any system zero sequence component affects both wyes equally, but a failed capacitor unit will appear as unbalanced in the neutral. Time coordination may be required to allow a fuse, in or on a failed capacitor unit, to blow. If it is a fuseless design, the time delay may be set short because no fuse coordination is required. If the current through the string exceeds the continuous current capability of the capacitor unit, more strings shall be added in parallel.

### 7.4.5 Delta Connected

Delta connected capacitor banks are used only at low voltages. A typical scheme is shown in Figure 7.18. Delta connection

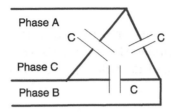

**Figure 7.18** The delta connected capacitors.

arrangement of capacitors requires two bushing units with a grounded rack or single bushing unit with isolated rack. With only one series group, overvoltage of a capacitor unit cannot occur due to unbalance, and therefore the unbalance detection is not required. The third harmonic currents can flow in the delta circuit of the capacitor bank. The individual capacitor fuse must be used for interrupting the system phase-to-phase, short-circuit current. This necessitates an expensive current limiting fuse rather than the expulsion fuse.

### 7.4.6 The H-Configuration

A schematic of the H-configuration of capacitor banks is shown in Figure 7.19 [6]. In this configuration, the capacitors per phase are divided into four quadrants and a current transformer sensor is installed across the bridge. The unbalance current through the bridge is used to monitor the change in the capacitance value. Any change in the capacitance value of the H will produce a current flow through the unbalance sensor. Such a configuration is used in Europe and Australia.

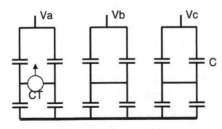

**Figure 7.19** The H-configuration of capacitors for a three-phase system.

**Example 7.2**

Select all possible capacitor combinations for power factor correction of a load supplied through a 13.8 kV, 60 Hz, three-phase system. The total load is (1,800 + j 1,400) kVA.

Solution

The solution is presented in Table 7.5. The resultant kVAR, the current, and the power factor are listed for various unit sizes.

## 7.5 CABLES FOR POWER CAPACITORS

Usually three-phase cables are used to connect the power factor correction capacitors to the power supply. The power factor capacitors draw a constant current from the power supply. Due to the various tolerances, the cable may be forced to carry a larger current. The tolerances include:

- Increase in supply voltage, up to 1.1 times the nominal.
- Increase in the system supply frequency, ±0.5%.
- Increase in the capacitor manufacturing tolerance of +10%.
- Effect of harmonic currents.

A safety factor of 1.25–1.30 can provide adequate current rating for the cable circuit.

### 7.5.1 Selection of Cables for Capacitor Circuit

There are several cable configurations available for the capacitor circuit interconnection application. The cables are classified according to voltage level, type of conductor material (copper or aluminum), type of insulation, type of shielding, and type of installation. In order to demonstrate the basic principles of cable selection, consider a 5,000 V, unshielded cable with copper conductor. The typical parameters for such a cable for installing on a nonmagnetic tray are presented in Table 7.6. The cable size, conductor area, the resistance per 1,000 ft, the reactance per 1,000 ft, and the nominal current are useful in the voltage drop and short-circuit current calcula-

**TABLE 7.5** Solution to Example 7.2

| Unit kVAR | kV | Numbers | Total kVAR | Delivered kVAR | Resultant kVAR | kVA | Current A | Power Factor |
|---|---|---|---|---|---|---|---|---|
| 0 | 0 | 0 | 0 | 0 | 1400.00 | 2280.35 | 95.41 | 0.79 |
| 50 | 15.125 | 3 | 150 | 124.87 | 1275.13 | 2205.89 | 92.29 | 0.82 |
| 100 | 15.125 | 3 | 300 | 249.74 | 1150.26 | 2136.14 | 89.37 | 0.84 |
| 150 | 15.125 | 3 | 450 | 374.61 | 1025.39 | 2071.58 | 86.67 | 0.87 |
| 200 | 15.125 | 3 | 600 | 499.48 | 900.52 | 2012.69 | 84.21 | 0.89 |
| 300 | 15.125 | 3 | 900 | 749.22 | 650.78 | 1914.03 | 80.08 | 0.94 |
| 50 | 15.125 | 6 | 300 | 249.74 | 1150.26 | 2136.14 | 89.37 | 0.84 |
| 100 | 15.125 | 6 | 600 | 499.48 | 900.52 | 2012.69 | 84.21 | 0.89 |
| 150 | 15.125 | 6 | 900 | 749.22 | 650.78 | 1914.03 | 80.08 | 0.94 |
| 200 | 15.125 | 6 | 1200 | 998.96 | 401.04 | 1844.13 | 77.16 | 0.98 |

TABLE 7.6  Typical Cable Data for Three 5,000 V Unshielded Cables

| KCMIL | mm² | R/1000 ft | X/1000 ft | Current A |
|---|---|---|---|---|
| 8 | 8.37 | 0.8170 | 0.0374 | 55 |
| 6 | 13.20 | 0.5140 | 0.0353 | 84 |
| 4 | 21.15 | 0.3230 | 0.0334 | 109 |
| 2 | 33.62 | 0.2030 | 0.0317 | 140 |
| 1 | 44.21 | 0.1610 | 0.0317 | 160 |
| 1/0 | 53.49 | 0.1280 | 0.0308 | 182 |
| 2/0 | 67.43 | 0.1020 | 0.0302 | 207 |
| 3/0 | 85.01 | 0.0809 | 0.0294 | 235 |
| 4/0 | 107.20 | 0.0644 | 0.0288 | 267 |
| 250 | 127.00 | 0.0546 | 0.0293 | 291 |
| 350 | 177.00 | 0.0395 | 0.0279 | 350 |
| 500 | 253.00 | 0.0286 | 0.0270 | 419 |
| 750 | 380.00 | 0.0202 | 0.0267 | 505 |
| 1000 | 507.00 | 0.0164 | 0.0260 | 568 |

tions. The voltage drop per phase $(V_d)$ due to the current $I$ through the cable is given by:

$$V_d = IR\cos(\theta) + IX\sin(\theta) \tag{7.2}$$

where $\theta$ is the power factor angle and $R$ and $X$ are the resistance and reactance of the cable, respectively. The voltage drop has to be within acceptable limits. The cable also should be able to carry the short-circuit current during fault conditions. The cable short-circuit current $(I_{SC})$ for copper conductor can be calculated as:

$$I_{SC} = A\sqrt{\frac{0.0297}{t}\log_{10}\left[\frac{(T_2 + 234.5)}{(T_1 + 234.5)}\right]} \tag{7.3}$$

where $t$ is the fault clearing time in seconds and $A$ is the cable size in millimeters. $T_1$ and $T_2$ are the Centigrade temperatures before and after the fault.

## Example 7.3

Consider a three-phase, 1,800 kVA capacitor bank, 4.16 kV, 60 Hz connected to the power supply through a 300 ft cable.

The capacitor bank can be switched on and off through a circuit breaker. Select a suitable cable for this application. State the assumptions made.

Solution

> kVA $= 1,800$
> kV $= 4.160$ V
> V/phase $= 2,401.8$ V
> $\quad\quad$ I $= 249.8$ A/phase
> Assume a power factor of 0.9, angle $= 25.8°$
> Safety factor $= 1.3$
> Required cable ampacity $(1.3 \times 249.8\,\text{A}) = 324.8$ A/phase

Select cable 350 kcmil from Table 7.6, with $I = 350$ A

> $Z$ per 1000 ft $= (0.0395 + j\ 0.0279)$ Ω/phase
> $Z$ (for 300 ft) $= (0.01185 + j\ 0.00837)$ Ω/phase
> $V_d [324.8(0.01185 \cos 25.8 + 0.0084 \sin 25.8) = 46.3$ V/phase
> Percentage voltage drop $(46.3 \times 100/2401.8) = 1.9\%$
> Temperature of the cable before fault $= 90°$C (assumed)
> Temperature of the cable after fault $= 250°$C (assumed)
> Conductor, $A = 350,000$ cmil
> Fault clearing time, $t = 0.05$ s (3 cycles, assumed)
> Using Equation (7.3), $I_{sc} = 112,521$ A
> Substation symmetrical short-circuit current $= 40,000$ A
> Asymmetrical $I_{SC}$ $(1.6 \times 40,000\,\text{A}) = 64,000$ A

The symmetrical and asymmetrical short-circuit current values are less than the cable short-circuit current. Therefore, the cable rating is acceptable.

## PROBLEMS

7.1. Power factor correction is performed at distribution, medium voltage, high voltage, and extra-high voltage levels. Why is the power factor correction applied at all voltage levels? Explain.

7.2. What are the advantages of pole-mounted capacitor banks?

7.3. A 13.8 kV/4.16 kV transformer is supplying loads at seven locations. The power factor range is 0.50–0.75. Compare the power factor correction on the high voltage side of the transformer versus the correction at individual load locations.

7.4. What are the trends in the installation of capacitor banks on the low voltage side versus the high voltage side of the transformer?

7.5. What are the commonly used connections for capacitor banks? Is there a specific relation between the transformer connection and the capacitor connection? If so, what is it?

7.6. Why is ferroresonance a factor in the capacitor bank connections?

7.7. Delta connected banks are recommended for low voltage applications. What will be the technical problems in using delta connected banks for a large MVAR rating with several series and parallel combinations?

7.8. A capacitor bank installation is needed in a 230 kV, three-phase, 60 Hz high voltage substation. The short-circuit rating at the substation is 46 kA. The total capacitor bank requirement is 100 MVAR, wye connected. How many steps are recommended from the switching point of view?

7.9. Select a three-phase cable for the interconnection of a 3,000 kVA capacitor bank, 4.16 kV, 60 Hz connected to the power supply through a 200 ft cable. The capacitor bank can be switched on and off through a circuit breaker. Use a fault clearing time of 0.1 s. State the assumptions made.

## REFERENCES

1. IEEE Standard 1036 (1992), IEEE Guide for Application of Shunt Capacitors.

2. ANSI/IEEE Standard C37.99 (1980), IEEE Guide for Protection of Shunt Banks.

3. Hammond, P. W. (1988). A Harmonic Filter Installation to Reduce Voltage Distortion from Solid State Converters, *IEEE Transactions on Industry Applications*, 24(1), 53–58.

4. Mendis, S. R., Bishop, M. T., Blooming, T. M., Moore, R. T. (1995). Improving the System Operations with Installation of Capacitor Filter Banks in a Paper Facility with Multiple Generating Units, IEEE Paper No. 0-7803-9418.

5. Hirakami, M., Jenkins, W. P., Ward, D. J. (1983). Capacitors: A Unique Opportunity for Application Reassessment, *Electric Forum*, 9(2), 14–19.

6. Illing, K. J. (2003). Capacitor Fuse Fail Detector, B.E. Thesis, The University of Queensland, Victoria, Australia.

7. U.S. Department of Labor, OSHA website, www.osha.gov/sltc/etools/electric_power/glossary.html.

8. S&C Electric Company (2005). Pad-Mounted Switchgear: Pad-Mounted Capacitor Bank, Photo Sheet, 661-700, S&C Electric Company, Chicago.

# 8

## POWER FACTOR IMPROVEMENT

### 8.1 INTRODUCTION

Most industrial loads such as induction motors operate at moderately low power factors. Around 60% of the utility load consists of motors and hence the overall power factor of the power system is low. Depending on the level of the load, these motors are inherently low power factor devices. The power factor of these motors varies from 0.30 to 0.95, depending on the size of the motor and other operating conditions. Therefore, the power factor level is always a concern for industrial power systems, utilities, and the user. The system performance can be improved by correcting the power factor. The system power factor is given by (see Figure 8.1):

$$\text{Power factor} = P/\text{kVA} \qquad (8.1)$$

where $P$ and kVA are the real and apparent power, respectively. The relation between the power factor and the $Q/P$ ratio is shown in Table 8.1. From Table 8.1, it can be seen that even at 90% power factor, the reactive power requirement is 48% of the real power. At low power factors, the reactive power demand is much higher. Therefore, some form of

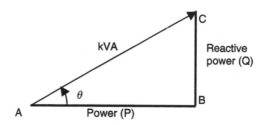

**Figure 8.1** Relationship between real, reactive, and apparent power (kVA).

TABLE 8.1 Power Factor and $Q/P$ Ratio

| Power Factor % | Angle Degree | $Q/P$ Ratio |
|---|---|---|
| 100 | 0 | 0.00 |
| 95 | 11.4 | 0.20 |
| 90 | 26.8 | 0.48 |
| 85 | 31.8 | 0.62 |
| 80 | 36.8 | 0.75 |
| 70.7 | 45.0 | 1.00 |
| 60 | 53.1 | 1.33 |
| 50 | 60.0 | 1.73 |

power factor correction is required in all the industrial facilities. The power factor of any operating system can be lagging or leading. The direction of active and reactive power can be used to determine the nature of the power factor. If both the real and reactive power flow are in the same direction, then the power factor is lagging. If the reactive power flows in the direction opposite to that of the real power, then the power factor is leading. A typical lagging power factor load is an induction motor [1–3]. A typical leading power factor load is a capacitor. Some typical power factors of industrial plants are presented in Table 8.2.

**Example 8.1**

The power factor of a 100 kVA load is 0.8. It is necessary to improve the power factor to 0.95. What is the rating of the shunt capacitor bank?

TABLE 8.2  Typical Power Factor of Some Industrial Plants

| Industry | % Power Factor | Industry | % Power Factor |
|---|---|---|---|
| Chemical | 80–85 | Arc welding | 35–60 |
| Coal mine | 65–80 | Machine shop | 45–60 |
| Electroplating | 65–70 | Arc furnace | 75–90 |
| Hospital | 75–80 | Spraying | 60–65 |
| Office building | 80–90 | Weaving | 60–75 |
| Cement | 80–85 | Clothing | 30–60 |
| Textile | 65–75 | Machining | 40–65 |
| Foundry | 75–80 | Plastic | 75–80 |

Solution

$$kVA = 100$$
$$P \ (100 \times 0.8) = 80 \, kW$$
$$\theta_1 \ (\cos^{-1}(0.8)) = 36.8°$$
$$Q_1 \ (100 \times \sin 36.8°) = 60 \, kVAR$$

Improved power factor condition:

$$kVA = 100$$
$$\theta_2 \ (\cos^{-1}(0.95)) = 18.2°$$
$$Q_2 \ (100 \times \sin 18.2°) = 31 \, kVAR$$

Required shunt capacitors, $(60 - 31) = 29 \, kVAR$

The power factor correction capacitors can be installed at high voltage bus, distribution, or at the load [1–3]. The power factor correction capacitors can be installed for a group of loads, at the branch location, or for a local load. The benefits due to the power factor correction for the utility are release in system generation capacity, savings in transformer capacity, reduction in line loss, and improved voltage profile. The benefits due to power factor correction to the customer are reduced rate associated with power factor improvement, reduced loss causing lower peak demand, reduced energy consumption, and increased short-circuit rating for the system.

## 8.2    FIXED VERSUS SWITCHED CAPACITORS

Shunt capacitors applied to distribution systems are generally located on the distribution lines or in the substations. The distribution capacitors may be in pole-mounted racks, pad-mounted banks, or submersible installations. The distribution banks often include three to nine capacitor units connected in three-phase grounded wye, ungrounded wye, or in delta configuration. The distribution capacitors are intended for local power factor correction by supplying reactive power and minimizing the system losses. The distribution capacitors can be fixed or switched depending on the load conditions. The following guidelines apply:

- Fixed capacitors for minimum load condition.
- Switched capacitors for load levels above the minimum load and up to the peak load.

Figure 8.2 shows the reactive power requirements of a distribution system are shown for a period of 24 hours. Such base load and peak load conditions are common in most utilities. Usually, the fixed capacitors satisfy the reactive power requirements for the base load and the switched capacitors compensate the inductive kVAR requirements of the peak load.

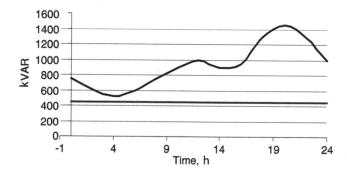

**Figure 8.2**    Distribution curve showing the base reactive power requirement (for fixed capacitors) and peak kVAR needs (for switched capacitors).

## 8.2.1 Sizing and Location of Capacitors

To obtain the best results, shunt capacitors should be located where they produce maximum loss reduction, provide better voltage profile, and are close to the load. When this is not practical, the following approaches can be used.

- For uniformly distributed loads, the capacitor can be placed at two thirds of the distance from the substation.
- For uniformly decreasing distributed loads, the capacitor can be placed at half the distance from the substation.
- For maximum voltage rise, the capacitor should be placed near the load.

Usually, the capacitor banks are placed at the location of minimum power factor by measuring the voltage, current, kW, kVAR, and kVA on the feeder to determine the maximum and minimum load conditions. Many utilities prefer a power factor of 0.95. The peaks and valleys in the kVAR demand curve make it difficult to use a single fixed capacitor bank to correct the power factor to the desired level. If a unity power factor is achieved during the peak load, then there would be leading kVAR on the line during off-peak condition, resulting in an over-corrected condition. Over-correction of power factor can produce excess loss in the system, similar to the lagging power factor condition. Overvoltage condition may occur during leading power factor condition causing damage to the equipment. Therefore, a leading power factor is not an advantageous condition. In order to handle such conditions, fixed capacitors are used to supply the constant kVAR requirements and switched capacitors are used for supplying the kVAR for the peak load conditions. Specifically, this will prevent over-correction of the power factor. Figure 8.2 shows the selection approach for the fixed and switched capacitors. The capacitive kVAR required to correct the given power factor to the desired level can be calculated as shown below.

$\theta_1 =$ Power factor angle of the given load
$\theta_2 =$ Desired power factor angle

kW = Three-phase real power

kVAR from shunt capacitors, $(Q_1 - Q_2)$

$\quad$ = kVA$_1$ sin $\theta_1$ − kVA$_2$ sin $\theta_2$

The above relation can be expressed in the form of a chart as shown in Table 8.3. To determine the needed capacitive reactive power, select the multiplying factor that corresponds to the present power factor and the desired power factor. Then multiply this factor by the kW load of the system. Select a kVAR bank close to the required kVAR.

## Example 8.2

The power factor of a system is 0.60. The desired power factor of the system is 0.95. The kW demand of the system is 100. Calculate the required shunt capacitors for power factor correction.

Solution

> The given power factor of the system = 0.60
> Desired power factor = 0.95
> Multiplying factor from Table 8.3 = 1.005
> kW demand = 100
> Capacitive kVAR required (1.005) (100 kW) = 100.5 kVAR
> Select 100 kVAR shunt capacitors.

### 8.2.2 Effect of Shunt Capacitors on Radial Feeders

Fixed capacitors can be used to improve the power factor on radial feeders [3]. The capacitors can be located at the source or at the load. In a radial system, the capacitor can be located very close to the load as shown in Figure 8.3. The voltage profile during light load on the radial system without and with shunt capacitors is shown in Figure 8.4. The voltage drop effects are dominant in the radial system when shunt capacitors are not present. The voltage rise effects are seen when the capacitor is present and during light load conditions. The voltage profile along the heavily loaded condition is shown

**TABLE 8.3**  Power Factor Correction Selection Table

| PF | \multicolumn{10}{c}{Desired Power Factor} |||||||||| |
|---|---|---|---|---|---|---|---|---|---|---|
|  | 0.80 | 0.81 | 0.82 | 0.83 | 0.84 | 0.85 | 0.86 | 0.87 | 0.88 | 0.89 |
| 0.55 | 0.768 | 0.794 | 0.820 | 0.846 | 0.873 | 0.899 | 0.925 | 0.952 | 0.979 | 1.006 |
| 0.56 | 0.729 | 0.755 | 0.781 | 0.807 | 0.834 | 0.860 | 0.886 | 0.913 | 0.940 | 0.967 |
| 0.57 | 0.691 | 0.717 | 0.743 | 0.769 | 0.796 | 0.822 | 0.848 | 0.875 | 0.902 | 0.929 |
| 0.58 | 0.655 | 0.681 | 0.707 | 0.733 | 0.759 | 0.785 | 0.811 | 0.838 | 0.865 | 0.892 |
| 0.59 | 0.618 | 0.644 | 0.670 | 0.696 | 0.723 | 0.749 | 0.775 | 0.802 | 0.829 | 0.856 |
| 0.60 | 0.583 | 0.609 | 0.635 | 0.661 | 0.687 | 0.714 | 0.740 | 0.767 | 0.794 | 0.821 |
| 0.61 | 0.549 | 0.575 | 0.601 | 0.627 | 0.653 | 0.679 | 0.706 | 0.732 | 0.759 | 0.787 |
| 0.62 | 0.515 | 0.541 | 0.567 | 0.593 | 0.620 | 0.646 | 0.672 | 0.699 | 0.726 | 0.753 |
| 0.63 | 0.483 | 0.509 | 0.535 | 0.561 | 0.587 | 0.613 | 0.639 | 0.666 | 0.693 | 0.720 |
| 0.64 | 0.451 | 0.477 | 0.503 | 0.529 | 0.555 | 0.581 | 0.607 | 0.634 | 0.661 | 0.688 |
| 0.65 | 0.419 | 0.445 | 0.471 | 0.497 | 0.523 | 0.549 | 0.576 | 0.602 | 0.629 | 0.657 |
| 0.66 | 0.388 | 0.414 | 0.440 | 0.466 | 0.492 | 0.519 | 0.545 | 0.572 | 0.599 | 0.626 |
| 0.67 | 0.358 | 0.384 | 0.410 | 0.436 | 0.462 | 0.488 | 0.515 | 0.541 | 0.568 | 0.596 |
| 0.68 | 0.328 | 0.354 | 0.380 | 0.406 | 0.432 | 0.459 | 0.485 | 0.512 | 0.539 | 0.566 |
| 0.69 | 0.299 | 0.325 | 0.351 | 0.377 | 0.403 | 0.429 | 0.456 | 0.482 | 0.509 | 0.537 |
| 0.70 | 0.270 | 0.296 | 0.322 | 0.348 | 0.374 | 0.400 | 0.427 | 0.453 | 0.480 | 0.508 |
| 0.71 | 0.242 | 0.268 | 0.294 | 0.320 | 0.346 | 0.372 | 0.398 | 0.425 | 0.452 | 0.480 |
| 0.72 | 0.214 | 0.240 | 0.266 | 0.292 | 0.318 | 0.344 | 0.370 | 0.397 | 0.424 | 0.452 |
| 0.73 | 0.186 | 0.212 | 0.238 | 0.264 | 0.290 | 0.316 | 0.343 | 0.370 | 0.396 | 0.424 |
| 0.74 | 0.159 | 0.185 | 0.211 | 0.237 | 0.263 | 0.289 | 0.316 | 0.342 | 0.369 | 0.397 |
| 0.75 | 0.132 | 0.158 | 0.184 | 0.210 | 0.236 | 0.262 | 0.289 | 0.315 | 0.342 | 0.370 |
| 0.76 | 0.105 | 0.131 | 0.157 | 0.183 | 0.209 | 0.235 | 0.262 | 0.288 | 0.315 | 0.343 |
| 0.77 | 0.079 | 0.105 | 0.131 | 0.157 | 0.183 | 0.209 | 0.235 | 0.262 | 0.289 | 0.316 |
| 0.78 | 0.052 | 0.078 | 0.104 | 0.130 | 0.156 | 0.183 | 0.209 | 0.236 | 0.263 | 0.290 |
| 0.79 | 0.026 | 0.052 | 0.078 | 0.104 | 0.130 | 0.156 | 0.183 | 0.209 | 0.236 | 0.264 |
| 0.80 |  | 0.026 | 0.052 | 0.078 | 0.104 | 0.130 | 0.157 | 0.183 | 0.210 | 0.238 |
| 0.81 |  |  | 0.026 | 0.052 | 0.078 | 0.104 | 0.131 | 0.157 | 0.184 | 0.212 |
| 0.82 |  |  |  | 0.026 | 0.052 | 0.078 | 0.105 | 0.131 | 0.158 | 0.186 |
| 0.83 |  |  |  |  | 0.026 | 0.052 | 0.079 | 0.105 | 0.132 | 0.160 |
| 0.84 |  |  |  |  |  | 0.026 | 0.053 | 0.079 | 0.106 | 0.134 |
| 0.85 |  |  |  |  |  |  | 0.026 | 0.053 | 0.080 | 0.107 |
| 0.86 |  |  |  |  |  |  |  | 0.027 | 0.054 | 0.081 |
| 0.87 |  |  |  |  |  |  |  |  | 0.027 | 0.054 |
| 0.88 |  |  |  |  |  |  |  |  |  | 0.027 |

*(Continued)*

**TABLE 8.3**   (CONTINUED) Power Factor Correction Selection Table

| PF | Desired Power Factor | | | | | | | | | |
|------|-------|-------|-------|-------|-------|-------|-------|-------|-------|-------|
|    | 0.90  | 0.91  | 0.92  | 0.93  | 0.94  | 0.95  | 0.96  | 0.97  | 0.98  | 0.99  |
| 0.55 | 1.034 | 1.063 | 1.092 | 1.123 | 1.156 | 1.190 | 1.227 | 1.268 | 1.315 | 1.376 |
| 0.56 | 0.995 | 1.024 | 1.053 | 1.084 | 1.116 | 1.151 | 1.188 | 1.229 | 1.276 | 1.337 |
| 0.57 | 0.957 | 0.986 | 1.015 | 1.046 | 1.079 | 1.113 | 1.150 | 1.191 | 1.238 | 1.299 |
| 0.58 | 0.920 | 0.949 | 0.979 | 1.009 | 1.042 | 1.076 | 1.113 | 1.154 | 1.201 | 1.262 |
| 0.59 | 0.884 | 0.913 | 0.942 | 0.973 | 1.006 | 1.040 | 1.077 | 1.118 | 1.165 | 1.226 |
| 0.60 | 0.849 | 0.878 | 0.907 | 0.938 | 0.970 | 1.005 | 1.042 | 1.083 | 1.130 | 1.191 |
| 0.61 | 0.815 | 0.843 | 0.873 | 0.904 | 0.936 | 0.970 | 1.007 | 1.048 | 1.096 | 1.157 |
| 0.62 | 0.781 | 0.810 | 0.839 | 0.870 | 0.903 | 0.937 | 0.974 | 1.015 | 1.062 | 1.123 |
| 0.63 | 0.748 | 0.777 | 0.807 | 0.837 | 0.870 | 0.904 | 0.941 | 0.982 | 1.030 | 1.090 |
| 0.64 | 0.716 | 0.745 | 0.775 | 0.805 | 0.838 | 0.872 | 0.909 | 0.950 | 0.998 | 1.058 |
| 0.65 | 0.685 | 0.714 | 0.743 | 0.774 | 0.806 | 0.840 | 0.877 | 0.919 | 0.966 | 1.027 |
| 0.66 | 0.654 | 0.683 | 0.712 | 0.743 | 0.755 | 0.810 | 0.847 | 0.888 | 0.935 | 0.996 |
| 0.67 | 0.624 | 0.652 | 0.682 | 0.713 | 0.745 | 0.779 | 0.816 | .857  | 0.905 | 0.966 |
| 0.68 | 0.594 | 0.623 | 0.652 | 0.683 | 0.715 | 0.750 | 0.787 | 0.828 | 0.875 | 0.936 |
| 0.69 | 0.565 | 0.593 | 0.623 | 0.654 | 0.686 | 0.720 | 0.757 | 0.798 | 0.846 | 0.907 |
| 0.70 | 0.536 | 0.565 | 0.594 | 0.625 | 0.657 | 0.692 | 0.729 | 0.770 | 0.817 | 0.878 |
| 0.71 | 0.508 | 0.836 | 0.566 | 0.597 | 0.629 | 0.663 | 0.700 | 0.741 | 0.789 | 0.849 |
| 0.72 | 0.480 | 0.508 | 0.538 | 0.569 | 0.601 | 0.635 | 0.672 | 0.713 | 0.761 | 0.821 |
| 0.73 | 0.452 | 0.481 | 0.510 | 0.541 | 0.573 | 0.608 | 0.645 | 0.686 | 0.733 | 0.794 |
| 0.74 | 0.425 | 0.453 | 0.483 | 0.514 | 0.546 | 0.580 | 0.617 | 0.658 | 0.706 | 0.766 |
| 0.75 | 0.398 | 0.426 | 0.456 | 0.487 | 0.519 | 0.553 | 0.590 | 0.631 | 0.679 | 0.739 |
| 0.76 | 0.371 | 0.400 | 0.429 | 0.460 | 0.492 | 0.526 | 0.563 | 0.605 | 0.652 | 0.713 |
| 0.77 | 0.344 | 0.373 | 0.403 | 0.433 | 0.466 | 0.500 | 0.537 | 0.578 | 0.626 | 0.686 |
| 0.78 | 0.318 | 0.347 | 0.376 | 0.407 | 0.439 | 0.474 | 0.511 | 0.552 | 0.599 | 0.660 |
| 0.79 | 0.292 | 0.320 | 0.350 | 0.381 | 0.413 | 0.447 | 0.484 | 0.525 | 0.573 | 0.634 |
| 0.80 | 0.266 | 0.294 | 0.324 | 0.355 | 0.387 | 0.421 | 0.458 | 0.499 | 0.547 | 0.608 |
| 0.81 | 0.240 | 0.268 | 0.298 | 0.329 | 0.361 | 0.395 | 0.432 | 0.473 | 0.521 | 0.581 |
| 0.82 | 0.214 | 0.242 | 0.272 | 0.303 | 0.335 | 0.369 | 0.406 | 0.447 | 0.495 | 0.556 |
| 0.83 | 0.188 | 0.216 | 0.246 | 0.277 | 0.309 | 0.343 | 0.380 | 0.421 | 0.469 | 0.530 |
| 0.84 | 0.162 | 0.190 | 0.220 | 0.251 | 0.283 | 0.317 | 0.354 | 0.395 | 0.443 | 0.503 |
| 0.85 | 0.135 | 0.164 | 0.194 | 0.225 | 0.257 | 0.291 | 0.328 | 0.369 | 0.417 | 0.477 |
| 0.86 | 0.109 | 0.138 | 0.167 | 0.198 | 0.230 | 0.265 | 0.302 | 0.343 | 0.390 | 0.451 |
| 0.87 | 0.082 | 0.111 | 0.141 | 0.172 | 0.204 | 0.238 | 0.275 | 0.316 | 0.364 | 0.424 |
| 0.88 | 0.055 | 0.084 | 0.114 | 0.145 | 0.177 | 0.211 | 0.248 | 0.289 | 0.337 | 0.397 |
| 0.89 | 0.028 | 0.057 | 0.086 | 0.117 | 0.149 | 0.184 | 0.221 | 0.262 | 0.309 | 0.370 |
| 0.90 |       | 0.029 | 0.058 | 0.089 | 0.121 | 0.156 | 0.193 | 0.234 | 0.281 | 0.342 |
| 0.91 |       |       | 0.030 | 0.060 | 0.093 | 0.127 | 0.164 | 0.205 | 0.253 | 0.313 |

*(Continued)*

TABLE 8.3 (CONTINUED) Power Factor Correction Selection Table

| | Desired Power Factor | | | | | | | | |
|---|---|---|---|---|---|---|---|---|---|
| PF | 0.90 0.91 | 0.92 | 0.93 | 0.94 | 0.95 | 0.96 | 0.97 | 0.98 | 0.99 |
| 0.92 | | 0.000 | 0.031 | 0.063 | 0.097 | 0.134 | 0.175 | 0.223 | 0.284 |
| 0.93 | | | 0.000 | 0.032 | 0.067 | 0.104 | 0.145 | 0.192 | 0.253 |
| 0.94 | | | | 0.000 | 0.034 | 0.071 | 0.112 | 0.160 | 0.220 |
| 0.95 | | | | | 0.000 | 0.037 | 0.078 | 0.126 | 0.186 |
| 0.96 | | | | | | | 0.041 | 0.089 | 0.149 |
| 0.97 | | | | | | | | 0.048 | 0.108 |
| 0.98 | | | | | | | | | 0.061 |
| 0.99 | | | | | | | | | |

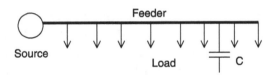

**Figure 8.3** Example radial system.

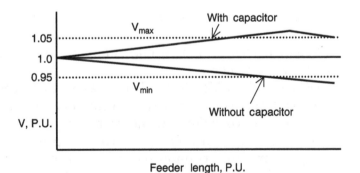

**Figure 8.4** Voltage profile along the feeder during light load.

in Figure 8.5. The voltage profile is within allowed limits with the shunt capacitors. Therefore, there is always a need to find the optimum location for the installation of fixed shunt capacitors.

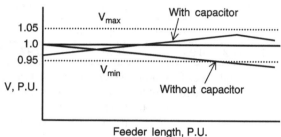

**Figure 8.5** Voltage profile along the feeder during heavy load.

### 8.2.3 Switched Capacitors

Switched capacitors provide additional flexibility to control system voltage, power factor, and losses. Switched capacitors are usually applied with some type of automatic switching control. A sensor detects a particular condition and then initiates a close or trip signal to the circuit breaker connected to the capacitor bank. A typical automatic capacitor control includes the following:

- Voltage: Control of the voltage regulation is a major consideration.
- Current: If current magnitude is directly proportional to the VAR demand.
- VAR control: VAR demand increases with certain loads and decreases when the specific load is off.
- Time switch: To switch on the capacitors during peak hours and to switch them off during off-peak hours.
- Temperature: In certain loads such as air conditioners, the VAR demand goes up when temperature increases.

The fixed capacitor banks are usually left energized on a continuous basis. In certain loading conditions, selected capacitor banks can be switched on and off based on seasonal conditions. Remote switching of capacitor banks is used in some areas. The capacitor switching is generally performed using radio signals, a power line carrier, or through telephone signals.

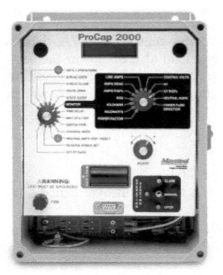

**Figure 8.6** A typical automatic capacitor controller. (Courtesy of Maysteel LLC, Menomonee Falls, WI.)

A typical controller for switching the capacitors automatically is shown in Figure 8.6. Such a controller is suitable for pole- and pad-mounted capacitor stations. Switching is accomplished by using VAR control with voltage override. This type of controller provides precise and economical control of capacitor banks as well as dependable operation over years of service. Such controllers have reliable data recorder capability because they feature high-level graphics and Windows software. Their intuitive front panel design and LCD display make reading from any angle easy. Selection of manual or automatic settings can be made with the turn of a dial.

### 8.2.4   Released System Capacity

The power factor correction reduces the kVA demand through the line, cable, transformer, or generator. This means capacitors can be used to reduce the equipment overloading of existing facilities. The released capacity due to power factor correction is shown by means of an example.

**Example 8.3**

A 4.16 kV/440 V, three-phase, 300 kVA, delta/wye transformer is used to supply a load at 0.65 power factor. The transformer is overloaded by 10%. Analyze this condition and suggest a suitable mitigation approach through power factor correction.

Solution

kVA = 300
Power factor = 0.65
Delivered kVA load when overloaded $(300 \times 1.1) =$
330 kVA
$\cos \theta = 0.65$,     $\sin \theta = 0.7599$
Before shunt compensation, kW $(330 \text{ kVA} \times 0.65) =$
214.5
kVAR $(330 \text{ kVA} \times 0.7599) = 250.8$
After power factor correction, kVA $= 300$, kW $= 214.5$

$$kVAR = \sqrt{300^2 - 214.5^2} = 210 \text{ kVAR}$$

Select a 60 kVAR, three-phase shunt capacitor bank.

$$\text{New kVA} = \sqrt{214.5^2 + (250.8 - 60)^2} = 287 \text{ kVA}$$

$$\text{New power factor} = \frac{214.5}{287} = 0.75$$

By using a 60 kVAR shunt capacitor on the 4.16 kV side of the transformer, the power factor can be improved from 0.65 to 0.75. The load on the transformer will be 287 kVA and hence the load on the transformer remains within the nominal rating of 300 kVA.

## 8.3  FIXED AND SWITCHED CAPACITOR APPLICATIONS

To improve the voltage profile, fixed capacitors are used in high voltage circuits at a point two thirds of the distance from the source. The shunt capacitor size will be determined by the voltage rise during light load conditions. If too much shunt compensation is connected, the losses in the system will increase during light load conditions.

After adding the fixed capacitors according to the above guideline, switched capacitors are added to about two thirds of the peak load reactive power requirements. The switched capacitors should be added to a location from a voltage correction point of view, but generally in the last one third of the circuit, close to the load. The switching can be performed using remote control mechanisms.

When shunt capacitors are used for improving the voltage profile, the fixed capacitors should not increase the voltage at light load above the allowed values. Additional capacitors can be switched to improve the power factor to 0.95 at rated load. The following examples illustrates the use of fixed and switched shunt capacitors.

## Example 8.4

The maximum load of a high voltage circuit is 5,000 kVA. The initial power factor is 0.85. The minimal load in the circuit is 25% of the maximum load, with a power factor of 0.80. The minimum kVAR demand is 0.6 of the maximum reactive power. The allowable power factor at light load is 0.95. Select the required fixed and switched capacitors.

Solution

      Maximum load $= 5{,}000$ kVA
      $\cos \theta = 0.85$,     $\sin \theta = 0.527$
      Maximum kW $(5{,}000 \text{ kVA} \times 0.85) = 4{,}250$ kW
      Maximum kVAR $(5{,}000 \text{ kVA} \times 0.527) = 2{,}635$ kVAR
      Minimum load $(5{,}000 \text{ kVA} \times 0.25) = 1{,}250$ kVA
      $\cos \theta = 0.80$     $\sin \theta = 0.60$
      Minimum kW $(1{,}250 \text{ kVA} \times 0.80) = 1{,}000$ kW
      Minimum kVAR $(1{,}250 \text{ kVA} \times 0.60) = 750$ kVAR

Selection of fixed capacitor: Select a 600 kVAR, fixed shunt capacitor.

$$\text{KVA with fixed capacitor} = \sqrt{1{,}000^2 + 150^2} = 1{,}011\,\text{kVA}$$

Power factor during light load $(1{,}000/1{,}011) = 0.989$
Allowable power factor $= 0.95$
Switched capacitor to meet the peak demand $(2{,}635 - 600) = 2{,}035\,\text{kVAR}$

Select a 1,800 kVAR switched capacitor.

$$\text{KVA with fixed and switched capacitor} = \sqrt{4{,}250^2 + 235^2}$$
$$= 4{,}257\,\text{kVA}$$

Power factor with fixed and switched capacitors $(4{,}250/4{,}257) = 0.998$

A smaller fixed and switched capacitor bank can meet the power factor correction requirement to achieve a power factor of 0.95. To minimize the losses, the fixed capacitor can be located at a point two thirds on the line from the source. The switched capacitor can be located at one third of the distance from the load.

## PROBLEMS

8.1   Why is the power factor of many industrial loads poor?

8.2   When would you recommend a fixed capacitor bank for power factor correction? When is a switched capacitor considered for power factor correction?

8.3   What are the different methods used to switch power factor capacitors remotely?

8.4   The power factor of a 200 kW induction motor load is 0.87. It is desired to improve the power factor to 0.95. What is the required kVAR of the shunt capacitor bank? What types of technical issues are expected when applying power factor correction to an induction motor?

8.5 A 4.16 kV overhead line is rated at 150 A but is carrying a load of 200 A at 0.67 power factor. Is it possible to reduce the load current to the rated value using power factor improvement? What is the value of reactive power to be supplied through shunt capacitors? What is the new power factor?

8.6 A load consists of a 50 hp, three-phase induction motor operating at 0.75 lagging power factor. There is an auxiliary load of $(15 + j\,0.0)$ kVA. What is the power factor of the combined load? If the power factor has to be improved to 0.95, what is the value of capacitor kVAR required to perform this function?

8.7 What will happen if a large capacitor bank is installed at the end of a long feeder in order to correct the power factor during peak load?

# REFERENCES

1. Hammond, P. W. (1988). A Harmonic Filter Installation to Reduce Voltage Distortion from Solid State Converters, *IEEE Transactions on Industry Applications*, 24(1), 53–58.

2. Phillips, K. J. (1994). Conducting a power factor study, *Consulting–Specifying Engineer*, 10, 54–58.

3. Porter, G. A., McCall, J. C. (1990). Application and Protection Considerations in Applying Distribution Capacitors, Pennsylvania Electrical Association.

# 9

## SYSTEM BENEFITS

### 9.1  INTRODUCTION

Using shunt capacitors to supply the leading currents required by the load relieves the generator from supplying that part of the inductive current. The system benefits due to the application of shunt capacitors include [1,2]:

- Reactive power support.
- Voltage profile improvements.
- Line and transformer loss reductions.
- Release of power system capacity.
- Savings due to increased energy loss.

These benefits apply for both distribution and transmission systems.

### 9.2  REACTIVE POWER SUPPORT

In distribution systems, the voltage at the load end tends to get lower due to the lack of reactive power. In such cases, local VAR support is offered using shunt capacitors. In the case of long transmission lines, the reactive power available at the end of the line during peak load conditions is small and hence needs

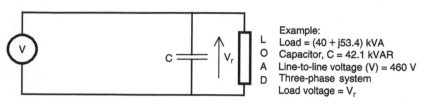

**Figure 9.1**   Per phase diagram of three-phase radial system.

**TABLE 9.1**   Example Load Flows in a Three-Phase Radial System

| Description | Without Shunt Capacitor | With Shunt Capacitor |
|---|---|---|
| Power, kW | 40 | 40 |
| Reactive power, kVAR | 53.4 | 13.1 |
| kVA | 66.7 | 42.1 |
| Power factor | 0.6 | 0.95 |
| Line current, A | 83.7 | 52.8 |

to be supplied using shunt capacitors. The advantage of providing local reactive power can be demonstrated by an example. Consider a 460 V, three-phase radial system without and with shunt capacitors as shown in Figure 9.1. The load at the end of the radial feeder is (40 + j 53.4) kVA. It can be seen that the load requires significant reactive power, which can be supplied using shunt capacitors as illustrated in Figure 9.1. Let the reactive power supplied by the shunt capacitors be 42.1 kVAR. Now compare the load flows from the source without and with shunt capacitors as listed in Table 9.1. From the table it can be seen that the reactive power drawn from the supply is substantially less, and the kVA and the current flows are less. The power factor at the load is improved.

## 9.3   VOLTAGE PROFILE IMPROVEMENTS

The shunt capacitors reduce the amount of inductive current in an electric circuit. The reduction in the line current

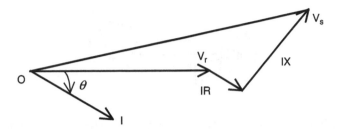

**Figure 9.2** Phasor diagram of the system without shunt capacitor.

decreases the *IR* and *IX* voltage drops, thereby improving the voltage level of the system from the capacitor location back to the source. In both the distribution and transmission systems, there is a need to maintain a voltage in the range 0.95–1.05 P.U. A lower system voltage will cause induction motors to operate with a larger than nominal current. With lower voltages, the recovery voltages after fault clearing will be slow. Therefore, maintaining acceptable voltage levels in the power system is an important objective. A one-line diagram of a power system for the voltage drop analysis is shown in Figure 9.1. The phasor diagram of the system without shunt capacitors is shown in Figure 9.2. The corresponding voltage relations are:

$$V_R = V_S - I \ (\cos \theta \pm j \ \sin \theta) \ (R + jX) \qquad (9.1)$$

where  $V_S$ = Sending end voltage/phase
$V_R$ = Receiving end voltage/phase
$I$ = Current, A
$R$ = Resistance, Ω/phase
$X$ = Reactance, Ω/phase
$\theta$ = Power factor angle, degrees

The phasor diagram of the system with shunt capacitors is shown in Figure 9.3. The corresponding current relations are:

$$I' = I(\cos \theta \pm j \ \sin \theta) - jI_C \qquad (9.2)$$

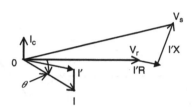

**Figure 9.3**  Phasor diagram of the system with shunt capacitor.

where the capacitor current is

$$I_C = \frac{V}{X_C} \tag{9.3}$$

The improved voltage profile at the load is due to the decrease in the line current and reduced voltage drop. This is illustrated by the following example.

**Example 9.1**

Consider a one-line diagram of the system as shown in Figure 9.1. The three-phase load is 40 kW, 460 V, and 0.6 power factor lagging. The line impedance is $(0.1 + j\ 0.3)\ \Omega/$phase. If the power factor is improved to 0.95 (lag) from the existing level using shunt capacitors, what is the reduction in the voltage drop?

Solution

Before power factor correction

$$I = \frac{40,000}{\sqrt{3} \times 460 \times 0.6} = 83.7\,\text{A/phase}$$

$I = 83.7\ (0.6 - j\,0.8)\,\text{A/phase}$
$Z = (0.1 + j\ 0.3)\,\Omega/\text{Phase}$
Voltage drop, $V_1 = 83.7\ (0.6 - j\ 0.8)\ (0.1 + j\ 0.3)$
$\qquad\qquad\qquad = (25.1 + j\ 8.4)$

After power factor correction

$$I = \frac{40,000}{\sqrt{3} \times 460 \times 0.95} = 52.9\,\text{A/phase}$$

$I = 52.9 \ (0.95 - j \ 0.31) \ \text{A}$

Voltage drop, $V_2 = 52.9 \ (0.95 - j \ 0.31) \ (0.1 + j \ 0.3) =$
$(10 + j \ 13.4)$

Improvement in the voltage drop ($V_1 - V_2$):

Reduction in the voltage drop $= 15.1 - j \ 5.0$

There is a 15.2 V reduction in the resistive voltage drop, which is responsible for the $I^2R$ loss. Overall reduction in the voltage drop is 16.0 V.

## 9.3.1 Voltage Rise due to the Addition of Shunt Capacitors

The addition of a shunt capacitor bank raises the voltage at the point of installation. The voltage drop equations without shunt capacitors ($VD_1$) and with shunt capacitors ($VD_2$) are:

$$VD_1 = \frac{kVA_1}{(10)(kV)^2} (R \cos \theta_1 - jX \sin \theta_1) \qquad (9.4)$$

$$VD_2 = \frac{kVA_2}{(10)(kV)^2} (R \cos \theta_2 - jX \sin \theta_2) \qquad (9.5)$$

$$VD_1 - VD_2 = \frac{1}{(10)(kV)^2} \{[R(kVA_1 \cos \theta_1 - kVA_2 \cos \theta_2)]$$
$$+ jX[(kVA_1 \sin \theta_1 - kVA_2 \sin \theta_2)]\} \qquad (9.6)$$

$$VD_1 - VD_2 = \frac{kVA_C}{(10)(kV)^2} X \qquad (9.7)$$

where ($kVA_1 \cos \theta_1 - kVA_2 \cos \theta_2$) is the change in the real power, which is equal to zero. The other component, ($kVA_1 \sin \theta_1 - kVA_2 \sin \theta_2$), is the change in the reactive power due to the addition of reactive power supplied through shunt capacitors.

### 9.3.2 Voltage Rise at the Transformer

Every transformer on the power system from the location of the capacitor bank to the generator will experience a voltage rise. This is an important component of the voltage rise due to the shunt capacitor. Since the transformer impedance is always a significant value, the voltage drop due to the transformer reactance in the presence of the shunt capacitor is given by:

$$VD_1 - VD_2 = \frac{kVA_C}{(10)(kV)^2} X_t \qquad (9.8)$$

$$kV\,(\text{base}) = (\text{MVA base})(X_b) \qquad (9.9)$$

$$VD_1 - VD_2 = \frac{kVAR_C}{(10)(\text{MVA base})(X_b)} = \frac{kVAR_C}{10(MVA_b)} X_t \frac{1}{X_b} \qquad (9.10)$$

$$\text{Voltage at the transformer location} = \frac{kVAR_C(X_t)}{(10)(kVA_t)} \qquad (9.11)$$

where $X_t$ is the transformer reactance per unit and $kVA_t$ is the base kVA of the transformer. If the voltage rise is excessive, then tap changers can be used to keep the voltage within acceptable levels.

### Example 9.2

The rating of a three-phase transformer is 500 kVA, 7% impedance, and delta/wye connected. A 100 kVAR capacitor bank is used on the high voltage side of the transformer. Calculate the voltage rise due to the installation of the capacitor bank.

Solution

Transformer $kVA = 500$, $X_t = 0.07$ P.U., $kVAR_C = 100$

$$\text{Voltage rise} = \frac{kVAR_C(X_t)}{kVA_t} \times 100 = \frac{100 \times 0.07}{500} \times 100 = 1.4\%$$

## Example 9.3

Consider a three-phase, 345 kV, 60 Hz system with a shunt capacitor bank of 135 MVAR. The short-circuit duty at the three-phase source is 2,000 A. Calculate the voltage rise due to the installation of the capacitor bank.

Solution

$$kV = 345 \text{ kV}, \quad MVAR_C = 135$$

$$X_{SC} = \frac{345 \text{ kV}}{\sqrt{3} \times 2 \text{ kA}} = 100 \, \Omega$$

$$Z \text{ (base)} = \frac{(345 \text{ kV})^2}{100 \text{ MVA}} = 1,190 \, \Omega$$

$$X_{SC} \text{ (P.U.)} = \frac{100 \, \Omega}{1,190 \, \Omega} = 0.084 \, \text{P.U.}$$

$$X_C = \frac{kV^2}{MVA} = \frac{(345 \text{ kV})^2}{135 \text{ MVA}} = 881 \, \Omega$$

$$X_C \text{ (P.U.)} = 881 \, \Omega / 1190 \, \Omega = 0.74 \text{ P.U.}$$

$$V \text{ at the transformer} = V_S \left[ 1 + \frac{X_{SC}}{(X_C - X_{SC})} \right]$$

$$= 1.0 \left[ 1 + \frac{0.084}{0.74 - 0.084} \right] = 1.128 \text{ P.U.}$$

## 9.4 LINE AND TRANSFORMER LOSS REDUCTIONS

When shunt capacitors are installed for power factor correction, the line current magnitude is decreased. Therefore, both $I^2R$ and $I^2X$ losses are reduced. In industrial power systems, the $I^2R$ losses vary from 3–8% of the rated load current depending on the hours of full load operation, conductor size, length of the feeder circuit, and the transformer impedance. The load on most electric circuits will vary depending on the time of the day. In order to account for the average load, the load factor is defined as:

$$\text{Load factor (LDF)} = \frac{\text{Average kW demand}}{\text{Peak kW demand}} \qquad (9.12)$$

The current also varies with time. The loss factor is defined in order to estimate the loss in the system. The loss factor is a function of the load factor and is given by:

$$\text{Loss factor (LF)} = 0.15\,\text{LDF} + 0.85\,\text{LDF}^2 \qquad (9.13)$$

$$\text{Average loss} = 3\left(I^2 R\right)\,(\text{Loss factor}) \qquad (9.14)$$

## Example 9.4

Consider a 4.16 kV line with a total impedance of $(1.5 + j\,7.0)\ \Omega$. The peak load on this line is 5 MW at 0.8 power factor. The load factor is 0.6. The customer wants to improve the power factor to 0.95. What will be the annual savings with the improved power factor? How much capacitive kVARs are to be installed in order to achieve this objective?

## Solution

Without capacitor bank:

    kV = 4.16,   PF = 0.6,   MW = 5
    MVA (5/0.6) = 6.25

$$I = \frac{5{,}000\,\text{kW}}{\sqrt{3} \times 4.16\,\text{kV} \times 0.8} = 857.4\,\text{A/phase}$$

Loss factor $= (0.15)(0.6) + (0.85)(0.6)^2 = 0.396$
Average loss $= (857.4^2 \times 1.5 \times 0.396)\,10^{-3} = 436.7\,\text{kW/phase}$

With capacitor bank:

The new power factor $= 0.95$

$$I = \frac{5,000\,\text{kW}}{\sqrt{3} \times 4.16\,\text{kV} \times 0.95} = 730.5\,\text{A/phase}$$

Average loss $(730.5^2 \times 1.5 \times 0.396 \times 10^{-3}) = 316.9\,\text{kW/phase}$

Total savings $[3(436.7 - 316.9)] = 359.4\,\text{kW}$

### 9.4.1 Savings due to Reduced Energy Losses

If the reactive compensation is provided in a feeder circuit, then the current through the feeder and the transformer circuit is reduced. If $I_1$ and $I_2$ are the currents through the feeder before and after compensation and if $R$ is the resistance of the circuit, then the cost of energy savings $(C_L)$ due to reduced losses per year is given by [1]:

$$C_L = 3\,(C_W)\,(I_1^2 - I_2^2)\,R\,(8,760)\,(\text{LF})\,10^{-3} \qquad (9.15)$$

where LF is the load factor, 8,760 is the number of hours in a year, and $C_W$ is the cost of the electric energy per kWh.

### Example 9.5

Consider a power system with shunt capacitors. The phase currents in the line without and with shunt capacitors are 80 A and 50 A, respectively. The feeder resistance is $0.2\,\Omega$. The cost of electrical energy is \$0.06 per kWh. The loss factor of the system is 0.6. Calculate the savings per year due to the reduced current as a result of power factor correction.

Solution

Current before power factor correction $(I_1) = 80\,\text{A}$
Current after power factor correction $(I_2) = 50\,\text{A}$
Feeder resistance/phase $= 0.2\,\Omega$
Loss factor $(\text{LF}) = 0.6$
Cost of electrical energy/kWh $= \$0.06$
Savings due to reduced energy savings $= S$

$$S = 3 \times 0.60\,(80^2 - 50^2)\,(0.2)\,(8,760)\,(0.6)\,(10^{-3}) = \$738$$

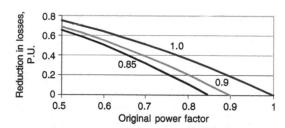

**Figure 9.4**   Reduction in losses due to improved power factor.


### 9.4.2   Loss Reduction due to Improved Power Factor

In industrial plants, the $I^2R$ losses vary from 2–7% of the load kWh depending on the duration of the rated load and the light load. Further, the losses are proportional to the circuit resistance and feeder length. When the shunt capacitors are installed, the line current is reduced. The relation can be expressed as:

$$\text{kW losses} \propto \left( \frac{\text{Original PF}}{\text{Improved PF}} \right)^2 \qquad (9.16)$$

$$\text{Loss reduction} = 1 - \left( \frac{\text{Original PF}}{\text{Improved PF}} \right)^2 \qquad (9.17)$$

This relation plotted in Figure 9.4 is based on the assumption that the kW load remains the same.

### Example 9.6

The power is supplied through a substation for a group of loads and the electric consumption per year is 120,000 kWh. The power factor at the substation is 0.72. It is decided to improve the power factor to 0.87 using shunt capacitors. Assuming the losses are 5% of the total kWh, what will be the annual savings in kWh if the power factor is improved?

Solution

> Total kWh/year $= 120{,}000$
> Loss per year $(120{,}000 \times 0.05) = 6{,}000$ kWh
> Original power factor $= 0.72$
> Improved power factor $= 0.87$
> Loss reduction, $1 - (0.72/0.87)^2 = 0.315$
> Reduction in the loss $(6{,}000 \text{ kWh} \times 0.315) = 1{,}890$ kWh

## 9.5 RELEASE OF POWER SYSTEM CAPACITY

Power factor correction capacitors provide the reactive current requirements locally and reduce the line current. Reduced line current means less kVA for transformers and feeder circuits. Thus, the shunt capacitor compensation helps to reduce the thermal overloads on transformers, transmission lines, generator, and cables.

### Example 9.7

A 4.16 kV, three-phase cable circuit is rated for 200 A, but it is carrying a load current of 250 A at a power factor of 0.6 lagging. How much reactive power compensation can reduce the load current to rated value?

Solution

> $kV = 4.16, \quad I = 250$ A
> Existing load demand $(1.732 \times 4.16 \times 250) = 1{,}801$ kVA
> Rated power of the cable $(1{,}801 \text{ kVA} \times 0.6) = 1{,}081$ kW
> Rated kVA of the cable $(1.732 \times 4.16 \times 200) = 1{,}441$ kVA
> Power factor at rated load $(1081/1{,}441) = 0.75$
> Capacitive kVAR required $(kVAR_C) = kW(\tan \theta_1 - \tan \theta_2)$
> $kVAR_C \ [1{,}081(\tan 53.1 - \tan 41.4)] = 488.7$

Use 500 kVAR, three-phase capacitors, with the nearest rating. With this power factor correction, the kVA demand on the cable circuit will be less than the allowable rating.

### 9.5.1 Release of Generator Capacity

The synchronous generator has kW limit as well as kVA limit. The kVA limit of the generator may correspond to unity power factor operation. If the load is a low power factor apparatus, then the generator has to deliver a kW at a lower power factor. But the generator output cannot exceed the nominal kVA rating. The power factor correction at the generator terminals can release the kVA capability.

**Example 9.8**

A 500 kW, 0.8 power factor, 625 kVA, three-phase generator is operating at the rated load. An emergency load of 100 kW at a power factor of 0.9 is to be added. Calculate the kVAR of the shunt capacitors to be connected at the terminals in order to keep the generator output within the rated value?

Solution

|                        | kW  | kVAR | kVA | Power Factor Angle |
|------------------------|-----|------|-----|--------------------|
| Original load degree   | 500 | 375  | 625 | 38.6               |
| Additional load degree | 100 | 48   | 111 | 25.8               |
| Total load degree      | 600 | 423  | 734 | 35.2               |

Allowable power factor $(600/625) = 0.96$
Allowable kVAR for this condition $(625 \times \sin 16.3) =$
  175 kVAR
Capacitive kVAR required $(423 - 175) = 248$ kVAR

**9.6   CONCLUSIONS**

In this chapter, the benefits of power factor correction were identified. The local reactive power support, voltage profile improvements, line and transformer loss reductions, release of power system capacity, and increased power flow capability can be achieved by the use of shunt capacitors. Specific examples were presented to illustrate the advantages of power factor correction.

## PROBLEMS

9.1. What are the benefits of power factor correction in a distribution system?

9.2. In Example 9.2, if the shunt capacitor rating is 270 KVAR, estimate the voltage rise due to the power factor correction.

9.3. A 13.8 kV cable circuit is rated for 150 A, but it is carrying a load current of 180 A at a power factor of 0.67 lagging. How much reactive power compensation can reduce the load current to rated value? What is the location of the shunt capacitors?

9.4. In a three-phase, 480 V circuit, the total impedance is $(0.5 + j\ 4.0)\ \Omega$ per phase. The peak load on this line is 4.5 kW at 0.72 power factor. The load factor is 0.62. The customer wants to improve the power factor to 0.95. What will be the annual savings with the improved power factor? How much capacitive kVARs are to be installed in order to achieve this objective?

9.5. The power consumption per year at a group load is 180,000 kWh. The power factor at the substation is 0.77. It is decided to improve the power factor to 0.95 using shunt capacitors. Assume the losses are 5.5% of the total kWh. What will be the annual savings in kWh if the power factor is improved?

## REFERENCES

1. Natarajan, R., Venkata, S. S., El-Sharkawi, M. A., Butler, N. G. (1987). Economic feasibility analysis of intermediate-size wind electric energy conversion systems employing induction generators, *International Journal of Energy Systems*, 7(2), 90–94.

2. Porter, G. A., McCall, J. G. (1990). Application and Protection Considerations in Applying Distribution Capacitors, Pennsylvania Electrical Association, 1990.

# 10

# SERIES CAPACITORS

## 10.1  INTRODUCTION

Shunt capacitor units are connected from phase to neutral or across the load. Series capacitors are connected in series in the circuit and hence carry the full line current. Therefore, the voltage across the shunt capacitor remains constant, and the drop across the series bank changes with load. It is this characteristic that produces an effect that is dependent on load and makes the series capacitor a valuable item in transmission and distribution line compensation.

The series capacitor is a negative reactance in series with a transmission line. The voltage rise across the capacitor is a function of circuit current and acts like a voltage regulator. The negative voltage drop across the series capacitor opposes the voltage drop due to the inductive reactance. Such effects are very valuable in radial feeders to reduce voltage drop and flicker effects. In the tie lines, the power transfer capability is significantly increased if the series compensation is applied.

A typical shunt-compensated circuit and the corresponding phasor diagrams are shown in Figure 10.1. The out-of-phase current component provides the reactive compensation in this arrangement. An example series-compensated circuit is

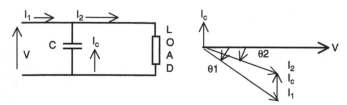

**Figure 10.1** Shunt compensation.

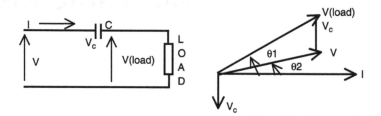

**Figure 10.2** Series compensation.

shown in Figure 10.2. In the series compensation circuit, the power factor improvement is achieved by using an out-of-phase component of the voltage. In practical applications, the reactive power provided by the series capacitor may be too low to improve the power factor.

## 10.2  SERIES CAPACITORS ON RADIAL FEEDERS

Consider a radial feeder shown in Figure 10.3 and the corresponding phasor diagram. The feeder impedance is given by:

$$Z = (R + j\,X_{\mathrm{L}}) \qquad\qquad (10.1)$$

The approximate voltage drop (VD) in this circuit is given by:

$$\mathrm{VD} = IR\cos\theta + X_{\mathrm{L}}\sin\theta \qquad\qquad (10.2)$$

where  $I$ = Total current, A
        $R$ = Resistance, $\Omega$

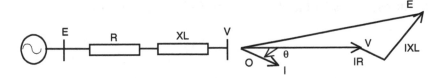

**Figure 10.3**  Radial feeder circuit.

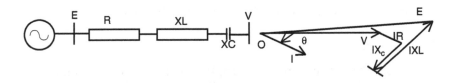

**Figure 10.4**  Radial feeder circuit with series capacitor.

$X_\mathrm{L}$ = Reactance, $\Omega$
$\theta$ = Power factor angle

Now consider the line with series compensation using a capacitive reactance ($X_\mathrm{C}$) as shown in Figure 10.4. The impedance ($Z'$) of the circuit is given by:

$$Z' = R + j(X_\mathrm{L} - X_\mathrm{C}) \qquad (10.3)$$

The corresponding approximate voltage drop (VD')is given by:

$$\mathrm{VD}' = IR\cos\theta + I\,(X_\mathrm{L} - X_\mathrm{C})\sin\theta \qquad (10.4)$$

If $X_\mathrm{C} = X_\mathrm{L}$, then the voltage drop is $IR\cos\theta$. In practical applications, $X_\mathrm{C}$ is chosen to be smaller than $X_\mathrm{L}$ in order to avoid overcompensation. Using kVA of the sending and receiving ends, the voltage relation can be calculated as:

$$\frac{\text{kVA (Receiving end)}}{\text{kVA (Sending end)}} = \frac{V_\mathrm{r}}{V_\mathrm{s}} = \frac{\sqrt{(P_\mathrm{r}^2 + Q_\mathrm{r}^2)}}{\sqrt{P_\mathrm{r}^2 + (Q_\mathrm{r} - Q_\mathrm{c})^2}} \qquad (10.5)$$

where $Q_\mathrm{c}$ is the reactive power supplied by the series capacitor bank. The reactive power supplied by the series capacitor bank can be calculated using the following relation:

$$Q_c = |I_C|^2 X_C \tag{10.6}$$

where $|I_C|$ is the magnitude of the current through the capacitance and $X_C$ is the reactance of the series capacitor. The total reactive power supplied by a three-phase series capacitor bank is obtained by multiplying Equation (10.6) by 3.

### Example 10.1

The total three-phase power supplied at the end of a radial line is 1,000 kVA at 0.6 power factor. The supply voltage is 13.8 kV at 60 Hz. The line impedance is (0.03 + j 7.2) Ω/phase. A series capacitor with a reactance of 6 Ω/phase was included at the end of the line to improve the voltage profile. Calculate the improvement in the voltage profile. What is the reactive power rating of the series capacitor bank?

Solution

> $V$ (line-to-line) = 13.8 kV, $V$ (phase) = 7.967 kV
> $I$ (load), [1,000 kVA/(1.732 × 13.8 kV × 0.6)] = 69.74 A
> Voltage drop, [(69.74(0.6 − j0.8)(0.03 + j 7.2)] =
>     (402.9 − j 299.6) V/phase
> Voltage at the load, [7,967 − (402.9 − j 288.6)] =
>     (7,564.1 − j 299.6) = 7,570.7 V/phase
> Voltage regulation, (7,967 − 7,570.7) × 100/7,967 = 4.98%
> With the − j6 Ω/phase capacitive reactance at the end
>     of the line:
> Line impedance (0.03 + j 7.2 − j 6.0) = (0.03 + j 1.2) Ω/phase
> Voltage drop, IZ [(69.74(0.6 − j 0.8)(0.03 + j 1.2)] =
>     (69.78 + j 68.2) V/phase
> Voltage at the load, [7,967 − (69.8 + j 68.2)] =
>     (7,897.2 − j 68.2) = 7,899.6 V/phase
> Voltage regulation (7,967 − 7,899.6) × 100/7,967 = 0.85%

There is a significant improvement in the voltage profile due to the series compensation using a − j 6.0 Ω/phase capacitive reactance.

$Q$ of the series capacitor, $(3 \times I^2 \times X_C) = 3 \times (69.74)^2$ (6)/ $1000 = 87.5$ kVAR

## 10.2.1 Series Capacitor on Distribution Systems

The starting of a motor draws a large current with a low power factor and causes a momentary voltage dip along the feeder. The voltage dip is sudden and lasts for a few seconds until the motor reaches the rated speed. The voltage flicker due to motor starting is objectionable in certain cases. Such a problem exists with large motors, welding equipment, furnaces, and other fluctuating loads. The traditional solutions to the voltage flicker problem are re-conductoring the feeder, upgrading the system voltage, providing a new feeder, constructing a new substation, or installing other compensating equipment across the load. The series capacitor is a viable solution to the flicker problem [1–5]. For the 60 Hz component of the motor starting current, the capacitive reactance of the series capacitor nullifies the inductive reactance of the feeder. Therefore, the series capacitor reduces the flicker level significantly at the load side. The effect is instantaneous and self-regulating. In order to have better results, the series capacitor location has to be carefully chosen. Further, the reactance of the series capacitor should be smaller than the total source reactance in order to avoid overcompensation. A one-line diagram of such a system is shown in Figure 10.5, along with the corresponding block diagram.

$$\text{Flicker} = \left[1 - \left|\frac{Z_2}{Z_2 + Z_1}\right|\right] \times 100\% \qquad (10.5)$$

$$Z_1 = R + j(X_1 - X_C) \qquad (10.6)$$

$$Z_2 = R_2 + jX_2 \qquad (10.7)$$

where $Z_1 =$ Impedance of the transformer, feeder, and series capacitor

$Z_2 =$ Impedance of the feeder and starting motor

Such a flicker reduction approach is discussed in Reference [3] for a utility distribution system. The calculated

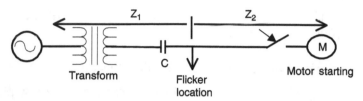

**Figure 10.5** One-line diagram to analyze flicker reduction technique.

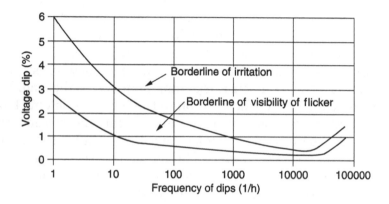

**Figure 10.6** The flicker curve.

flicker value can be compared with the acceptable flicker levels available in Figure 10.6.

### 10.2.2 Protection of Series Capacitor

The series capacitor installations include capacitors, overvoltage protection equipment, a bypass switch, and a control system. There are several schemes available for overvoltage protection. Some of the schemes include [4,5], self-sparking gap protection, triggered gap and varistor protection, varistor protection, and thyristor protected series capacitor.

#### 10.2.2.1   Self-Sparking Gap Protection

In this scheme, the overvoltage protection of the series capacitors is provided by a spark gap connected in parallel as shown

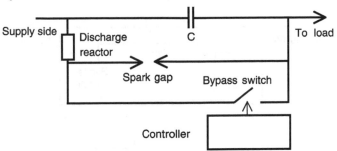

**Figure 10.7** Series capacitor scheme with spark gap protection.

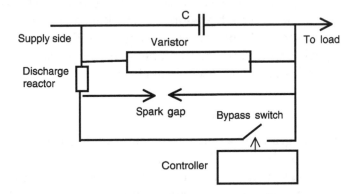

**Figure 10.8** Series capacitor scheme with spark gap and varistor protection.

in Figure 10.7. The spark gap is set to a predetermined over-voltage level at which it sparks and the bypass switch is closed across the series capacitors. The spark overvoltage is sensitive to weather conditions, dirt, and insects. Further, if the control fails to close the bypass switch, there will be prolonged arc-over.

### 10.2.2.2  Triggered Gap and Varistor Protection

A schematic of this protection scheme is shown in Figure 10.8. In this scheme, the spark gap is triggered as soon as an overvoltage condition occurs across the series capacitors. The

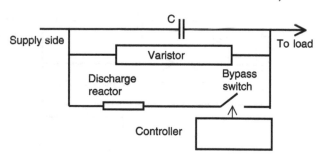

**Figure 10.9**   Distribution series capacitor with varistor protection.

trigger voltage level is set by the varistor across the series capacitor. Most of the transient overvoltages are controlled by the varistor and this scheme is more efficient than the simple spark gap arrangement. The sparking remains across the gaps until the bypass switch is closed.

### 10.2.2.3   Varistor Protection

A one-line diagram of a series capacitor with varistor protection scheme is shown in Figure 10.9. The distribution series capacitor design consists of a capacitor assembly, high energy varistor, bypass switch, discharge reactor, inrush bypass device, and a master control. The distribution series capacitor contains a varistor connected in parallel with the capacitor. The varistor limits the voltage across the capacitor in the event of power system faults. In the event of a single line-to-ground fault, ferroresonance occurs between the series capacitors and the distribution transformers (see Figure 10.13). Such a problem leads to the failure of the distribution transformers.

### 10.2.2.4   Thyristor Protected Series Capacitor

A typical thyristor protected series capacitor scheme is shown in Figure 10.10. Using this scheme, the thyristor can be triggered as soon as the overvoltage condition occurs due to a power system fault. The firing angle of a thyristor can be controlled closely and the protection is much smoother than

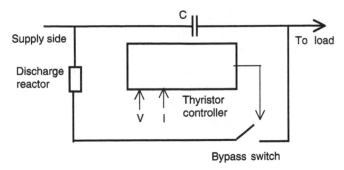

**Figure 10.10** Series capacitor with thyristor controller protection.

**Figure 10.11** Photographic view of one phase of the thyristor protected series capacitor project. (Courtesy of IEEE, Reproduced from Reference [5].)

the other schemes. It should be noted that in varistor protection there may be differences in the phase voltages at which the conduction occurs due to the mismatch in the devices. The thyristor protected scheme is gaining acceptance in utility applications. A typical thyristor protected scheme is shown in Figure 10.11 [5].

### The capacitor bank

The capacitor assembly is made up of a number of standard capacitor units connected in parallel. Each capacitor unit is equipped with an expulsion type fuse. Standard capacitor units can withstand slightly more than twice the rated voltage for a second; accordingly, the varistor protective level is selected at around twice the rated voltage of capacitor units. The bypass switches are closed as soon as the varistor conduction is detected. Consequently, the protective level voltage appears across the capacitor units for only a fraction of a second.

### Varistors

This is a surge arrester with nonlinear volt-ampere character-istics which will conduct in the event of a temporary over-voltage, switching surge, or an overvoltage due to a system fault. The selected varistor should have adequate energy capability to withstand the maximum fault current for the time required to operate the bypass device.

### Bypass switch

The bypass switch is closed by the control system as soon as a fault is identified and the varistor starts conduction. Therefore, in addition to the discharge current, the switch must carry the fault current from the series capacitor equipment. The bypass switch is closed when there are problems in the series capacitor circuit and is therefore expected to carry rated line current.

### Current limiting reactor

The purpose of the current limiting reactor is to control the capacitor discharge current when the bypass switch is closed. The inductance of the current limiting reactor is chosen to control the peak amplitude of the discharge current and the frequency of oscillation. Further, the resistance of the reactor provides damping of the capacitor discharge current.

### Control systems

In earlier versions, electromechanical devices performed the switching of the bypass switches. Now the electronic feedback control type of circuit is employed using current and voltage sensing signals.

The electronic controller is more flexible in setting the necessary time delays and accurate closing times. In the thyristor controlled capacitor schemes, the entire control function can be microprocessor based. A controller monitors the series capacitor through a voltage transformer connected to all three phases. The bypass switch is automatically closed if the voltage across the capacitor indicates the presence of a system fault.

## 10.3 SERIES CAPACITORS FOR TRANSMISSION LINES

Thermal considerations and transient or steady-state stability restrict the maximum power capability of a transmission system. The series capacitors are used on long transmission lines to increase the power transfer capability and to improve system stability.

The power transfer through a transmission line is given by:

$$P = \frac{EV\sin\delta}{X_{\mathrm{L}}} \tag{10.8}$$

where $\delta$ is the angle between the sending end voltage $(E)$ and receiving end voltage $(V)$. With a series capacitor, the expression for power transfer is:

$$P = \frac{EV\sin\delta}{(X_{\mathrm{L}} - X_{\mathrm{C}})} \tag{10.9}$$

Therefore, for a given phase angle difference between the voltages, the power transfer is greater with a series capacitor. Thus, by making a greater interchange of power possible, the normal load transfer and the synchronizing power flowing during transient conditions are increased, thereby improving stability.

Figure 10.12 shows the power transfer for a typical line without and with series capacitors. With the series capacitors, the maximum power transfer is increased significantly. Further, to transfer the same amount of power, the power angle $\delta$ is smaller, thus improving the stability.

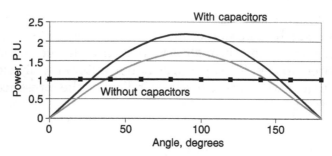

**Figure 10.12** Power transfer of a line without and with series capacitors.


## 10.4 LIMITATIONS OF SERIES CAPACITOR APPLICATIONS

Most of the series capacitor applications have been successful in improving the power system performance. However, in a few circumstances, unforeseen difficulties have been encountered, including ferroresonance, hunting of synchronous machines, and subsynchronous resonance.

### 10.4.1 Ferroresonance

Usually an unloaded transformer draws a significant inrush current during energization. The series capacitor may interact with the transformer inductance, producing a resonant condition that causes oscillatory or undamped currents to flow. This is known as ferroresonance. Also, in the case of series capacitor installations without the protection of MOV, any single line-to-ground fault can introduce ferroresonance problems. One such circuit reported in Reference [3] is shown in Figure 10.13. Such a condition leads to the failure of the distribution transformer. In order to avoid such a condition, the MOV arresters, bypass switches, damping resistance, and suitable controls are employed. A series capacitor, when installed in a long circuit supplying a transformer of very high no-load current, may resonate during normal operation at a frequency corresponding to a harmonic component of the exciting current. Sometimes a

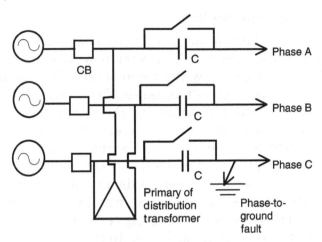

**Figure 10.13** Equivalent circuit showing ferroresonance followed by a single line-to-ground fault between a series capacitor and transformer.

fluctuating load may cause resonance when the transformer is energized. In many instances, a damping resistor can be used across the series capacitors. Typical waveforms of such oscillatory cases are shown in Reference [3].

### 10.4.2 Hunting of Synchronous Motors

Power systems are equipped with synchronous machines either for generation or motoring. Lightly loaded synchronous machines can go into hunting due to system disturbances such as switching operations or changes in the load or excitation. The presence of series capacitors in the feeder circuits reduces the effective reactance and hence violent hunting is experienced in some motors. A synchronous motor, when fed through a long line with overcompensation by a series capacitor, may hunt if started during periods of light load. Series capacitors should not be applied to circuits supplying either synchronous or induction motors driving reciprocating loads such as pumps or compressors. In such cases, both hunting and voltage flicker problems will be encountered.

### 10.4.3  Subsynchronous Resonance

Consider an induction or a synchronous machine started through a series capacitor. The rotor may lock and continue to rotate at a speed below the rated speed. This condition is known as subsynchronous resonance. The speed will be due to a resonant frequency dictated by the series capacitor and the circuit inductance. Such a resonant circuit conducts a large current and may damage the motor due to excessive heating and vibrations. This frequency is usually 20–30 cycles for a 60 cycle motor. In order to damp out such resonance, suitable damping resistance can be installed across the series capacitors. The value of the damping resistance has to be high in order to reduce continuous losses. In general, the possibility of subsynchronous resonance has to be checked for all the larger motors where the series capacitor is installed. Some utilities use subsynchronous relays at the generators to sense sustained low frequency operation and trip the unit.

### 10.4.4  Self-Excitation of Induction Motors

Consider an induction motor supplied through a line containing a series capacitor bank. Under certain circumstances, the motor may act as an induction generator producing current of lower than normal frequency. This low frequency current is limited only by the impedance of the supply circuit and may reach relatively large values. These large low frequency currents manifest not only as current surges and voltage swings, but also as strong oscillations of the rotor, producing large pulsating torques. This phenomenon of self-excitation will not always take place and may be eliminated by sufficient line or shunt resistance, or by locating the capacitor at a suitable distance. Such a self-excitation trend can also occur with synchronous machines [2].

### Example 10.2

A simplified diagram of a 360 mile, double-circuit, two-section, 525 kV AC transmission link connecting two areas is shown in Figure 10.14. For simplicity of calculations, neglect the

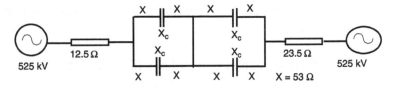

**Figure 10.14**  One-line diagram for Example 10.2.

resistance and shunt capacitor effects. Determine the maximum power that can be transmitted between the two areas under the following conditions.

1. $X=53\ \Omega$ and $X_C=0\ \Omega$ (No series compensation)
2. $X=53\ \Omega$ and $X_C=53\ \Omega$
3. $X=53\ \Omega$ and $X_C=0\ \Omega$ (One section of the line is out of service)
4. $X=53\ \Omega$ and $X_C=53\ \Omega$ (One section of the line is out of service)

Solution

1. $X=53\ \Omega$ and $X_C=0\ \Omega$ (No series compensation)
   $X=(53\times4)/2+12.5+23.5=142\ \Omega$

$$P_{\max}=\frac{kV^2}{X}=\frac{525^2}{142}=1{,}941\ \text{MW}$$

2. $X=53\ \Omega$ and $X_C=53\ \Omega$

$$\text{Inductive reactance}=12.5+2\times\frac{106}{2}+23.5$$
$$=142\ \Omega$$

Capacitive reactance $=2\times(53/2)=53\ \Omega$
Net reactance, $X=142-53=89\ \Omega$

$$P_{\max}=\frac{kV^2}{X}=\frac{525^2}{89}=3{,}097\ \text{MW}$$

3. $X = 53\ \Omega$ and $X_C = 0\ \Omega$ (One section of the line is out of service)

$$\text{Inductive reactance} = 12.5 + (53 \times 2) + 2$$
$$\times \frac{2 \times 53}{2} + 23.5 = 195\ \Omega$$

Capacitive reactance $= 0\ \Omega$
Net reactance, $X = 195 - 0 = 195\ \Omega$

$$P_{\text{max}} = \frac{\text{kV}^2}{X} = \frac{525^2}{195} = 1{,}413\ \text{MW}$$

4. $X = 53\ \Omega$ and $X_C = 53\ \Omega$ (One section of the line is out of service)

$$\text{Inductive reactance} = 12.5 + (53 \times 2) + 2$$
$$\times \frac{2 \times 53}{2} + 23.5 = 195\ \Omega$$

Capacitive reactance $= 53 + 53/2 = 79.5\ \Omega$
Net reactance, $X = 195 - 79.5 = 115.5\ \Omega$

$$P_{\text{max}} = \frac{\text{kV}^2}{X} = \frac{525^2}{115.5} = 2{,}387\ \text{MW}$$

The benefits due to the use of series capacitors can be seen from the MWs transmitted.

## Example 10.3

A series capacitor scheme is required to control the flicker problem due to a 500 hp motor starting. A line diagram of the system is shown in Figure 10.15. The kVA of the motor during starting was 2,200 at a power factor of 0.32. Assume the nominal kVA of the motor to be 555 at a power factor of 0.9. Use a flicker limit of 7.5%.

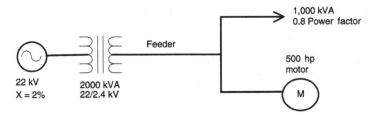

**Figure 10.15** One-line diagram for Example 10.3.

Solution

Impedance data on 2,000 MVA base:

> Source impedance $= 0.0 + j\, 0.02$ P.U.
> Feeder impedance $= 0.04 + j\, 0.09$ P.U.
> Transformer impedance $= 0.009 + j\, 0.06$ P.U.
> Total impedance $= 0.049 + j\, 0.17$ P.U.

Load kW, kVAR, and kVA; load is 1000 kVA at a power factor of 0.8:

> kW $(1000 \times 0.8) = 800$ kW
> kVAR $(1000 \times 0.6) = 600$ kVAR

During motor starting, kVA $= 2,200$ at a power factor of 0.32.

> kW $(2,200 \times 0.342) = 704$ kW
> kVAR $(2,200 \times 0.947) = 2083$ kVAR

|  | kW | kVAR |
|---|---|---|
| Load | 800 | 600 |
| Motor starting | 704 | 2083 |
| Total | 1504 | 2683 |

kVA of the load $(1504^2 + 2683^2)^{0.5} = 3076$

$\cos \theta = 0.4889, \quad \sin \theta = 0.8722$

Per unit load $= \dfrac{3{,}076}{2{,}000} = 1.538$

Normal load, motor operating at rated power factor:

$\cos \theta = 0.90, \quad \sin \theta = 0.436$
kW $(555 \times 0.90) = 500$ kW
kVAR $(555 \times 0.436) = 242$ kVAR

|                | kW   | kVAR |
|----------------|------|------|
| Load           | 800  | 600  |
| Motor starting | 500  | 242  |
| Total          | 1300 | 842  |

kVA of the load $(1{,}300^2 + 842^2)^{0.5} = 1{,}549$
$\cos \theta = 0.839, \quad \sin \theta = 0.544$

Per unit load $= \dfrac{1{,}549}{2{,}000} = 0.775$

To limit the flicker level to 7.5%, calculate the capacitive reactance $(X_C)$:

$e = IR \cos \theta + I(X_L - X_C) \sin \theta$
$7.5 = (1.538)(4.9)(0.4889) + 1.538\,(17 - X_C)(0.8772)$

Solving the above equation, $X_C = 14.566\%$

$$\text{Base impedance, } Z = \frac{2.2^2}{2{,}000} = 242 \ \Omega$$

$X_C = (242 \ \Omega)(0.14566 \ \text{P.U.}) = 35.23 \ \Omega$

$$\text{Motor starting current} = \frac{3{,}076 \ \text{kVA}}{\sqrt{3} \times 22 \ \text{kV}} = 80.7 \ \text{A}$$

This current magnitude corresponds to 150% factor for the momentary load of a series capacitor, and the current through the series capacitor is:

$$I_C = \frac{80.7}{1.5} = 54 \ \text{A}$$

The voltage across the series capacitor is:

$E_C = (54\,\text{A})\,(35.23\,\Omega) = 1{,}902\,\text{V}$
Say $E_C = 1{,}920\,\text{V}$
$X_C$ at $60\,\text{Hz} = 246\,\Omega$
Rated current $= 7.81\,\text{A}$

$$\text{Number of capacitor units} = \frac{246\,\Omega}{35.23\,\Omega} = 7$$

Select the next whole number, i.e., 8 units.

The actual capacitive reactance of 8 units in parallel $(246/8) = 30.75\,\Omega$
The series capacitor bank will consist of 8 units in parallel, each rated $1{,}920\,\text{V}$, $246\,\Omega$, and $7.81\,\text{A}$
The rated current of the series capacitor bank $(8 \times 7.81\,\text{A}) = 62.48\,\text{A}$
Momentary current rating $(1.5 \times 62.48\,\text{A}) = 93.72\,\text{A}$
kVAR rating of the bank $(1.732 \times 22\,\text{kV} \times 93.72\,\text{A}) = 3{,}571\,\text{kVAR}$

$$X_C = \frac{(30.75)(2{,}000)}{(10)(22\,\text{kV})^2} = 12.71\%$$

The capacitive reactance is less than the value required (14.566%) and the actual voltage drop will be less than 7.5%.

*Location of the capacitor*

The series capacitor can be located by considering the voltage drop without and with series capacitors. As a first approximation, assume that the capacitor is located at the middle of the feeder. The voltages at various locations without and with the series capacitor for the motor starting condition are listed in Table 10.1. From this table, it can be seen that the voltage drop without series capacitor during the motor starting condition is not acceptable. With a series capacitor, the voltage drop during the motor starting is acceptable. From Table 10.2, it can be seen that the voltage drop without

**TABLE 10.1** Voltages without and with Series Capacitor for Motor Starting

| Location | Without Series Capacitor Voltage Drop % | With Series Capacitor Voltage Drop % |
|---|---|---|
| Utility source | 2.68 | 2.68 |
| 1/2 Line | 7.42 | 7.42 |
| Series Capacitor | | −10.49 |
| Line | 14.83 | −3.07 |
| Transformer | 26.19 | 8.28 |

**TABLE 10.2** Voltages without and with Series Capacitor for Normal Condition

| Location | Without Series Capacitor Voltage Drop % | With Series Capacitor Voltage Drop % |
|---|---|---|
| Utility source | 0.84 | 0.84 |
| 1/2 Line | 2.59 | 2.59 |
| Series Capacitor | | −3.04 |
| Line | 5.19 | −0.44 |
| Transformer | 8.87 | 3.24 |

a series capacitor during the normal motor operation is not acceptable. With a series capacitor, the voltage drop during the normal motor operation is acceptable.

## 10.5 OTHER SERIES CAPACITOR APPLICATIONS

Series capacitors are used in many applications where the load is fluctuating in nature. Such applications include resistance welding, arc furnaces, saw mills, rolling mills, crusher load, traction applications, induction generator-wind turbines, and arc welding. Some details of these applications are discussed below.

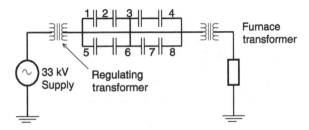

**Figure 10.16** One-line diagram of power supply for an arc furnace with series capacitors.

## 10.5.1 Series Capacitance in Arc Furnace Power Supply

A one-line diagram of an arc furnace installation with series capacitor scheme is shown in Figure 10.16 [7]. The system consists of a 30 MVA arc furnace, supplied through a 30 MVA furnace transformer. The series capacitor is connected in an H-configuration and is protected by a shunt bypass arrangement consisting of a bypass circuit breaker along with spark gaps. The principle of operation behind this protection system involves the instantaneous operation of the spark gap whenever the voltage across the spark gap exceeds the predetermined safe value. Thus, the series capacitor voltages are maintained within acceptable value. The current required in the arc furnace during the melting process is fluctuating and the series capacitor offers a means of smoothing the current flow.

## 10.5.2 Series Capacitors for Induction Generator Applications

The induction machine is a very simple and robust device. As a motor, it performs very well and as a generator it has limitations due to the absence of excitation. The excitation requirements can be supplied through shunt capacitors but self-excitation limits the size of the capacitor equipment. Also, for better performance, the capacitor has to be switched.

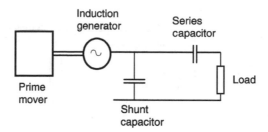

**Figure 10.17** Induction generator with shunt and series capacitors.

The shunt capacitor and series capacitor combination is considered to provide self-excitation and self-regulation. The prime mover can be diesel engine, water-turbine, or wind-turbine. The voltage regulation of a stand-alone generator can be achieved by using series capacitors. A one-line diagram of such a scheme is shown in Figure 10.17. The performance characteristic of such a scheme is discussed in Reference [8].

### 10.5.3 Series Capacitor for Traction Applications

In traction systems, the system supply voltages are limited due to the presence of tunnels, bridges, etc. As a rule of thumb, a 20 mile line is acceptable with 25 kV and a 40 mile line for 50 kV system voltages. If the distance between the substations is short and if the train loading is high, the voltage drop imposes a limit on the performance of the locomotive. With heavy loading on the long lines, the performance is again limited by the voltage drop. Some viable solutions to the voltage drop may be to add more substations, low impedance catenary, shunt capacitors, series capacitors, or higher catenary voltage.

In the coal-mining industry, the shunt and series capacitor compensation was applied in order to maintain the voltage at the remote end [9]. The series capacitor was used in the Black Mesa and Lake Powell Railroad at 50 kV system voltage. In this coal-haul railroad, two trains were to operate with three 6,000 hp locomotives per train. The catenary was fed radially for 78 miles with no intermediate substations. The load of 36 MVA was

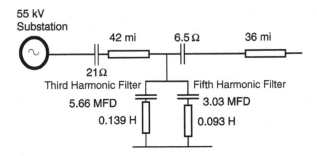

**Figure 10.18** Shunt and series capacitor compensated traction power system.

simply far too great to supply over 78 miles by one substation. The shunt and series capacitor scheme used to compensate the line is shown in Figure 10.18. The shunt capacitor-supported filter bank is installed at 42 miles from the substation. The harmonic filters are tuned to 3rd and 5th harmonics. The series capacitors are 21 Ω and 6.5 Ω, installed at the substation and at the 42 mile location. The performance of the traction system is described in Reference [9]. The system was installed in early 1980.

### 10.5.4 Control of Load Sharing Between Feeders

Consider an overhead transmission line or distribution line to be reinforced by a second circuit operating in parallel. If the second circuit is of different length with different impedance, then there may be a load sharing problem. This concept is illustrated in Figure 10.19. In this circuit arrangement, the circuit impedances are different and hence the load sharing will be different. In order to ensure equal load current through both the lines, a series capacitor can be used in the longer line with higher impedance (Figure 10.19b).

### Example 10.4

Consider an 11 kV, 60 Hz, three-phase distribution line between X and Y as shown in Figure 10.19. The reactance of line 1 is 4 Ω/phase. The reactance of the second line is 6 Ω/phase.

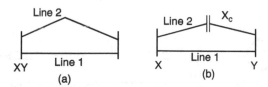

**Figure 10.19**  Load sharing between two lines: (a) without series capacitor; (b) with series capacitor.

The load supplied at substation Y is 5,000 kVA. Calculate the current through each line. In order to ensure equal load current through each feeder, a series capacitor bank is proposed at the middle of line 2. Calculate the reactance of the series capacitor. Ignore the effect of resistance and assume that the conductors in both circuits are the same.

Solution

>Reactance of line 1 = 4 $\Omega$/phase
>Reactance of line 2 = 6 $\Omega$/phase
>Total load current (5,000 kVA/1.732 × 11 kV) = 262 A/phase
>Current through line 1 (262 × 6/10) = 157.2 A/phase
>Current through line 2 (262 × 4/10) = 104.8 A/phase

Since both line conductors are the same, an equal current distribution is expected. In order to ensure equal current distribution through the conductors, a series capacitor is proposed, as shown in Figure 10.19b. The expected current in line 1 is equal to the current in line 2, 131 A.

>The new reactance of line 2 = 4 $\Omega$/phase
>Reactance of the series capacitor bank (6 − 4) $\Omega$ = 2 $\Omega$/phase

## 10.5.5  Capacitors for Induction Heaters

Induction heaters are designed to heat ferrous or nonferrous materials. A typical one-line diagram of a power supply for an induction heater is shown in Figure 10.20. A motor generator set can be used to generate the required high frequency voltage. The induction coil can be supplied through a series

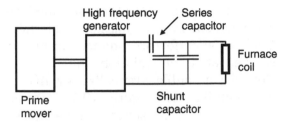

**Figure 10.20** A induction heater supplied from a motor generator set through a series capacitor.

capacitor. The power factor of the induction furnace is always low and can be improved by using shunt capacitors. The series capacitor is used to improve the voltage regulation. A typical voltage range for a high voltage capacitor for induction heating is 2.5–6 kV single-phase or three-phase. The heater ratings are 1.5–4 MW.

## CONCLUSIONS

The fundamentals of the shunt and series capacitor were presented using phasor diagrams. The series capacitors in the radial system can be used to improve the voltage profile on distribution systems. The basic components used in a series capacitor scheme in a radial system were explained. The theory of application of series capacitors for the transmission line to increase the transfer capability is shown. The numerical examples show how the power transfer is increased and how the flicker problem can be controlled. Other applications of the series capacitor, such as the arc furnace, induction generator, and the traction system performance improvement, are illustrated.

## PROBLEMS

10.1. Explain series compensation.
10.2. What is the difference between series and shunt compensation?

10.3. Why are series capacitors provided with bypass arrangement in transmission projects?

10.4. Explain why ferroresonance occurs in a series compensated transmission line.

10.5. Consider a double-circuit 345 kV AC transmission line (similar to Figure 10.14) connecting two areas. Use the source impedance given in Example 10.2. Determine the maximum power that can be transmitted between the two areas under the following conditions:

1. $X = 51$ $\Omega$ and $X_C = 0$ $\Omega$ (No series compensation)
2. $X = 51$ $\Omega$ and $X_C = 41$ $\Omega$
3. $X = 51$ $\Omega$ and $X_C = 0$ $\Omega$ (One section of the line is out of service)
4. $X = 51$ $\Omega$ and $X_C = 41$ $\Omega$ (One section of the line is out of service)

10.6. Define hunting in the context of series compensation.

10.7. Repeat Example 10.3 using a 100 hp motor.

10.8. Consider a 6.6 kV, 60 Hz, three-phase distribution line between X and Y as shown in Figure 10.19. The reactance of line 1 is 3.5 $\Omega$/phase. The reactance of line 2 is 6.5 $\Omega$/phase. The load supplied at the substation Y is 4,000 kVA. Calculate the current through each line. In order to ensure equal load current through each feeder, a series capacitor bank is proposed at the middle of line 2. Calculate the reactance of the series capacitor. Ignore the effect of resistance and assume that the conductors in both circuits are the same.

## REFERENCES

1. Butler, J. W., Concordia, C. (1937). Analysis of series capacitor application problems, *AIEE Transactions*, 56, 975–988.

2. Kimbark, E. W. (1966). Improvement of system stability by switched series capacitors, *IEEE Transactions on Power Apparatus and Systems*, PAS-85, 180–188.

3. Morgan, L., Barcus, J. M., Ihara, S. (1993). Distribution series capacitors with high energy varistor protection, *IEEE Transactions on Power Delivery*, 8(3), 1413–1419.

4. Miske, S. A. (2001). Considerations for the application of series capacitors to radial power distribution circuits, *IEEE Transactions on Power Delivery*, 16(2), 306–318.

5. Bhargava, B., Haas, R. G., Thyristor protected series capacitors project, IEEE Power Engineering Society, Summer Meeting, Vol. 1, 2002, pp. 241–246.

6. IEEE Standard 824 (1994), IEEE Standard for Series Capacitors in the Power System.

7. Pienaar, J., Stewart, P.H. (2002). Ferroresonance in a series capacitor system supplying a submerged-arc furnace, *Proceedings of the Sixth IEEE African Conference*, Vol. 2, October 2002, pp. 785–789.

8. Murthy, S. (1996). Analysis of series capacitor compensated self-excited induction generators for autonomous power generation, *Proceedings of the 1996 International Conference on Power Electronics, Drives and Energy Growth*, Vol. 2, 687–693.

9. Burke, J. J. (1985). Optimizing performance of commercial frequency electrified railroads, IEEE Transportation Division, New York Section, 22 pages.

# 11

## SURGE CAPACITORS

### 11.1 INTRODUCTION

Oil is used on circuit breakers and transformers as an insulation to smooth the stress that results from electric voltage. In the case of rotating machines such as motors and generators, there is no oil insulation and the windings are equipped with dry insulation. Therefore, rotating machines are more vulnerable to high stress and failure compared to circuit breakers or transformers. With the fast front transients, the winding structure can cause surge reflections and oscillations that can damage winding insulation. There are rotating machines connected directly to overhead lines and exposed to lightning surges. Other electrical machines are connected through transformers and are exposed to traveling waves produced by lightning strikes. There are overvoltages generated due to insulation failure of the electric motor. Such effects are important in the reliable operation of the motor. Almost all rotating machines are exposed to lightning surges, switching surges, and internally generated surge voltages [1–5].

TABLE 11.1   BIL Levels of Motors, Transformers, and Switchgear and Acceptable Rate of Rise of Surge Voltage [1]

|  | | Basic Insulation Level kV | | |
| :---: | :---: | :---: | :---: | :---: |
| Voltage Rating V | Voltage Class kV | AC Motor kV | Transformer kV | Enclosed Switchgear |
| 480 | 0.6 | 3.4 | 45 | 30 |
| 2,300 | 2.5 | 10.0 | 60 | 60 |
| 4,000 | 5.0 | 16.0 | 75 | 60 |
| 12,000 | 15.0 | 44.0 | 110 | 95 |

## 11.2   INSULATION WITHSTAND STRENGTH

The air gaps in electrical machines are very small and hence the built-in impulse insulation strength is low. With multi-turn coils in the same slot, the inter-turn insulation level is much lower than the coil slot insulation. With series-connected stator coils, there is a larger capacitance coupling between the conductors of each coil and the grounded slots into which they are fitted. Table 11.1 compares the BIL of AC motors, transformers, and switchgears in order to illustrate that the rotating machines are weak from the insulation point of view [1]. The fast front transients can cause voltage gradients among the turns, and hence there is a need to reduce the magnitude and rate of rise of the surge voltages traveling to the rotating machines.

A typical time-voltage envelope of the impulse strength of a rotating machine based on an IEEE working group report is shown in Figure 11.1. The insulation withstand capabilities of a rotating machine can be derived based on the following considerations. A generally accepted maximum rate is equal to one tenth of the maximum permitted line-to-ground impulse voltage per microsecond, which is a 10 μs surge, front to crest. The various voltage levels defined in Figure 11.1 are listed in Table 11.2. From the impulse withstand curve, the following observations can be made.

1. The initial fast rising or nearly instantaneous step voltage must not exceed the rated peak voltage of

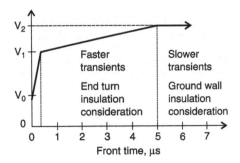

**Figure 11.1** Impulse strength of rotating machines.

**TABLE 11.2** Probable Impulse Withstand Volt-Time Impulse Voltage [2]

| Rated kV (L–L) | $V_0$ kV | $V_1$ kV | $V_2$ kV | $V_2$ P.U. |
|---|---|---|---|---|
| 0.46 | 0.375 | 0.75 | 3.4 | 9.0 |
| 0.575 | 0.470 | 0.94 | 3.8 | 8.1 |
| 2.3 | 1.9 | 3.8 | 9.9 | 5.27 |
| 4.0 | 3.3 | 6.5 | 15.9 | 4.87 |
| 4.6 | 3.8 | 7.5 | 18.1 | 4.80 |
| 6.6 | 5.4 | 10.8 | 25.1 | 4.66 |
| 13.2 | 10.8 | 21.6 | 48.4 | 4.49 |
| 18.0 | 14.7 | 29.4 | 65.4 | 4.45 |
| 20 | 16.3 | 32.7 | 72.5 | 4.44 |
| 25 | 20.4 | 40.8 | 90.2 | 4.42 |

the machine. Usually the rated peak voltage of a motor or generator can be assumed to be 110% of the nameplate rating.

2. For surge voltages of short duration, the maximum peak voltage must not exceed 200% of the rated instantaneous peak value. This value is shown as voltage $V_2$ in Figure 11.1.

3. The rate of rise of the fast front of the fast rising voltage surge, after the initial step voltage, must not exceed the rate based on a minimum rise time of 5 µs, relative to the short duration withstand strength. This value is shown as voltage $V_2$ in Figure 11.1.

**Table 11.3**  Impulse Insulation Strength and Permissible Surge Voltage Rate of Rise at the Motor Terminals [2]

| Machine Voltage, kV | Impulse Withstand, kV | Capacitance, $\mu F$ |
|---|---|---|
| 0.65 | 4 | 0.4 |
| 2.4 | 10 | 1.0 |
| 4.16 | 16 | 1.6 |
| 4.8 | 19 | 1.9 |
| 6.9 | 21 | 2.1 |
| 11.5 | 42 | 4.2 |
| 13.8 | 50 | 5.0 |

**Table 11.4**  Station Type Arresters for Rotating Machines [1]

| Machine kV | System Not Effectively Grounded | | | System Effectively Grounded | | |
|---|---|---|---|---|---|---|
| | Arrester kV rms | Impulse Sparkover kV, Peak | Discharge Voltage kV, Peak | Arrester kV rms | Impulse Sparkover kV, Peak | Discharge Voltage kV, Peak |
| 0.6 | 0.75 | 2.8 | 2.6 | 0.75 | 2.8 | 2.6 |
| 2.4 | 3.00 | 12.0 | 5.0 | 3.00 | 12.0 | 5.0 |
| 4.16 | 4.5 | 16 | 7.4 | 3 | 12 | 5.0 |
| 7.20 | 7.5 | 25 | 12.2 | | 20 | 9.8 |
| 12.0 | 12 | 39 | 19.4 | 9 | 30 | 14.6 |
| 13.8 | 15 | 48 | 24.2 | 12 | 39 | 19.4 |

The surge withstand strength of a rotating machine depends on the peak magnitude, duration, rise time, and repetition rate. The acceptable impulse insulation strength and permissible rate of rise at the machine terminals are presented in Table 11.3 [2]. These values differ slightly from the Table 11.1 values (which illustrate the BIL capabilities of AC machines, transformers, and switchgear). The station type surge arresters to be considered for various voltage classes are presented in Table 11.4 [1]. An effectively grounded system is defined as the system in which the reactance ratio $X_0/X_1$ is positive and less than 3. The ratio $R_0/X_1$ is positive and less than 1. These values are presented for guideline purposes. Detailed calculations are necessary to select a surge arrester

**TABLE 11.5** Station Type Surge Capacitors for Rotating Machines [1]

| Machine kV | System Not Effectively Grounded | | | System Effectively Grounded | | |
|---|---|---|---|---|---|---|
| | Arrester kV rms | Impulse Sparkover kV, Peak | Discharge Voltage kV, Peak | Arrester kV rms | Impulse Sparkover kV, Peak | Discharge Voltage kV, Peak |
| 0.6 | 0.75 | 2.8 | 2.6 | 0.75 | 2.8 | 2.6 |
| 2.4 | 3 | 12 | 5 | 3 | 12 | 5 |
| 4.16 | 4.5 | 16 | 7.4 | 3 | 12 | 5 |
| 7.2 | 7.5 | 25 | 12.2 | | 20 | 9.8 |
| 12 | 12 | 39 | 19.4 | 9 | 30 | 14.6 |
| 13.8 | 15 | 48 | 24.2 | 12 | 39 | 19.4 |

type. Surge capacitors recommended for rotating machines are presented in Table 11.5 [1]. Again, calculations are necessary to verify the suitability of the surge arrester and surge capacitor for the specific application.

## 11.3 SOURCES OF FAST FRONT SURGES

The sources of overvoltages traveling to the rotating machines may be lightning surges or switching surges. A lightning surge on an overhead conductor can travel through the line and can produce an overvoltage problem on a rotating machine. The lightning can be a direct stroke or a back flashover through an insulator. Whatever the nature of the surge, lightning surges are known to have very sharp rise times and thus the rotating machine needs protection. The characteristics of the various lightning strokes are discussed in Chapter 18.

Switching surges are another source of overvoltages in the power system. These are due to circuit breaker operations such as energizing, de-energizing, fault clearing, reclosing, backup fault clearing, capacitor switching, prestriking, restriking, and other similar circuit breaker operations. The nature of the overvoltages and the resulting waveforms are discussed in Chapter 22. Surge voltages may be generated within the rotating machine due to switching or a fault outside the machine. Steep wave fronts may be generated due to switching operations or breakdown of insulation. Sparkover at the surge arrester produces overvoltages that are harmful to the machine windings. Insulation failure can produce overvoltages in unfaulted phases. Motor starting can produce oscillatory voltages due to the inductance of the motor and capacitance of the cable connecting the motor and the starting device.

## 11.4   THE PROTECTION SYSTEM

In order to control the rate of voltage rise in an LC circuit, the impressed voltage has to be limited. Consider an LC circuit as shown in Figure 11.2, where a voltage $V$ is applied. The voltage across the capacitor $V_C$ will be oscillatory. The period of oscillation will be $T = 2\pi\sqrt{LC}$. The capacitor voltage reaches a peak value at a time $T/2$. By properly choosing the values of $L$ and $C$, the rate of rise of the voltage can be controlled. Still, the amplitude of the surge voltage can be higher. In this context, the three items required to control the rate of rise and the peak voltage are a capacitance, an inductance, and a means to limit the voltage magnitude. In practical circuits, the voltage can be limited by using a surge arrester. In Figure 11.3, the voltage across the capacitor $V_C$ and the applied voltage $V$ are shown for two conditions. The capacitor voltage rise can be twice the value of the applied voltage $V$ and is oscillatory. In practical circuits, there will be some damping and hence the voltage rise will be smaller. However, the crest of value of $V_C$ may exceed the applied voltage $V$ by a considerable amount. In order to limit the voltage across the capacitor, a surge arrester is needed. The surge arrester will clip the capacitor voltage $V_C$ if it exceeds the sparkover voltage.

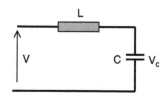

**Figure 11.2**   The LC oscillatory circuit.

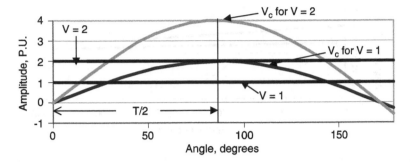

**Figure 11.3** Voltage across the capacitor in the circuit shown in Figure 11.2.

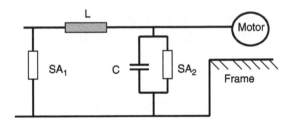

**Figure 11.4** The surge capacitor–surge arrester protection scheme.

The surge protection circuit based on the above principles is shown in Figure 11.4. The inductance may be distributed in the cable or line conductors. The surge arrester $SA_1$ is installed at the high voltage side of the transformer. The surge capacitor and the surge arrester $SA_2$ are installed at the motor terminals. The various voltage waveforms of the surge protection scheme are shown in Figure 11.5. The surge arrester limits the voltage at the motor terminals and the rate of rise of the voltage is controlled by the capacitor.

## Example 11.1

Consider a 0.125 microfarad/phase at the terminal of a 2 MVA, 60 Hz, three-phase, 4.2 kV motor along with a surge arrester to ground. Calculate the steady-state current through the surge capacitor. Also calculate the current through the surge

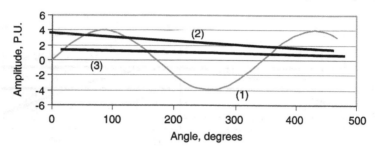

**Figure 11.5**   (1) = Voltage across the motor terminals without the surge arrester $SA_2$; (2) = Voltage across the surge arrester $SA_1$; (3) = Voltage across the motor terminals with $SA_2$ in service.

capacitor if a surge voltage arrives at the capacitor with a frequency of 10 kHz. State your observations.

Solution

$$C = 0.125 \text{ MFD/phase}$$

$$X_C = \left( \frac{1}{2\pi \times 60 \times 0.125 \times 10^{-6}} \right) = 21{,}231 \ \Omega/\text{phase}$$

$$\text{Steady-state current} = \left( \frac{4{,}200 \text{ V}}{1.732 \times 21{,}231} \right) = 0.11 \text{ A/phase}$$

Current through the capacitor at 10 kHz transient

$$X_C = \left( \frac{1}{2\pi \times 10 \times 10^3 \times 0.125 \times 10^{-6}} \right) = 127.4 \ \Omega/\text{phase}$$

$$\text{Current due to high frequency surge} = \left( \frac{4{,}200 \text{ V}}{1.732 \times 127.4} \right)$$

$$= 19 \text{ A/phase}$$

It is observed that the current through the surge capacitor during steady-state condition is very small. The current through the capacitor during the high frequency surge is significant.

### 11.4.1 Calculation of Peak Surge Voltage at the Motor

Figure 11.6 shows a motor supplied through a transformer. A surge arrester at the primary of the transformer protects the transformer from the incoming lightning and switching surges. In order to calculate the maximum overvoltages arriving at the motor terminals, consider the following parameters:

$$kVA = \text{Transformer kVA}$$

$V_1$ = Primary voltage of the transformer (line-to line), kV

$V_2$ = Secondary voltage of the transformer (line-to line), kV

$Z$ = Transformer impedance, %

$K$ = Transformer correction factor (see Figure 11.7)

$E_a$ = Surge arrester spark overvoltage, kV

$R$ = Surge impedance of the motor, $\Omega$

$A$ = Minimum number of running motors

$B$ = Length of the cable between circuit breaker and motor, ft

$C_f$ = Capacitance of the cable per 100 ft, MFD

$C_S$ = Surge capacitance, MFD; typical surge capacitance values are presented in Table 11.5

$L$ = Transformer inductance, µH

The time period $T$ of the wave, the surge voltage at the motor terminals $V_m$, and the surge voltage rate of rise are calculated in order to evaluate the need for surge capacitor protection.

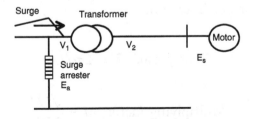

**Figure 11.6** Motor supplied through a transformer.

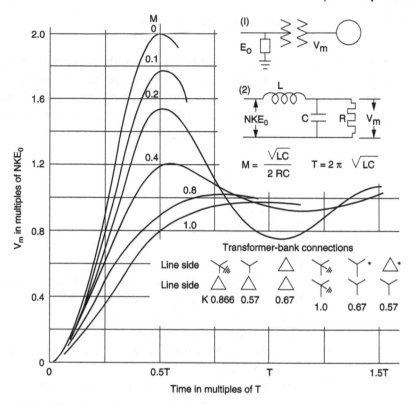

**Figure 11.7** The overvoltage factor curve from Reference [1].

The inductance of the transformer $L$ is calculated from the given parameters as:

$$\text{Transformer inductance, } L = \frac{26.5(Z)(V_2)^2 10^3}{\text{kVA}} \tag{11.1}$$

$$\text{Time period, } T = 2\pi\sqrt{LC} \tag{11.2}$$

$$\text{Multiplying factor, } M = \frac{\sqrt{LC}}{2RC} \tag{11.3}$$

The surge voltage at the transformer terminals ($E_s$) is given by:

$$E_s = \frac{V_2}{V_1} KE_a \qquad (11.4)$$

Knowing the factor $M$, the corresponding peak voltage factor can be read from the graph given in Figure 11.7. The peak voltage at the motor terminals ($V_m$) is given by:

$$V_m = (\text{Factor from Figure 11.7}) \, E_s \qquad (11.5)$$

$$\text{The surge voltage rate of rise} = \frac{V_m}{0.5 \, T} \qquad (11.6)$$

If the peak voltage is below the BIL of the motor and if the surge voltage rate of rise is less than the maximum value, then there is no need to have additional surge protection. If these parameters exceed the allowed values, then there is a need to have a surge capacitor and surge arrester protection.

## Example 11.2

Consider a motor supplied through a 5,000 kVA, 12 kV/2.4 kV, delta/delta connected transformer with 5.5% impedance. The cable length is 200 ft and the cable is 500 kcmil, unshielded. The surge arrester located at the primary of the transformer is 12 kV and has a sparkover voltage of 45 kV. Evaluate if a surge capacitor and surge arrester protection are needed for this motor. Make suitable assumptions.

## Solution

| | | |
|---|---|---|
| Transformer kVA | kVA | 5,000 |
| Primary voltage of the transformer (line-to-line), kV | $V_1$ | 12 |
| Secondary voltage of the transformer (line-to-line), kV | $V_2$ | 2.4 |
| Transformer impedance, % | $Z$ | 5.5 |
| Transformer correction factor | $K$ | 0.67 |
| Surge impedance of the motor | $R$ | 1,000 |
| Minimum number of running motors | $A$ | 2 |
| Length of the cable between CB and motor, ft | $B$ | 200 |

| | | |
|---|---|---|
| Capacitance of the cable per 100 ft, MFD | | 0.016 |
| Capacitance of the cable, MFD | C (Cable) | 0.032 |
| Surge capacitance, MFD | $C_S$ | 0 |
| Total capacitance, MFD | C | 0.032 |
| Transformer inductance, µH | L | 167.904 |
| Surge impedance motor, Ω | RA | 500 |
| Time period, µs | T | 14.557 |
| M | M | 0.072 |
| Surge arrester spark overvoltage, kV | $E_a$ | 45 |
| $E_s$ | $E_s$ | 6.03 |
| Factor from Figure 11.7 | $F_a$ | 1.85 |
| Maximum voltage | $V_m$ | 11.16 |
| BIL of the motor from Table 11.3, kV | BIL | 10 |

This voltage ($V_m = 11.6\,\text{kV}$) exceeds the BIL rating of the motor and hence a surge capacitor is required. With a surge capacitor of 0.125 MFD/phase and 2.4 kV surge arrester:

| | | |
|---|---|---|
| Capacitance of the cable, MFD | C (Cable) | 0.032 |
| Surge capacitance, MFD | $C_S$ | 0.125 |
| Total capacitance, MFD | C | 0.157 |
| Transformer inductance, µH | L | 167.9 |
| Surge impedance motor, Ω | RA | 500 |
| Time period, µs | T | 32.243 |
| M | M | 0.033 |
| Surge arrester spark overvoltage, kV | $E_a$ | 12 |
| $E_s$ | $E_s$ | 1.608 |
| Factor from Figure 11.7 | $F_a$ | 1.9 |
| Maximum voltage | $V_m$ | 3.06 |
| BIL of the motor from Table 11.3, kV | BIL | 10 |
| Surge voltage rate of rise, kV/µs | | 0.095 |

The maximum surge voltage is below the BIL of the motor. The surge voltage rate of rise is below the acceptable value of 1.0 kV/µs.

The surge arrester calculation procedure is presented in Chapter 18. It is necessary to have the correct surge arrester at the motor terminal along with the surge capacitor. The cable lead length in surge arrester applications has been discussed in various documents. It is better to have a short lead length and install the surge capacitor and the surge arrester close to the motor terminals.

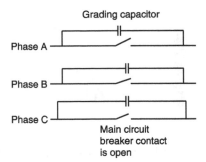

**Figure 11.8**  Grading capacitor in a circuit breaker.

## 11.5  CAPACITORS FOR CIRCUIT BREAKERS

### 11.5.1  Grading Capacitors for EHV Circuit Breakers

In EHV circuit breakers, the voltage across the open circuit breaker blades is significant. In order to distribute the voltage uniformly across the multi-break interrupters in an open circuit condition, grading capacitors are used as shown in Figure 11.8. The grading capacitors also provide control on the rate of rise of the transient recovery voltage. Such grading capacitors provide an indirect connection between the load side to the source side of the circuit breaker when the circuit breaker is in an open condition. Sometimes the grading capacitance and the magnetizing reactance of the transformer form a resonant circuit and produce ferroresonance. The resonant circuit is excited by the main supply source through the grading capacitance. A typical value of grading capacitance on a 500 kV circuit breaker is 700 pF/phase. A photograph of a grading capacitor is shown in Figure 11.9(a) [4]. The grading capacitor is usually installed in parallel to the porcelain insulator.

### 11.5.2  Surge Capacitors for Circuit Breakers

A close-in fault at the generating station will result in a very high short-circuit current. The circuit breakers used to clear such

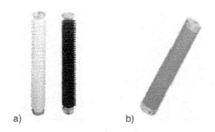

**Figure 11.9**  Photograph of (a) two designs of grading capacitor and (b) a grading capacitor for GIS circuit breaker. (Courtesy of Maxwell Technologies, San Diego, CA.)

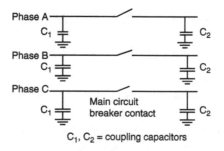

**Figure 11.10**  Diagram showing the coupling (surge) capacitor at the circuit breaker.

fault currents may experience significant Transient Recovery Voltage (TRV) and Rate of Rise of Recovery Voltage (RRRV). In order to control both the TRV and RRRV at the circuit breaker blades, surge capacitors are installed on both sides of the circuit breaker as shown in Figure 11.10. A typical surge capacitor used in a 13.8 kV generator circuit breaker is of the order of 0.13–0.25 µF/phase depending on short-circuit rating and other factors. Sometimes the surge capacitor is also called the coupling capacitor in circuit breaker terminology. This device is similar to the grading capacitor shown in Figure 11.9(a).

### 11.5.3  Gas Insulated Switchgear (GIS) Capacitors

The gas insulated switchgear (GIS) type of circuit breaker has been used extensively in utility and industrial applications in the past few years. This is due to the superior performance of GIS equipment and related technological advantages. GIS circuit breakers for tough environments with extreme temperature variations, dust, pollution, or confined space use grading capacitors to control the rate of rise in voltage during switching conditions. The minimum grading capacitor available for the 500 kV level GIS circuit breaker is 200 pF/phase. An example of a GIS grading capacitor is shown in Figure 11.9(b) [4].

## 11.6 CONCLUSIONS

Rotating machines are equipped with dry type insulation in the windings, and hence the impulse withstand capability is lower than the transformers or circuit breakers with oil insulation. The impulse withstand capability of the rotating machines are represented by a curve. The basis of surge protection on any rotating machine is to install a surge capacitor to reduce the rate of rise in voltage and to install a surge arrester to limit the amplitude of the voltage rise. A typical surge protection circuit is presented. The use of grading capacitors in the EHV circuit breakers is a common practice to distribute the voltage uniformly across the interrupting chambers. Grading capacitors are also used in GIS circuit breakers. The use of surge capacitors in circuit breakers for controlling the rate of rise in voltage during a close-in fault is a common practice. These concepts are illustrated using one-line diagrams and photographs.

## PROBLEMS

1. What are the sources of surge voltage in a motor or generator circuit?
2. Describe features of the impulse withstand curve of a motor or generator.
3. Describe functions of the surge protector circuit when installed at the motor terminal.
4. Consider a 0.25 µF/phase at the terminal of a 7.5 MVA, 60 Hz, three-phase, 6.6 kV circuit along with a surge arrester to ground. Calculate the steady-state current

through the surge capacitor. Also calculate the current through the surge capacitor if a short surge arrives at the capacitor with a frequency of 12 kHz. State your observations.

5. What are the advantages of coupling capacitors across the generator circuit breaker?

6. Discuss the grading capacitors used in EHV circuit breakers.

7. A three-phase motor is supplied through a 2,000 kVA, 4.16 kV/0.6 kV, delta/wye connected transformer with 7% impedance. The cable length is 150 ft and the cable is 500 kcmil, unshielded. The surge arrester located at the primary of the transformer is 4.5 kV and has a sparkover voltage of 16 kV. Evaluate if a surge capacitor and surge arrester protection are needed for this motor. State the assumptions.

## REFERENCES

1. Hoenigmann, W. F. (1983). Surge protection of AC motors – when are protective devices required? *IEEE Transactions on Industry Applications, Vol. IA*, 19(5), 836–843.

2. Jackson, D. W. (1979). Surge protection of rotating machines, *IEEE Tutorial Course*, 79 EH0144–6 PWR, 90–111.

3. Gupta, B. K., Lloyd, B. A., Dick, E. P., Narang, A. (1998). Switching surges at large AC machines, CIRGE, Paper No. 11–07.

4. www.maxwelltechnologies.com, Maxwell Technologies, San Diego, CA.

5. ANSI Standard 141 (1993), Recommended Practice for Electric Power Distribution for Industrial Plants (Red Book).

# 12

## CAPACITORS FOR MOTOR APPLICATIONS

### 12.1 INTRODUCTION

Motors are classified by the National Electrical Manufacturers Association (NEMA) as Designs A, B, C, D, F, and wound rotor machines. Design A motors usually have low resistance rotors that provide good running characteristics at the expense of high starting current. A reduced voltage starter may be required for starting this type of motor. Example loads are blowers, fans, machine tools, and centrifugal pumps [1].

Design B motors have a double cage motor and are used for full voltage starting. They have the same starting torque as Design A, but with only 75% of the starting current of a Design A motor. The applications are the same as Design A. Design B motors are more popular than Design A motors.

Design C motors have a double cage and deep bar construction, with higher rotor resistance than Design B. Design C motors have higher starting torque, but lower efficiency and somewhat greater slip than Design B motors. These motors are suitable for constant speed loads, requiring fairly high starting torque while drawing relatively low starting current. Typical loads are compressors, conveyors, crushers, and reciprocating pumps.

Design D motors have the highest starting torque of all the designs. They are single cage motors that provide high starting torque but also have high slip with correspondingly lower efficiency. Design D motors are used for high inertia loads such as bulldozers, die-stamping machines, punch press, and shears.

Design F motors are usually high speed drives directly connected to loads that require low starting torques such as fans or centrifugal pumps. The rotor has low resistance, which produces low slip and correspondingly high efficiency but also low starting torque.

The wound rotor machines are provided with slip rings in the rotor circuit. These machines can be controlled from the rotor circuit using variable resistance or solid state controllers.

### 12.1.1   Effect of Voltage Variation on Induction Motor Characteristics

The typical characteristics of an induction motor corresponding to the voltage variation are listed in Table 12.1. The most important effect of low voltage is the reduction in the starting torque and increase in temperature rise when delivering the rated load. The significant effects with increased voltage are increased starting torque, increased starting current, and decreased power factor. The increased starting torque may cause the couplings to shear off or damage the load equipment.

### 12.2   REACTIVE POWER REQUIREMENTS

The real, reactive, and total current components of a lagging current are shown in Figure 12.1. It can be seen that the reactive current is 90° out of phase with the real current. The magnitude of the current at zero voltage gives the maximum reactive current component of the given current. Induction motors are used in all types of drive applications such as pumps, fans, and industrial drives. More than 60% of the connected load in the electric utility is the induction motor.

**TABLE 12.1** Effect of Voltage Variation on Induction Motor Characteristics

| Item | Function of V | 90% of Voltage | 110% of Voltage |
|---|---|---|---|
| Torque | Voltage$^2$ | Decrease 18% | Increase 20% |
| Slip | 1/Voltage$^2$ | Increase 20% | Decrease 17% |
| Full load speed | Slip speed | Decrease 1% | Increase 1% |
| Efficiency | | | |
| Full load | – | Decrease 2% | Increase 1% |
| 1/2 Load | – | Increase 2% | Decrease 1% |
| Power | | | |
| Full load | – | Increase 1% | Decrease 4% |
| 1/2 Load | – | Increase 4% | Decrease 5% |
| Full load current | – | Increase 10% | Decrease 7% |
| Starting current | Voltage | Decrease 10% | Increase 10% |
| Temperature rise | – | Increase 6°C | Decrease 2°C |
| Full load | Voltage$^2$ | | |
| Max. overload | | Decrease 19% | Increase 20% |

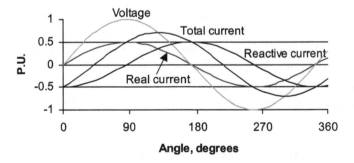

**Figure 12.1** The real, reactive, and total current components of a lagging current.

The typical reactive power characteristics of an induction motor are shown in Figure 12.2. It can be seen that the minimum demand for reactive power occurs at no-load, very close to the synchronous speed. As the load increases, the motor speed decreases and the reactive power consumption increases.

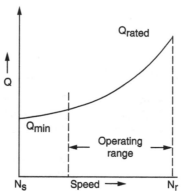

**Figure 12.2** Reactive power profile of an induction motor. $N_S$ = Synchronous speed. $N_r$ = Rated speed.

The power factor of an induction motor at rated load is between 25 and 90%, depending on the size and speed of the motor. At lighter loads, the power factor is poor. In fact, many induction motors operate below the nominal rating, resulting in a lower power factor. Further, the reactive power drawn by these motors is almost constant. These characteristics make the motor load a potential candidate for power factor improvement using shunt capacitors. Addition of significant shunt capacitors at the motor terminal may lead to overvoltages due to self-excitation when the motors are switched off with capacitors.

## 12.3  ADDITIONAL EFFECTS DUE TO CAPACITORS

### 12.3.1  Self-Excitation

The magnetizing current of an induction motor can vary significantly depending on the design. For example, the high efficiency motors operate at lower flux density, and hence the magnetizing current will be less. A capacitor can be used to supply part of the magnetizing current. When the contactor is open and disconnected from the power source, the shunt capacitor provides the magnetizing current and the motor will self-excite or act as a generator. The magnitude of the generated voltage will depend on the capacitor current and the motor speed. In Figure 12.3 the motor excitation curve and the self-excitation curves for various capacitor ratings are shown. The self-excitation is measured by using a voltmeter across the motor terminals. If the capacitor volt-ampere curve does not cross the motor magnetization curve, there can be no voltage due to self-excitation. This condition is represented in Figure 12.3 by capacitor $C_1$. Self-excitation occurs only if the capacitive reactance is equal to or less than the inductive reactance of the motor. The capacitor of curve $C_2$ is approximately the size required to correct the no-load power factor of the motor. Capacitor $C_3$ is approximately large enough to correct the full load power factor of the motor, and the voltage due to self-excitation will be of the order of 1.40 P.U. Such overvoltage magnitudes are not acceptable. The magnitude of the self-excitation voltage depends on the motor design, speed, and inertia of the rotor circuits. However, the actual motor slows down rapidly after the switch is opened and the induced voltage decreases rapidly. The voltage of self-excitation usually collapses in a few seconds as the motor slows down. With high inertia loads, the voltage due to self-excitation is sustained for several minutes.

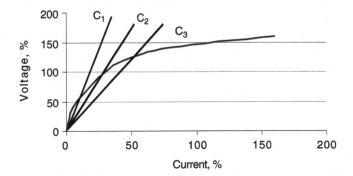

**Figure 12.3** Capacitor and motor magnetization curve for self-excitation.

### 12.3.2   Transient Torques

Even if the capacitor has been properly chosen from the over-voltage point of view, it may still be large enough to produce excessive torques in certain motor applications with high inertia loads such as large compressors and air conditioners. High stresses may be produced in the shaft and couplings if the motor is reconnected while still rotating and generating a voltage due to self-excitation. The magnitude of the line and motor voltages, the angle between them, and the impedance of the motor determine the peak value of electrical torque. In motors, the peak transient torque should not be permitted to exceed the allowable peak torque.

## 12.4   SELECTION OF CAPACITOR RATINGS

The capacitor ratings selected for the power factor correction of an induction motor should not produce self-excitation and undesirable transient torques. Usually the capacitor current should not exceed the motor no-load current. The desirable capacitor ratings for various motors are listed in Tables 12.2 through 12.5:

> Table 12.2: Maximum capacitor ratings for pre-U-frame NEMA Design B, 230 V, 460 V, and 575 V squirrel-case induction motors
> Table 12.3: Maximum capacitor ratings for U-frame NEMA Design B, 230 V, 460 V, and 575 V squirrel-cage motors
> Table 12.4: Maximum capacitor ratings for T-frame NEMA Design B, 230 V, 460 V, and 575 V squirrel-cage motors
> Table 12.5: Maximum capacitor ratings for NEMA Design C, D, and wound rotor motors

### Example 12.1

Consider a 100 hp, 460 V, 3,600 rpm, three-phase NEMA U-frame motor. Select a suitable capacitor bank from the table and calculate the reduction in the line current.

**TABLE 12.2** Maximum Capacitor Ratings for Pre-U Frame NEMA Design B Induction Motors

| Motor hp | kVAR for 3,600 rpm Motor | kVAR for 1,800 rpm Motor | kVAR for 1,200 rpm Motor | kVAR for 900 rpm Motor | kVAR for 720 rpm Motor | kVAR for 600 rpm Motor |
|---|---|---|---|---|---|---|
| 3 | 1.5 | 1.5 | 1.5 | 1.5 | 2.5 | 3.5 |
| 5 | 2 | 2 | 2 | 2 | 4 | 4.5 |
| 7.5 | 2.5 | 3 | 3 | 4 | 5.5 | 6 |
| 10 | 3 | 3.5 | 3.5 | 5 | 6.5 | 7.5 |
| 15 | 4 | 5 | 5 | 6.5 | 8 | 9.5 |
| 20 | 5 | 5 | 6.5 | 7.5 | 9 | 12 |
| 25 | 6 | 6 | 7.5 | 9 | 11 | 14 |
| 30 | 7 | 7 | 9 | 10 | 12 | 16 |
| 40 | 9 | 9 | 11 | 12 | 15 | 20 |
| 50 | 12 | 11 | 13 | 15 | 19 | 24 |
| 60 | 14 | 14 | 15 | 18 | 22 | 27 |
| 75 | 17 | 16 | 18 | 21 | 26 | 32.5 |
| 100 | 22 | 21 | 25 | 27 | 32.5 | 40 |
| 125 | 27 | 26 | 30 | 32.5 | 40 | 47.5 |
| 150 | 32.5 | 30 | 35 | 37.5 | 47.5 | 52.5 |
| 200 | 40 | 37.5 | 42.5 | 47.5 | 60 | 65 |
| 250 | 50 | 45 | 52.5 | 57.5 | 70 | 75 |
| 300 | 57.5 | 52.5 | 60 | 65 | 80 | 87.5 |
| 350 | 65 | 60 | 67.5 | 75 | 87.5 | 95 |
| 400 | 70 | 65 | 75 | 85 | 95 | 105 |
| 450 | 75 | 67.5 | 80 | 92.5 | 100 | 110 |
| 500 | 77.5 | 72.5 | 82.5 | 97.5 | 107.5 | 115 |

**TABLE 12.3**   Capacitor Ratings for U-Frame NEMA Design B Motors

| Motor hp | kVAR for 3,600 rpm Motor | kVAR for 1,800 rpm Motor | kVAR for 1,200 rpm Motor | kVAR for 900 rpm Motor | kVAR for 720 rpm Motor | kVAR for 600 rpm Motor |
|---|---|---|---|---|---|---|
| 2 | 1 | 1 | 1 | 1 | – | – |
| 3 | 1 | 1 | 1 | 2 | – | – |
| 5 | 2 | 2 | 2 | 2 | – | – |
| 7.5 | 1 | 2 | 4 | 4 | 5 | 5 |
| 10 | 2 | 2 | 4 | 5 | 5 | 5 |
| 15 | 4 | 4 | 4 | 5 | 10 | 10 |
| 20 | 4 | 5 | 5 | 5 | 10 | 10 |
| 25 | 5 | 5 | 5 | 5 | 10 | 10 |
| 30 | 5 | 5 | 5 | 10 | 10 | 10 |
| 40 | 5 | 10 | 10 | 10 | 15 | 15 |
| 50 | 5 | 10 | 10 | 15 | 20 | 20 |
| 60 | 10 | 10 | 10 | 15 | 25 | 30 |
| 75 | 15 | 15 | 15 | 20 | 40 | 35 |
| 100 | 15 | 20 | 25 | 25 | 45 | 45 |
| 125 | 20 | 25 | 30 | 30 | 45 | 45 |
| 150 | 25 | 30 | 30 | 40 | 55 | 50 |
| 200 | 35 | 40 | 60 | 55 | 60 | 60 |
| 250 | 40 | 40 | 60 | 80 | 80 | 100 |
| 300 | 45 | 45 | 80 | 80 | – | 120 |
| 350 | 60 | 70 | 80 | 80 | – | – |
| 400 | 60 | 80 | 80 | 160 | – | – |
| 450 | 70 | 100 | – | – | – | – |
| 500 | 70 | – | – | – | – | – |

TABLE 12.4 Capacitor Ratings for T-Frame NEMA Design B Motors

| Motor hp | kVAR for 3,600 rpm Motor | kVAR for 1,800 rpm Motor | kVAR for 1,200 rpm Motor | kVAR for 900 rpm Motor | kVAR for 720 rpm Motor | kVAR for 600 rpm Motor |
|---|---|---|---|---|---|---|
| 3 | 1.5 | 1.5 | 2.5 | 3 | 3 | 3 |
| 5 | 2 | 2.5 | 3 | 4 | 4 | 4 |
| 7.5 | 2.5 | 3 | 4 | 5 | 5 | 5 |
| 10 | 4 | 4 | 5 | 6 | 7.5 | 8 |
| 15 | 5 | 5 | 6 | 7.5 | 8 | 10 |
| 20 | 6 | 6 | 7.5 | 9 | 10 | 12 |
| 25 | 7.5 | 7.5 | 8 | 10 | 12 | 18 |
| 30 | 8 | 8 | 10 | 14 | 15 | 22.5 |
| 40 | 12 | 13 | 16 | 18 | 22.5 | 25 |
| 50 | 15 | 18 | 20 | 22.5 | 24 | 30 |
| 60 | 18 | 21 | 22.5 | 26 | 30 | 35 |
| 75 | 20 | 23 | 25 | 28 | 33 | 40 |
| 100 | 22.5 | 30 | 30 | 35 | 40 | 45 |
| 125 | 25 | 36 | 35 | 42 | 45 | 50 |
| 150 | 30 | 42 | 40 | 52.5 | 52.5 | 60 |
| 200 | 35 | 50 | 50 | 65 | 68 | 90 |
| 250 | 40 | 60 | 62.5 | 82 | 87.5 | 100 |
| 300 | 45 | 68 | 75 | 100 | 100 | 120 |
| 350 | 50 | 75 | 90 | 120 | 120 | 135 |
| 400 | 75 | 80 | 100 | 130 | 140 | 150 |
| 450 | 80 | 90 | 80 | 140 | 160 | 160 |
| 500 | 100 | 120 | 82.5 | 160 | 180 | 180 |

**TABLE 12.5**  Capacitor Ratings for NEMA Design C, D, and Wound Rotor Motors

| Motor hp | Design C 1800 rpm and 1200 rpm Motor | Design C 900 rpm Motor | Design D 1,200 rpm Motor | Wound Rotor Motor |
|---|---|---|---|---|
| 15 | 5 | 5 | 5 | 5.5 |
| 20 | 5 | 6 | 6 | 7 |
| 25 | 6 | 6 | 6 | 7 |
| 30 | 75 | 9 | 10 | 11 |
| 40 | 10 | 12 | 12 | 13 |
| 50 | 12 | 15 | 15 | 17.5 |
| 60 | 17.5 | 18 | 18 | 20 |
| 75 | 19 | 22.5 | 22.5 | 25 |
| 100 | 27 | 27 | 30 | 33 |
| 125 | 35 | 37.5 | 37.5 | 40 |
| 150 | 37.5 | 45 | 45 | 50 |
| 200 | 45 | 60 | 60 | 65 |
| 250 | 54 | 70 | 70 | 75 |
| 300 | 65 | 90 | 75 | 85 |

Solution

HP $= 100$,     V $= 460$

Capacitor bank for power factor correction from Table 12.3 $= 15$ kVAR

Assume a power factor of 0.8.

$$I = \frac{100\,\text{hp} \times 746}{\sqrt{3} \times 460\,\text{V} \times 0.8} = 117\,\text{A}$$

$$I = 117\,\text{A}\,(0.8 - j\,0.6) = (93.6 - j\,70.2)\,\text{A}$$

$$I_\text{C} = \frac{15{,}000}{\sqrt{3} \times 460\,\text{V}} = 18.8\,\text{A}$$

$$I_\text{r} = 93.6 - j\,70.2 + j\,18.8 = (93.6 - j\,51.4) = 106\,\text{A}$$

$$\text{Reduction in the phase current} = \frac{(117 - 106)}{117} \times 100 = 9.4\%$$

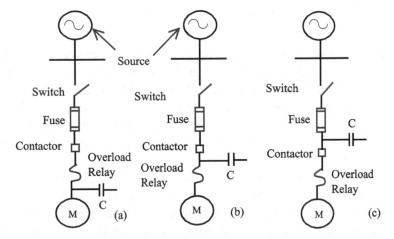

**Figure 12.4** Location of shunt capacitors for power factor correction.

### 12.4.1  Location of Shunt Capacitors

The shunt capacitors can be connected across the motor terminals in three different modes as shown in Figure 12.4. The three important schemes are described below.

#### 12.4.1.1  Shunt Capacitors at the Motor Terminal

The shunt capacitors can be connected at the motor terminals as shown in Figure 12.4(a). Usually the induction motors are equipped with thermal overload relay. If shunt capacitors are applied, the line current will be reduced. Therefore the thermal overload relays need careful resetting. This scheme is suitable for new motor installations; both the fuse and overload relay setting can be selected for reduced line current.

#### 12.4.1.2  Switching Motor and Capacitor as One Unit

In the case of an existing motor and overload relay, the capacitor bank can be installed after the overload relay, as shown in Figure 12.4(b). The line current seen by the overload relay will be same before and after the shunt capacitor installation.

12.4.1.3   Shunt Capacitor on the Line Side of
the Starter

In certain motor applications such as hoist and crane, the motor
is used for plugging, reversing, and different speeds with higher
inertia. In such applications, it is advisable to have the shunt
capacitors connected to the line side of the starter as shown
in Figure 12.4(c). In this scheme, the capacitor is connected
before the starting device and the capacitor can be left
permanently.

Each scheme has its own advantages and drawbacks. In
the first two cases, when the capacitor bank is switched along
with the motor as a unit, overvoltages can be produced due to
self-excitation and transient torques. In the third scheme, the
overvoltage problem can arise at the open end, once the motor
load is disconnected.

## 12.5   MOTOR STARTING
## AND RELATED ISSUES

Large squirrel-cage motors and industrial synchronous motors
draw several times their full load current from the supply
during starting. The power factor during the starting is usually
in the range of 0.15–0.30 lagging. A typical starting current
profile is illustrated in Figure 12.5. The actual shape and magnitude of the staring current curve depends on the motor
design, the voltage at the motor terminals, and the speed-
torque characteristic of the mechanical load connected to the
motor. The starting current through the system impedances

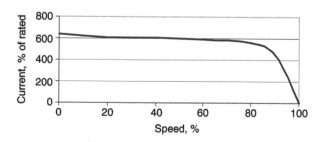

**Figure 12.5**   Starting current profile of a typical motor.

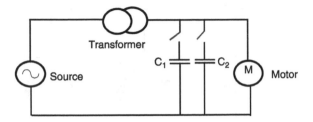

**Figure 12.6** Switched capacitors for motor starting; shown for one phase only.

can result in an unacceptable voltage drop that may be large enough to cause contactors to drop out and influence the ability of the motor to start [2,3].

Shunt capacitors are sometimes used to reduce the voltage dip when starting a large motor. Their effect is to reduce the reactive component of the input kVA. With this method, the high inductive component of the normal starting current is offset, at least partially, by the addition of capacitors to the motor bus during the starting period. The capacitor size needed for this purpose is usually 2–3 times the motor full load kVA rating. In order to control the voltage properly, the capacitor is usually switched out in steps as the motor accelerates. Due to the large kVAR size of these capacitors, they are usually in the circuit for only a few seconds during motor starting. Such a scheme with two steps of capacitor is shown in Figure 12.6.

## Example 12.2

Consider a 100 hp, three-phase, 60 Hz induction motor. The estimated starting current is six times the rated current and the power factor is 0.3. If the full load kVA is 100, calculate the capacitor rating required for starting.

Solution

Inrush kVA $(6 \times 100) = 600\,\text{kVA}$
Power factor $(\cos \theta) = 0.3$

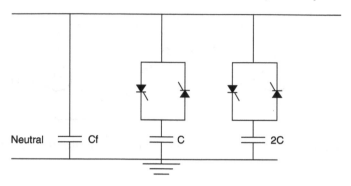

**Figure 12.7** Fixed and thyristor switched capacitors for motor starting.

$$\theta = \cos^{-1}(0.3) = 81°, \ \sin \theta = 0.954$$
$$\text{kVAR} \ (600 \times 0.954) = 572.4 \, \text{kVAR}$$

Consider kVAR sizes of 200, 200, and 100 and arrive at an optimum value. Use a switched capacitor controller.

### 12.5.1 Motor Starting Using Shunt Capacitors [2,3]

An experimental controller with fixed and thyristor switched capacitors was used to analyze the performance of motor starting. The per phase schematic of the controller is shown in Figure 12.7. The motor is a 2 hp, 220 V, 60 Hz, 7 A, 3600 rpm, and wye-connected induction machine. The minimum reactive power requirement of this motor is 1,000 VAR. A fixed capacitor of 50 μF/phase is required to provide this compensation. To avoid self-excitation, a fixed capacitor value of 40 μF was used. The corresponding reactive power profile is shown in Figure 12.8, curve B. The switched capacitors are selected to provide the varying load demand of the reactive power. The required switched capacitors are provided in two steps with values of 10 and 20 μF/phase, respectively. The measured reactive power of the motor with fixed and switched capacitors is shown in Figure 12.8, curve C. The power factor profile of the conventional motor, with 40 μF fixed capacitors and with both fixed and switched capacitors, is shown in Figure 12.9.

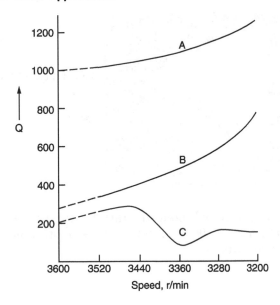

**Figure 12.8** Reactive power profile of the induction motor. A. Conventional motor. B. With fixed capacitors. C. With fixed and switched capacitors.

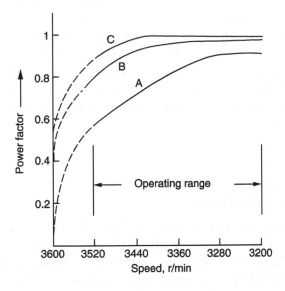

**Figure 12.9** Power factor profile of the induction motor.

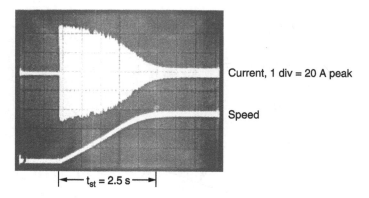

Current, 1 div = 20 A peak

Speed

$t_{st} = 2.5$ s

**Figure 12.10** Measured no-load starting current and speed of a 2 hp, 220 V, 7 A, 3,600 rpm, three-phase motor, no capacitors; X axis: 1 division = 0.5 s.

This experiment shows that the combination of fixed and switched capacitors can improve the power factor considerably throughout the operating range.

The starting current and the speed are recorded for the laboratory experimental motor without and with shunt capacitors. The starting time of the test motor is 2.5 and 3 s at no-load and rated-load, respectively. The starting current and the speed of the motor for the no-load without any shunt capacitors are shown in Figure 12.10. The starting current is 40 A and is 2.5 times the rated current. The steady-state, no-load current is 4.3 A and the rated-load current is 16 A.

### 12.5.2  Starting Transients with Fixed Capacitors

The starting current of the motor, the capacitor, and the motor speed of a motor with fixed capacitors are shown in Figure 12.11. The starting current reduces significantly. The steady-state line current increases, depending on the size of the capacitor. The capacitor current remains constant.

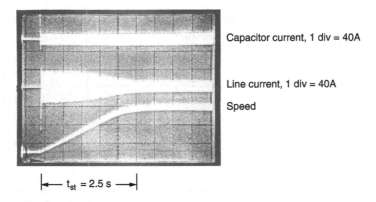

**Figure 12.11** Measured no-load starting current and speed of a 2 hp, 220 V, 7 A, 3,600 rpm, three-phase motor; C = 240 MFD/phase; X axis: 1 division = 0.5 s.

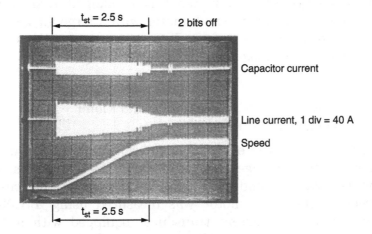

**Figure 12.12** Measured no-load starting current and speed of a 2 hp, 220 V, 7 A, 3,600 rpm, three-phase motor with switched capacitors; X axis: 1 division = 0.5 s.

### 12.5.3 Starting Transients with Switched Capacitors

The starting current of the motor, the capacitor, and the motor speed of a motor with switched capacitors are shown in Figure 12.12. The starting current of the motor reduces

**TABLE 12.6**  Summary of Measured Currents in the 2 hp Laboratory
Test Motor

| Condition | Starting Current, A | Steady-State Current, A |
|---|---|---|
| Without capacitors | 50.0 | 4.3 |
| With 240 MFD/phase | 40.0 | 6.0 |
| With 480 MFD/phase | 35.0 | 28.0 |
| With switched capacitors | 32.0 | 4.3 |

significantly with fixed capacitors. The steady-state line current decreases since the capacitor is switched off after the starting. The summary of the measured starting currents is shown in Table 12.6.

With switched capacitors, the starting current is reduced and the steady-state current remains unchanged. It should be noted that this is a very small motor with a large air gap and the no-load current is higher. This demonstrates the use of fixed or switched capacitors for motor starting.

### 12.5.4 Effect of Shunt Capacitors on the Motor Voltage and Current

In order to show the effect of shunt capacitors at the motor terminal, the voltage and the current waveforms are analyzed without and with shunt capacitors. The circuit current in the 2 hp motor was measured using a 1,000:1 ratio clamp on a current transducer equipped with a $10\,\Omega$ burden in the secondary. The output of the current sensor will produce 1 V for a line current of 100 A. A DC tacho-generator mounted on the motor shaft provides a signal proportional to the motor speed. The output of the speed signal is 140 V DC at a rated speed of 3,000 rpm. The voltage is input to a potential divider with $70\,k\Omega$ and $1\,k\Omega$ in series. The voltage across the $1\,k\Omega$ with respect to the ground is used as the speed signal, which is 2 V at rated 3,600 rpm. The steady-state and the transient waveforms are recorded using a storage oscilloscope. Polaroid pictures are taken for

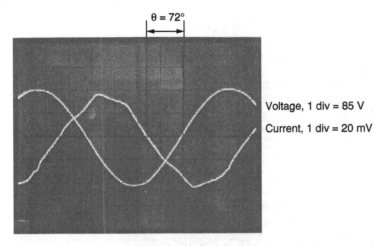

θ = 72°

Voltage, 1 div = 85 V
Current, 1 div = 20 mV

**Figure 12.13** Steady-state voltage and phase current at 120 V/phase, 3 A, 3,590 rpm and at no-load without capacitors; X axis: 1 division = 2 ms.

future use. The following waveforms are presented at no-load and at rated-load.

1. *No-load voltage and current without capacitors.* The steady-state voltage and the current of the motor without shunt capacitors are presented at no-load, 120 V/phase, 3 A, 3,590 rpm in Figure 12.13. The voltage is a sine wave and the current is distorted. The phase angle between the voltage and the current is 72° and the corresponding power factor is 0.31.

2. *Rated-load voltage and current without capacitors.* The steady-state voltage and the current of the motor without shunt capacitors are presented at rated-load, 120 V/phase, 7 A, 3,200 rpm in Figure 12.14. The voltage is a sine wave and the current is peaky and distorted. The phase angle between the voltage and the current is 31° and the corresponding power factor is 0.86.

3. *No-load voltage and current with 240 MFD/phase.* The steady-state voltage and the current of the motor with 240 MFD/phase shunt capacitors are presented at

θ = 31°

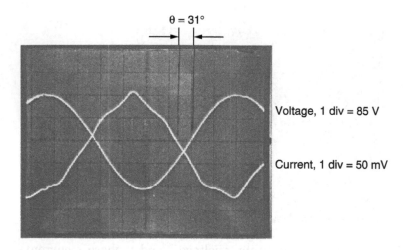

Voltage, 1 div = 85 V

Current, 1 div = 50 mV

**Figure 12.14** Steady-state voltage and phase current at 120 V/phase, 7 A, 3,200 rpm and at rated-load without capacitors; X axis: 1 division = 2 ms.

θ = 22°

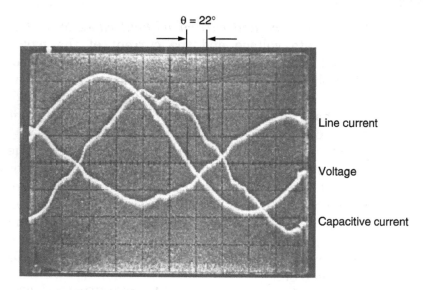

Line current

Voltage

Capacitive current

**Figure 12.15** Steady-state voltage and phase current at 120 V/phase, 3 A, 3,590 rpm and at no-load with 240 MFD/phase; X axis: 1 division = 2 ms.

θ = 10°

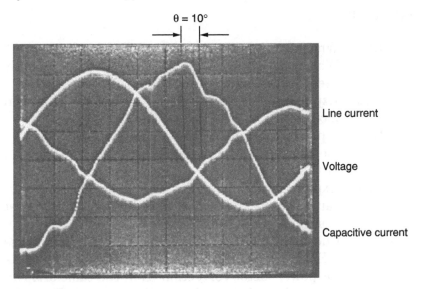

Line current

Voltage

Capacitive current

**Figure 12.16** Steady-state voltage and phase current at 120 V/phase, 7 A, 3,200 rpm and at no-load with 240 MFD/phase; X axis: 1 division = 2 ms.

no-load, 120 V/phase, 3 A, 3,590 rpm in Figure 12.15. The voltage is a sine wave and the current is distorted. The phase angle between the voltage and the current is 22° and the corresponding power factor is 0.93. The capacitor current is leading.

4. *Rated-load voltage and current with 240 MFD/phase.* The steady-state voltage and the current of the motor with 240 MFD/phase shunt capacitors are presented at rated-load, 120 V/phase in Figure 12.16. The phase angle between the voltage and the current is 10° and the corresponding power factor is 0.99. The capacitor current is leading. The capacitor current and the line are distorted to some extent.

From the waveforms it can be seen that the shunt capacitors at the motor terminal cause distortion in the line current to a certain level. The level of distortion depends on the loading level of the motor.

## PROBLEMS

12.1. What are the different types of induction motor? Which design of induction motor is commonly used in drive applications?

12.2. Explain self-excitation as it applies to induction motors.

12.3. What is the typical power factor of an induction motor? How is the power factor related to the load of the motor?

12.4. Explain where the shunt compensation capacitors can be connected in the motor? Which is the most desirable location?

12.5. Consider a 20 hp, 575 V, 3,600 rpm, three-phase NEMA Design D motor. Select a capacitor bank for power factor correction from the table and calculate the reduction in the line current. State the assumptions, if any.

12.6. A 400 hp, three-phase, 60 Hz induction motor is used for pumping application. The estimated starting current is 6.2 times the rated current and the power factor is 0.32. If the full load KVA is 400, calculate the capacitor rating required for starting. How can the required capacitive compensation be applied?

## REFERENCES

1. ANSI Standard 141 (1993), Recommended Practice for Electric Power Distribution for Industrial Plants (Red Book).

2. Natarajan, R., Misra, V. K. (1991). Starting transient current of induction motors without and with terminal capacitors, *IEEE Transactions on Energy Conversion*, EC–6(1), 1134–1139.

3. Natarajan, R. (1989). A solid state power factor controller for continuous miner application, *Mineral Resources Engineering*, 2(3), 239–248.

# 13

# CAPACITORS FOR MISCELLANEOUS APPLICATIONS

## 13.1 INTRODUCTION

Capacitors are used in many industrial applications other than utility installations for power factor correction purposes [1–11]. Applications are discussed for arc furnaces, resistive welding equipment, lighting, capacitor-start, capacitor-run motors, ferroresonance transformers, and inverters. Pulsed power supply applications are also discussed.

## 13.2 ELECTRIC ARC FURNANCE

Steel production with electric arc furnaces is vital to the infrastructure of industrialized countries. As is well known, there are three types of electric furnaces available: the resistance, induction, and arc types. The resistance furnace produces limited flicker due to the resistive nature of the load. Most induction furnaces operate at high frequency and are therefore connected to the power system through frequency converters and present a constant load. Three-phase electric arc furnaces are used extensively to make high quality steel with significant melting capabilities. During the melting period, pieces of steel between the electrodes in these furnaces produce short circuits on the secondary of the transformer to which the electrodes are connected. Therefore, the melting period is characterized by severe fluctuations of current at low power factor values. This melting process takes around 50–120 min, depending on the type of fur-nace. When the steel is melted down to a pool, the arc length can be maintained uniformly by regulating the electrodes. This is known as the refining mode and the electrical load is constant with fairly high power factor. During the refining period, the power is supplied to the furnace every 2 h for a period of 10–40 min, depending on the size of the furnace. The severe fluctuation during the melting process is responsible for significant voltage drop in the power system and causes flicker. Voltage flicker in an electric arc furnace corresponds to oscillations in the bus voltages with a frequency of 1–30 Hz. Therefore, suitable filters or static VAR controllers are used to reduce flicker and harmonics, and to improve the power factor.

A typical power supply system for an electric arc furnace is shown in Figure 13.1. In this case, the power system consists of

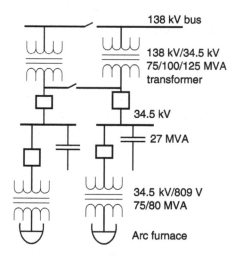

**Figure 13.1**  One-line diagram of the power supply for two arc furnaces.

two 138 kV lines with a normally open tie switch between them. The stepdown transformer is 75 MVA, 138 kV/34.5 kV, delta/ grounded wye, 8% impedance. The 34.5 kV circuit breakers are intended to open and close so that each arc furnace can operate independently. The arc furnace transformer is rated at 75 MVA, 34.5 kV/277 V, delta/wye connected. Each arc furnace has a 200 ton capacity. In order to improve the power factor of the arc furnace, each unit is provided with a 27 MVAR shunt capacitor bank. To protect the capacitor bank from energization transients, pre-insertion resistors are used in the circuit switcher. Vacuum circuit breakers are used to energize the arc furnaces. The power factor is 0.707 during the melting process without reactive compensation and the desired power factor is 0.90. In order to provide enough voltage margins on the capacitor rating, three series capacitors are used in each phase. This gives a voltage rating of $3 \times 7,200$ V/can $\times 1.732 = 37,400$ V/phase. The voltage rating of the system is 34.5 kV. The capacitor bank is connected in ungrounded wye with neutral unbalance detection scheme [1].

Figure 13.2 shows the typical real power and reactive power flow through an arc furnace for one cycle. It can be seen that the furnace is charged in three stages during the melting period. Toward the end of each charging, the power output is reduced since a continuous arc is established and the power input is decreased to prevent overheating of the refractory. During the refining period, both the real and reactive power requirements are very small. The power supply characteristics of a few practical arc furnace installations are presented in Table 13.1. It can be seen that if the furnace transformer rating to main transformer ratio is around 75% the furnace operates satisfactorily. Significant filter rating is needed in each application.

## 13.3 RESISTIVE SPOT WELDING

In the electric welding process, two pieces of metal are brought together and then fused by generating heat. Resistance welding (butt welding, flash butt welding, spot welding, projection

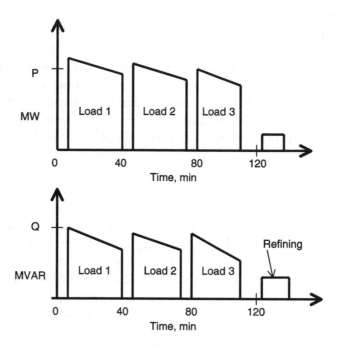

**Figure 13.2**  Typical P and Q profiles of an arc furnace.

**TABLE 13.1**  Power Supply Systems for a Few Arc Furnaces

| Description | Furnace A | Furnace B | Furnace C | Furnace D |
|---|---|---|---|---|
| Rating MW | 20 | 60 | 40 | 57 |
| Transformer MVA | 27.1 | 56 | 56 | 60 |
| Transformer kV | 230/22.8 | 230/66 | 230/66 | 230/35 |
| Furnace voltage | 22.8 | 33 | 33 | 34.5 |
| Furnace MVA/ Transformer MVA | 0.73 | 1.07 | 0.71 | 0.95 |
| Filter MVA | 15 (3rd) | 65, SVC | 65, SVC | 25 (3rd) |
| Operation | Well | No | Well | Not clear |

welding, seam welding, and energy storage welding), arc welding (metal arc welding and carbon arc welding), hydrogen welding, and helium or organ welding are commonly used welding methods [2,3].

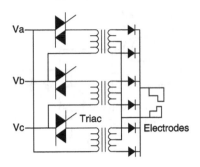

**Figure 13.3**  Power supply configuration for a pair of electrodes.

Resistive welding is used to weld two plates by placing them together between the electrodes. The welding current flows through the electrode tips, producing spot-weld. In a specific configuration, there are four electrode pairs (top and bottom), a total of eight electrodes. A speed control cam determines the mechanical timing of the welder. During each rotation of the cam, the electrodes move to the next position, squeeze the welding material, make the weld, and hold the material. Then the electrodes move to the next position. The front and back electrodes are 180° out of phase, so that when the front electrodes are welding, the back welding time is 1–4 cycles.

The power supply arrangement for a pair of electrodes is shown in Figure 13.3. The three-phase power supply uses triacs, which are equivalent to back-to-back thyristors with common gates to determine the weld time and heat. The triacs control the amount of current flow through the electrodes. The current flow through the electrodes is peaky, lasting only for a few cycles. The equivalent AC currents can be peaky and the power factor is very low. When the welder current magnitude reaches around 1,000 A, there will be voltage dips and hence flicker problems. The rating of a typical three-phase spot welder in Reference [2] is 1500 kVA. The power supply system for the resistive spot welder is shown in Figure 13.4. The power factor is compensated at the 12.47 kV level using a three-phase, 1,200 kVAR shunt capacitor bank. At 480 V, there is a 900 kVAR switched capacitor bank.

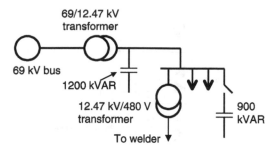

**Figure 13.4**  One-line diagram of the power supply circuit to the electric spot welder.

## 13.4  CAPACITORS FOR LIGHTING APPLICATIONS

The tungsten filament lamp is a resistive element, operating at unity power factor and sensitive to voltage changes. The electric discharge lamps produce light based on the phenomenon of excitation and ionization in a gas or vapor. The commonly used lamps for advertisement or decoration purposes are the sodium vapor lamp for yellow light, the mercury vapor lamp with low pressure for bluish green light, the mercury vapor lamp with high pressure for bluish white light, and the neon vapor lamp for red light. In these types of lamps, the electrons have to be emitted from the cathode for starting and maintaining the arc. In the cold cathode type of lamps, a voltage supply of 100–200 V is required between the electrodes. Therefore, a transformer is required to provide such a voltage. The hot cathode lamp can work from a standard distribution supply. The cathode needs to be heated using a power supply. Some of the power supplies and the role of capacitors in these circuits are presented below.

### 13.4.1  Neon Lamps for Red Light

The neon lamp is a cold cathode type and the electrodes are in the form of iron shells and coated on the inside. The neon gas is used in the tube and the color emitted is red. The power supply

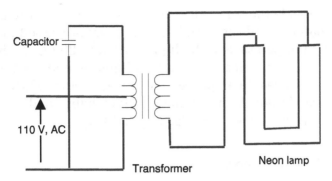

**Figure 13.5** Power supply for a neon lamp.

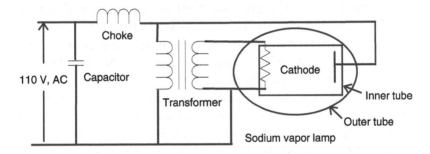

**Figure 13.6** Power supply for a sodium vapor lamp.

is provided through a transformer. The circuit is shown in Figure 13.5. The transformer has a high leakage reactance to stabilize the arc in the lamp. A capacitor is used to improve the power factor of the circuit. Helium and neon with color tubing produces a variety of color effects on the light.

### 13.4.2  Sodium Vapor Lamp

A typical electric circuit of a sodium vapor lamp is shown in Figure 13.6. The lamp consists of a discharge tube suitable for withstanding high temperatures. An outer tube for filtering certain parts of the spectrum of the emitted light surrounds the discharge tube. The cathodes are heated by a power supply from a transformer. The lamp is started using the neon gas inside the tube. The sodium is provided in the form of a solid material and the initial discharge vaporizes sodium. The operating temperature of the light is around 300°C. A choke is provided for stabilizing the electric discharge and the terminal capacitor is used to improve the power factor. These vapor lamps are designed to produce yellow light, and they are mainly used for street and highway lighting.

### 13.4.3 Mercury Vapor Lamp

A typical electric circuit of a mercury vapor lamp is shown in Figure 13.7. The lamp consists of a discharge tube suitable for withstanding high temperatures. An outer tube for filtering certain spectrum of the emitted light surrounds the discharge tube. The space between the two tubes is filled with an inert gas. The lamp is started using argon gas inside the tube. The mercury is provided in the form of a solid material and the initial discharge vaporizes mercury. The choke is provided for stabilizing the electric discharge and the terminal capacitor is used to improve the power factor. The pressure inside the discharge tube is of the order of 1–10 atmospheres used for lighting purposes. If the pressure is low in the discharge tube, most of the radiation is in the ultraviolet region. By coating the inside of the lamp with phosphor, the ultraviolet radiation is converted to visible light. Depending on the pressure inside the lamp, bluish green or bluish white type of light will be available.

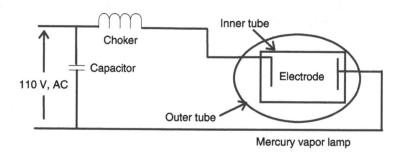

**Figure 13.7** Power supply for a mercury vapor lamp.

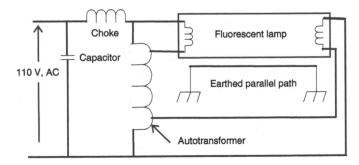

**Figure 13.8** Power supply for a fluorescent lamp.

### 13.4.4 Fluorescent Lamps

The power supply circuit for a typical fluorescent lamp is shown in Figure 13.8. The fluorescent lamp is a low intensity type of electric discharge lamp using mercury vapor. There are two electrodes at the end of the tube. The electrons are propelled at extremely high speeds at each end of the lamp. The energy resulting from the collisions between the electrons and the mercury atom because of the very low vapor pressure is emitted in the ultraviolet region. To convert the ultraviolet into visible light, the inside of the tube is coated with phosphors. An autotransformer is used to provide much of the voltage to start the lamp. The starting voltage directly strikes the arc. A small amount of voltage from the autotransformer heats the electrodes. Fluorescent lamps produce flicker or stroboscopic effect since the 60 Hz AC supply goes to voltage zero on a regular basis. Flicker corrections can be applied to a pair of lamps. There are many designs available based on the same principle. In this circuit, the capacitor is used to correct the power factor. The typical values of capacitor units used in some lamp circuits are summarized in Table 13.2.

### 13.5 CAPACITOR RUN SINGLE-PHASE MOTORS

A one-line diagram of a single-phase, capacitor start, capacitor run motor is shown in Figure 13.9. The specification of a

**TABLE 13.2**   Typical Capacitor Values Used in Lamp Circuits

| Lamp Type | Wattage | Capacitor μF |
|---|---|---|
| High pressure sodium lamp | 50 | 10 |
| High pressure sodium lamp | 200 | 32 |
| Metal halide lamp | 70, 100 | 12 |
| Metal halide lamp | 150 | 20 |
| High mercury vapor lamp | 50 | 7 |
| High mercury vapor lamp | 250 | 18 |
| Fluorescent lamp | 18, 36 | 4.5 |
| Fluorescent lamp | 58 | 7 |
| Compact fluorescent lamp | 26 | 4.5 |

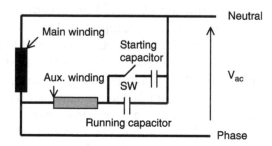

**Figure 13.9**   Single-phase capacitor motor.

typical single-phase capacitor motor is 1.5 kW, 220 V, 60-Hz, 1800 r/min, capacitor start capacitor run. These motors are produced for ratings of 0.25 hp, 0.33 hp, 0.5 hp, 0.75 hp, 1 hp, and 1.5 hp. The main winding is designed for 120 V operation. The auxiliary winding is connected to the starting capacitor and the run capacitor. As soon as the motor is started, the starting winding circuit is opened through the centrifugal switch. The run capacitor remains in service and provides power factor correction. The current through the auxiliary winding and the run capacitor is phase-shifted from the main winding and a rotating magnetic field is created. The capacitor produces phase-shifted currents and also improves the overall power factor of the motor. The typical values of capacitor units used in some single-phase capacitor motors are summarized in

TABLE **13.3**  Typical Capacitor Values Used in Capacitor Motors

| Motor Rating | Voltage | Capacitor μF |
|---|---|---|
| 0.25–hp, 1075 RPM | (208 to 230) V | 5 MFD, 370 V |
| 0.5–hp, 1075 RPM | (208 to 230) V | 5 MFD, 370 V |
| 0.75–hp, 1075 RPM | 277 V | 15 MFD, 370 V |

Table 13.3. A brief description of the electrolytic capacitors used for starting of the AC induction motors is given in Chapter 5.

## 13.6  FERRORESONANT TRANSFORMER APPLICATIONS

### 13.6.1  Ferroresonant Transformer

In certain AC and DC applications, there is a need to maintain constant input voltage. Such loads include computers, televisions, and microprocessor-supported equipment. Ferroresonant transformers have been used to provide constant voltage supply over the past several decades. This is a robust type of power supply with no maintenance requirements. A ferroresonant transformer is a nonlinear device that is designed to provide passive voltage regulation, using magnetics. There are no complex feedback circuits to monitor and adjust the output voltage level. This transformer is designed to operate within a predetermined regulation band, typically 1–4%. A nonlinear transformer differs from a linear transformer in that the output voltage will not deviate outside this regulation band, regardless of what happens on the input. On the other hand, as the term implies, the output of a linear transformer is directly proportional to the input. That is, whatever happens at the input will directly affect the output. Ferroresonant transformers also have the unique characteristic of being able to store energy up to one half cycle because of their "tank circuit" design. A typical ferroresonant transformer is shown in Figure 13.10. It consists of a primary winding,

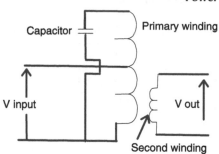

**Figure 13.10** A typical ferroresonant transformer for constant voltage output.

a secondary winding, and a capacitor. For a typical input variation of 180–250 V, the output will be maintained at ±1%. The capacitor is used to produce resonance in the circuit and the output voltage is maintained constant. The advantages of the ferroresonant transformer are that it provides isolation, it has no moving parts, and it responds within one cycle. The disadvantages of the ferroresonant transformer are that the output contains some harmonics, it is less efficient, and it can be noisy.

### 13.6.2   Ferroresonant Inverter

The basic function of an inverter is to convert DC voltage into AC voltage at the desired frequency. A simplified block diagram of a ferroresonant inverter is shown in Figure 13.11. The inverter consists of a DC filter, a thyristor controlled inverter bridge, a triggering circuit to the thyristor circuit, and a ferroresonant transformer. The DC filter protects the inverter circuit from the voltage transients of the DC source. The thyristor controlled circuit produces square wave output at the desired frequency. The triggering circuit provides the firing signal to the inverter circuit and the frequency control. The ferroresonant transformer takes the square wave input voltage and produces sine wave output voltage at the desired frequency.

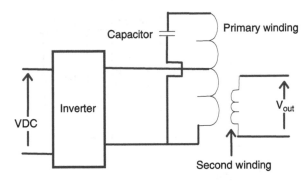

**Figure 13.11** A typical ferroresonant inverter for constant voltage output.

## 13.7 CAPACITORS FOR PULSED POWER SUPPLY

Capacitors are used in pulsed power supplies to store energy for a short duration and rapid discharge. The main difficulty in a pulsed power supply is the commutation of currents within the short duration. A conventional pulse generator uses an active commutator in which flows the same current as the load. Consequently, the output power is limited by the capability of the commutation switch device. Plasma switches such as thyratrons have limited repetition rate and lifetime. The magnetic compression network can go beyond this limit and generate high power pulses in tens of nanoseconds. The principle is to generate a long pulse with a classical switching device such as a fast thyristor and apply it at the input of a multi-stage magnetic compression circuit shown in Figure 13.12. Each stage of magnetic compression will reduce the time duration of the pulse, increasing the power of the pulse. A compression stage is built with a capacitor and a saturable inductor that acts as a switch off when it is unsaturated and as a switch on when it is saturated [4,5].

### 13.7.1 Transformer Pulse Generator

A transformer pulse generator consists of a high efficiency low leakage inductance pulse transformer, a charging circuit

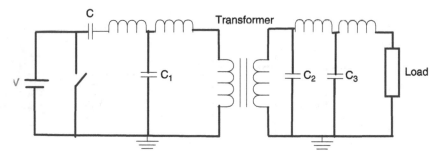

**Figure 13.12**   Magnetic compression circuit.

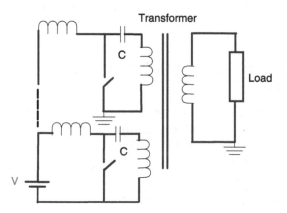

**Figure 13.13**   A typical pulse circuit.

(inductance and capacitance), and a switching device such as an IGBT. The primary coils are supplied by the charging circuit. The pulse is produced by alternate switching on and off by the switching device. It is well suited for pulse power up to 10 MW with a duty cycle up to 5% and a mean power up to 20 kW. Pulse transformers have an exceptionally low leakage inductance that permits the obtaining of a rise time of less than 1 µs for 100 kV/100 A pulses. A typical pulse generator circuit is shown in Figure 13.13.

### 13.7.2   Resonant Charge Transformer Circuit

A resonant charge transformer generator consists of a resonant circuit for storing the energy, a switching device, and a

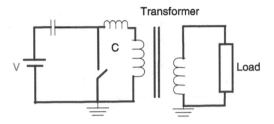

**Figure 13.14**   Resonant charge transformer circuit.

transformer for transferring the pulse to the load circuit. This makes it particularly suitable for nonlinear loads. A typical resonant charge transformer circuit is shown in Figure 13.14. The high voltage pulsed power-based technologies are rapidly emerging as key to the efficient and flexible use of electrical power for many industrial applications. Some of the applications include laser power supply, pulsed electric field sterilization of liquids, corona plasma discharge oxidation for sterilization and de-pollution, electromagnetic forming, and particle accelerators for fruit and vegetable juice extraction.

### 13.7.3   Pulsed Power Supply for Varistor Testing

Power system equipment such as circuit breakers, power transformers, surge arresters, varistors, current transformers, and voltage transformers are tested for short time rating and insulation withstand capability in the laboratory setup. Such testing requires power supplies with high voltages and specific waveforms to simulate the expected impedance loading. To demonstrate this concept, consider a high voltage power supply used to evaluate the transient immunity of a varistor. There are three different tests involved in this application. One of them is the 0.5 µs, 10 kHz ring wave, suitable for medium impedance load. The typical load impedance used is of the order of 100 Ω. The circuit diagram to generate such high voltage power supply is shown in Figure 13.15. The output of the power supply to test the varistor to the industry specifications is shown in Figure 13.16. It can be seen that the

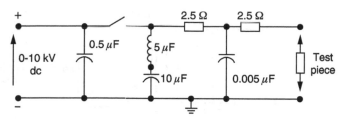

**Figure 13.15**  Circuit diagram to produce 0.5 μs, 100 kHz ring wave.

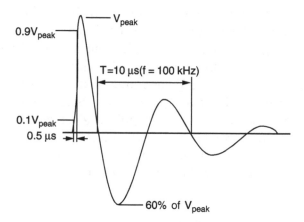

**Figure 13.16**  Open circuit waveform from the Figure 13.15 circuit.

capacitors are used in this power supply to generate such specific waveforms [6,7].

### 13.7.4   High Voltage Power Supply for Transformer Testing

All transformers are tested and evaluated to meet the industry standards in order to ensure adequate performance in the power system. Sometimes the transformers are tested at the substation locations. A typical no-load test of a high voltage transformer requires power supplies with rated voltage and adequate MVA output. A portable power supply for testing

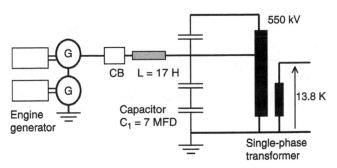

**Figure 13.17** No-load test arrangement of an EHV transformer [8].

a high voltage power transformer is discussed in Reference [8]. The capacitors are used as feedback elements. A one-line diagram of the power supply is shown in Figure 13.17. A twin-engine generator set is used to produce a 10 kV output. A single-phase three winding transformer is to be tested for no-load performance. A series inductance $L$ is used and the power supply is connected at a suitable tap level. A capacitance bank is connected across the main winding that can excite the entire transformer. The tertiary winding is left open. The parameters of the various components used in testing a 550 kV/230 kV/13.8 kV, 60 Hz, single-phase autotransformer are shown in Figure 13.17. It can be seen that the capacitors are important items in this test setup.

### 13.7.5 Power Supply for Circuit Breaker Testing

The circuit breaker interrupter capabilities are evaluated to meet the industry standard requirements and include both direct and indirect tests. In the direct test, the rated voltage and full load conditions are applied and the performance is evaluated. The indirect tests are performed in the field or in a laboratory setup. The laboratory tests are performed using specialized equipment. The basic diagram of a typical high power laboratory setup for circuit breaker testing is shown in Figure 13.18 [9]. The main equipment used in the test circuit are the main power supply source from the utility or generators with adequate short-circuit capability.

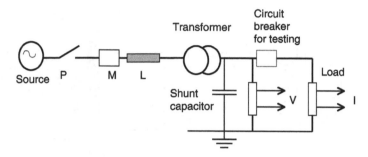

**Figure 13.18**   A high voltage test circuit for circuit breakers.

A protective circuit breaker $P$ is used that can clear the fault
if the test circuit breaker fails during the testing conditions. A
circuit breaker $M$, with independent pole closing capabilities to
simulate various test conditions such as single pole closing and
two pole closing, is used. A current limiting reactor $L$ is used to
control the short-circuit current to the desired values. A transformer is used to vary the secondary voltage accordingly. A
shunt capacitor bank is used to control the transient recovery
voltage to the desired test values. The circuit voltage and current are recorded in the time domain. Although the shunt
capacitor value is not specified, the requirements are based
on the size of the circuit breaker to be tested.

## 13.8   CAPACITORS IN EHV APPLICATIONS

### 13.8.1   Carrier Coupling Capacitor

In power systems, overhead lines are used to transmit
60 Hz power and communication signals. The circuits used to
transmit signals through the power circuits are called
carrier communication. The frequency spectrum used in this
type of communication is 30–500 kHz. Since the signal is
applied directly to the power line, this type of communication
is very reliable for a few channels. Long distance communication can be handled without the use of repeaters. Since
the power line is maintained for continuous operation,
the maintenance cost of the carrier communication is low.

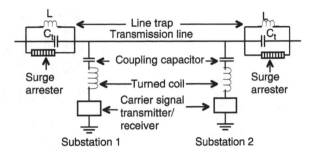

**Figure 13.19** Carrier communication arrangement.

The communication signals are susceptible to line noise and disturbances, if any, on the power line. A typical carrier communication arrangement between two substations is shown in Figure 13.19. In this arrangement, the communication signal from the transmitter is passed through a tuned circuit to the overhead line. The tuned circuit contains a coupling capacitor and an inductor. The coupling capacitor is a high frequency element in porcelain housing with cast metal ends, containing several capacitor elements in series. A typical coupling capacitor is shown in Figure 13.20 along with a substation line trap. The cylindrical device is the line trap inductor [10]. Each capacitor unit is made up of paper/ foil, noninductively wound and impregnated. The capacitor units are mounted on a metal base that contains a grounding arrangement, protection, and a drain coil. The drain coil connects the capacitor to the ground and provides high impedance at the carrier frequency. The typical values of coupling capacitors are shown in Table 13.4.

### 13.8.2 Line Trap Capacitors

A line trap is a tuned circuit connected in series with the transmission line, as shown in Figure 13.19. A line trap circuit is used to break the carrier current from one section to the other. This is a parallel resonant circuit tuned to provide high impedance at a specific carrier frequency. This is a very low impedance device and does not produce significant voltage

**Figure 13.20** Photograph of a coupling capacitor for carrier communication (left) and a substation line trap (right). (Courtesy of U.S. Department of Labor, OSHA website [10].)

TABLE 13.4   Ratings of Capacitive Voltage Transformers

| kV | Maximum C pF | BIL |
|---|---|---|
| 72.5 | 23,000 | 350 |
| 110 | 12,000 | 550 |
| 145 | 10,000 | 650 |
| 170 | 100,000 | 750 |
| 220 | 9,000 | 1,050 |
| 300 | 7,000 | 1,050 |
| 362 | 5,000 | 1,175 |
| 420 | 4,500 | 1,425 |
| 500 | 4,500 | 1,425 |
| 750 | 3,000 | 2,250 |

(Courtesy of Maxwell Technologies, San Diego, CA.)

drop at the power frequency. The main coil is rated to carry the full power current of the line conductor. The coil is wound on a porcelain cylinder, which is the housing for the adjustable capacitor unit. The coil and the capacitor unit are tuned to specific frequency. Usually, a surge arrester is connected across the line trap to protect the tuned circuit from lightning surges and other traveling waves.

### 13.8.3  Capacitive Coupling Voltage Transformer (CCVT)

At extra high voltage levels such as 500 kV, conventional voltage transformers are expensive. Also, such transformers will have many turns of very small conductors and thus introduce weakness into the high voltage insulation system. An effective alternative is to use a CCVT for the voltage measurement. The CCVT output is used for monitoring, protection relays, and control applications. A typical one-line diagram of a CCVT is shown in Figure 13.21. A photograph of a CCVT is shown in Figure 13.22.

### 13.8.4  Capacitors for Laboratory Applications

High voltage capacitors are used in laboratory applications as coupling capacitors for potential discharge measurement, high voltage potential dividers for 60 Hz measurement, as high voltage RC dividers for impulse measurement, in capacitive loads for series resonance installation, and as power supplies for equipment testing. The capacitor range for such applications can be 150 pF–500 nF. The voltage range can be from 20 to 1,000 kV. Some examples of these special capacitors are shown in Figure 13.23.

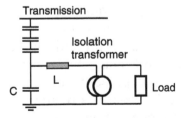

**Figure 13.21** One-line diagram of a capacitive coupling voltage transformer (CCVT).

**Figure 13.22**  Photograph of a capacitive coupling voltage transformer (CCVT). (Courtesy of Maxwell Technologies, San Diego, CA.)

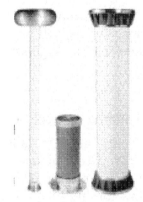

**Figure 13.23**  Photograph of laboratory capacitors. (Courtesy of Maxwell Technologies, San Diego, CA.)

## 13.9  CAPACITORS FOR ENERGY STORAGE

Standby power generators are used in order to achieve a system with high reliability in telecommunication circuits. Battery backup, DC-DC converters, and parallel type UPS are examples of highly reliable power systems. The other approach is to use engine generators or fuel cells to achieve both high reliability and better efficiency. In these systems, the standby power unit does not operate in normal mode when the line voltage is present but starts to operate in the event of line outage. These units require certain time to start. To overcome this problem, it is necessary to hold the output voltage during the starting time. Ultra capacitors provide an interesting and simple solution to this problem. The ultra

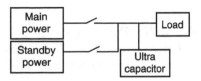

**Figure 13.24** A typical standby power system with ultra capacitor.

capacitors are available in hundreds of farads in very small size with significant energy, are maintenance free, and have a long lifetime. Consequently, there is a possibility that the output voltage of the system is backed up during the starting time. A block diagram of such a standby power system is shown in Figure 13.24. Usually the voltage rating of an ultra capacitor is very small (2–3 V) and hence several units are to be connected in series–parallel combination. Several such applications are discussed in Reference [11], including power quality ride through applications, power stabilization, adjustable speed drive support, and voltage flicker mitigation.

## PROBLEMS

13.1. Draw the power supply circuit of an electric arc furnace installation and identify the power factor values in different locations without and with power factor correction.

13.2. How is the low power factor condition created in the resistive arc welding units? How do you improve the power factor in the resistive arc welding circuit?

13.3. Discuss the basic principles behind vapor lamps. Show the circuit diagram of a neon lamp. What is the purpose of the capacitor?

13.4. What are the main differences between the sodium vapor lamp and the fluorescent lamp?

13.5. Draw the power supply circuit suitable for a mercury vapor lamp.

13.6. How are capacitor-run motors different from other single-phase induction motors?

13.7. Discuss the use of capacitors in pulsed power supplies.

## REFERENCES

1. Yehling, T. O. (1977). Power systems design for a large arc furnace shop, *Proceedings of the IEEE Industry Applications Society Conference Meeting*, October 2–6. IEEE Publication 77, Ch. 1246-8-IA.

2. Key, T. (1977). Light flicker caused by resistive spot welders, *PQTN Case Study*, Journal from EPRI Electronics Applications Center, No. 1, May, 1–8.

3. Choe, Y. M., Gho, J. S., Hok, H. S., Choe, G. H., Shin, W. H. (1999). A new instantaneous output control method for inverter arc welding machine, *Power Electronics Specialist Conference*, 1, 521–526.

4. Riberro, P. F., Johnson, B. K., Crow, M. L., Arsoy, A., Liu, Y. (2001). Energy storage systems for advanced power applications, *Proceedings of the IEEE*, 89(12), 1744–1756.

5. Rodewald, A. (1972). A New Triggered Multiple Gap System for Any Kind of Voltages, International Symposium High Voltage Technology, Munich.

6. IEEE Standard 587 (1980), IEEE Guide for Surge Voltages in Low-Voltage AC Power Circuit.

7. Natarajan, R., Croskey, C. L. (1989). A microcomputer-based data acquisition system for power systems transient analysis, *Laboratory Microcomputer*, 8(2), 53–58.

8. Gerloch, H. G. (1991). Resonant power supply kit system for high voltage testing, *IEEE Transactions on Power Delivery*, 6(1), 1–7.

9. Panek, J. (1975). Test procedures, *IEEE Tutorial on Application of Circuit Breakers, Course Text*, 75CH0975-3-PWR, 65–71.

10. www.osha.gov/sltc/etools/electric_power/glossary.html., website of the Occupational Safety and Health Administration, U.S. Department of Labor.

11. Harada, K., Sakai, K., Kutake, H., Ariyashi, G., Yamasaki, K. (1998). Power system with cold-standby using ultra capacitors, *Twentieth Telecommunications Energy Conference*, San Francisco, CA, October 4–8, 1998, 498–504.

# 14

---

# STATIC VAR COMPENSATORS

## 14.1  INTRODUCTION

The need to control reactive power through transmission and distribution lines has been recognized since the emergence of the AC power system. Fixed and switched shunt capacitors are used to ensure desirable voltage profile along the transmission and distribution lines. To handle dynamic disturbances such as line switching, loss of generation, load rejection, and system faults, the reactive power has to be supplied quickly to keep the system stable. With fluctuating loads such as arc furnaces, crushing mills, paper mills, and continuous miners in the mining industry, there is a need for adaptive power factor correcting devices. The thyristor controlled static VAR compensators (SVC) offer such opportunities for dynamic power control.

## 14.2  COMPENSATION CONCEPTS

Consider the simple two-machine model shown in Figure 14.1 for the basic power transfer. The sending end voltage ($V_S$) and

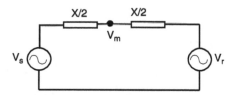

**Figure 14.1** Two-machine system for the analysis.

receiving end voltages ($V_R$) are given by [1]:

$$V_S = V\left(\cos\frac{\delta}{2} + j\sin\frac{\delta}{2}\right) \tag{14.1}$$

$$V_R = V\left(\cos\frac{\delta}{2} - j\sin\frac{\delta}{2}\right) \tag{14.2}$$

The voltage at the midpoint is:

$$V_M = \left(\frac{V_S + V_R}{2}\right) = V\cos\left(\frac{\delta}{2}\right) \tag{14.3}$$

The current through the line is given by:

$$I = \left(\frac{V_S - V_R}{jX}\right) = \frac{2V}{X}\sin\frac{\delta}{2} \tag{14.4}$$

The relation between $V_S$, $V_R$, and $I$ is shown in the phasor diagram, Figure 14.2. If a line without any loss is assumed, the power delivered is:

$$P = \frac{V^2}{X}\sin\delta \tag{14.5}$$

The reactive power entering at each end of the line is:

$$Q_S = Q_R = \frac{V^2}{X}(1 - \cos\delta) \tag{14.6}$$

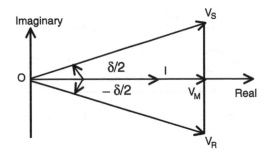

**Figure 14.2** Phasor diagram showing the relationship between $V_S$, $V_R$, and $I$.

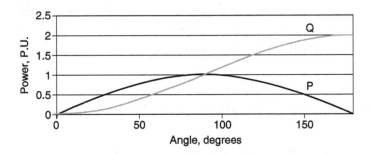

**Figure 14.3** Power angle diagram.

The total reactive power is:

$$Q = 2Q_S = \frac{2V^2}{X}(1 - \cos \delta) \qquad (14.7)$$

The relationship between $P$, $Q$, and $\delta$ is plotted in Figure 14.3. The maximum real power limit is 1.0 P.U. and the maximum reactive power limit is 2.0 P.U.

## 14.3 EFFECT OF SHUNT COMPENSATION

If a controllable synchronous voltage source is connected at the midpoint, then the reactive power loss can be compensated. Such reactive compensation will increase the real power

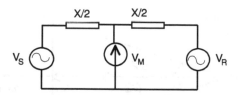

**Figure 14.4**  Two-machine system with compensation at the midpoint.

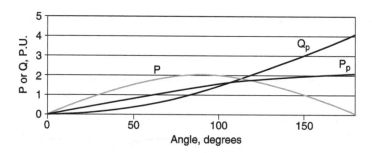

**Figure 14.5**  $P$ and $Q$ of the compensated scheme.

transmission capability as well. This concept is shown in Figure 14.4. If the voltage at the midpoint is kept the same as the sending end voltage then the real power $P_c$ is:

$$P_c = \frac{2V^2}{X} \sin \frac{\delta}{2} \qquad (14.8)$$

The corresponding reactive power relation is:

$$Q_c = \frac{4V^2}{X} \left( 1 - \cos \frac{\delta}{2} \right) \qquad (14.9)$$

The power and the reactive power plots are shown in Figure 14.5. This example demonstrates the effect of shunt compensation. In lines with losses, the maximum $P$ and $Q$ will be significantly less. The maximum power transmission in the uncompensated line $P$ occurs at 90°. The maximum power and reactive power of the compensated line occurs at the 180° position.

Synchronous condenser and shunt capacitors provide shunt compensation. The following SVCs are used to provide dynamic shunt compensation.

- Thyristor controlled reactor and fixed capacitor
- Thyristor controlled reactor and thyristor switched capacitor
- Thyristor controlled reactor and mechanically switched capacitor
- Thyristor switched capacitor
- Microprocessor-based SVC
- STATCOM type of static compensator

The required type of SVC is selected [3] based on application requirements and cost.

## 14.4  DESCRIPTION OF AN SVC

A typical SVC includes thyristors with control circuits, a cooling system for thyristor heat sinks, electronic control equipment, capacitor banks, filter reactors, power circuit breakers, and mounting racks. The thyristors are still low or medium voltage devices and hence the SVCs are manufactured at low or medium voltage levels. In the microprocessor-based SVCs, the thyristor switched capacitor and thyristor controlled reactor are present along with intelligent control. Usually an interfacing transformer connects the device to a high voltage power system.

### 14.4.1  Thyristor Controlled Reactor (TCR) and Fixed Capacitor (FC)

In this scheme, the capacitors are usually selected to provide the maximum reactive power needed at the point of installation. The required inductive power is dynamically controlled to maintain the desired voltage profile when the demand for reactive power is less than the maximum. The control is performed through phase angle variation. A typical scheme is shown in Figure 14.6. The result of this compensation is a power factor of nearly unity, with minimum voltage changes due to

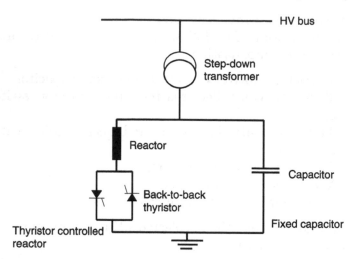

**Figure 14.6** Thyristor controlled reactor (TCR) and fixed capacitor (FC).

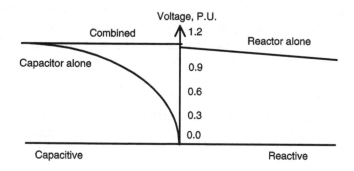

**Figure 14.7** Voltage characteristics with a fixed capacitor and a controlled reactor.

continuous phase control on the reactor. At maximum leading VAR, the switch is open and the current in the reactor is zero. As the firing angle increases, the harmonic content increases. A 10 MVAR unit typically consists of 10 MVAR of capacitor bank and 10 MVAR of reactor in addition to the thyristor controls. The variation of inductive, capacitive VARS with the system voltage is shown in Figure 14.7.

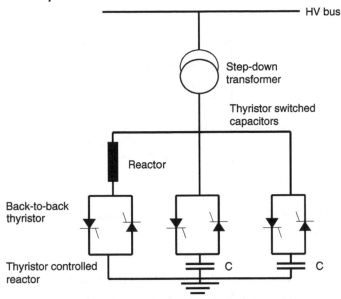

**Figure 14.8** Thyristor controlled reactor (TCR) and thyristor switched capacitor (TSC).

### 14.4.2 Thyristor Controlled Reactor (TCR) and Thyristor Switched Capacitor (TSC)

Another variation of the above-mentioned scheme is shown in Figure 14.8. It contains several capacitor sections operating in parallel with a phase controlled reactor. The task of controlling both the reactor and capacitor requires an electronic controller. The overall efficiency of the scheme is less due to losses in the reactor. The number of capacitor branches depends on the amount of kVAR, thyristor ratings, etc. The variation of reactive power due to a two-switched capacitor and one reactor per phase is shown in Figure 14.8. For a 10 MVAR SVC, the capacitor bank will be two 5 MVAR (or in some other combination) and the reactor will be 10 MVAR. A typical VAR demand versus VAR output profile of a scheme is shown in Figure 14.9.

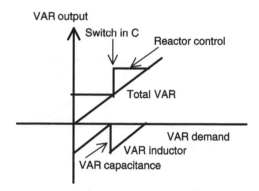

**Figure 14.9** VAR output versus VAR demand characteristics of TSC and TCR.

### 14.4.3 Thyristor Controlled Reactor (TCR) and Mechanically Switched Capacitor (MC)

Such a scheme is shown in Figure 14.10 where the capacitors are switched through mechanical switches. This method is suitable for steady load conditions, where the reactive power requirements are predictable and the capacitors can be switched using circuit breakers. Depending on the need, the three capacitor banks can be selected in various combinations. The reactor provides a smooth control. This scheme is lower in cost but has a slower response. For a 10 MVAR SVC, the capacitor bank will be in three steps, 3.33 MVAR each, and the reactor will be 10 MVAR.

### 14.4.4 Thyristor Switched Capacitor (TSC)

In this case, all the required capacitors are switched in and out using SCRs. To reduce the number of capacitors, binary grouping is sometimes employed. This scheme is suitable for both balanced and unbalanced loads. A typical scheme is shown in Figure 14.11. For a 10 MVAR size, bank 1 is 4 MVAR, bank 2 is 3 MVAR, and bank 3 is 3 MVAR. Some designs use binary-based steps. In this scheme, only the capacitor banks are used. In the previous schemes, a controlled reactor

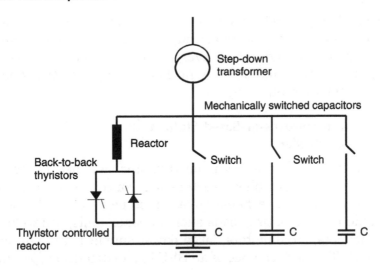

**Figure 14.10** Thyristor controlled reactor (TCR) and mechanically switched capacitors (MC).

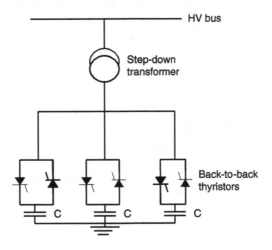

**Figure 14.11** Thyristor switched capacitors (TSC).

of equivalent capacity is used. It can be noted that the cost of this kind of scheme will be considerably less than the previous schemes. With this scheme, the reactive compensation is corrected on a cycle-by-cycle basis. Each phase is compensated independently and correction to unbalance is made. Ferroresonance conditions are suppressed since the capacitors are switched in and out on every cycle. The phase controlled reactor is absent and there is no harmonic generation due to the reactor currents.

### 14.4.5 Microprocessor-Based Static VAR Controller

In this type of device, both the thyristor switched capacitors and thryristor controlled reactors are controlled by a microprocessor. Such a scheme gives a stepless control and is suitable for fluctuating loads such as paper mills, arc welders, arc furnaces, large motors, large power supplies, and pumping stations. The performance of an SVC in a distribution system is demonstrated from a practical standpoint [8]. The one-line diagram of the microprocessor-based controller is shown in Figure 14.12. The photographic view of the pole-mounted installation is shown in Figure 14.13. The controller has a range of 50 kVAR lagging to 750 kVAR leading per phase. The other features of this controller are:

- For a paper mill application, where the motor load may cause flicker.

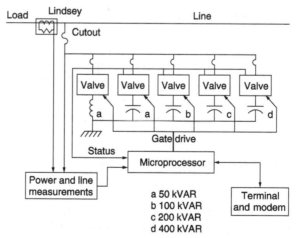

**Figure 14.12** One-line diagram of a microprocessor-based SVC. (Reproduced from Reference [5], with permission from IEEE.)

**Figure 14.13** Photographic view of the SVC installed at the paper mill. (Reproduced from Reference [8], with permission from IEEE.)

- 7,200 V line to neutral rating.
- 50 kVAR inductive to 750 kVAR capacitive per phase, stepless control.
- Power factor ranging from 0.70 lagging to 0.866 leading.

The measured voltage profile before and after the installation of the SVC is shown in Figure 14.14. It is claimed that before installation of the SVC there were noticeable voltage fluctuations. After the installation of the SVC, the voltage profile was smoother and it was claimed that the flicker problem was corrected.

### 14.4.6 STATCOM

The STATic synchronous COMpensator (STATCOM) consists of a solid-state voltage source inverter with several Gate Turn Off thyristor switch-based valves, a DC link capacitor, magnetic circuit, and a controller. The number of thyristors and the various configurations of the magnetic circuit depend on the desired quality of AC waveforms generated by the controller. A one-line diagram of a typical STATCOM is shown in Figure 14.15. The DC link capacitor is switched through the inverter circuit and the required reactive power is injected to the transmission circuit. The injected current is almost

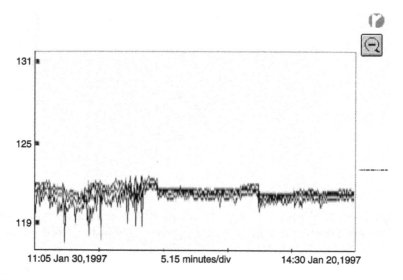

131 ■
125 ■
119 ■

11:05 Jan 30,1997          5.15 minutes/div          14:30 Jan 20,1997

**Figure 14.14**   Voltage profile without and with SVC. (Reproduced from Reference [8], with permission from IEEE.)

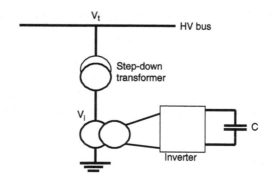

$V_t$
HV bus

Step-down transformer

$V_i$

C

Inverter

**Figure 14.15**   A typical STATCOM (one phase only).

in quadrature with the line voltage, thereby emulating an inductive or capacitive reactance at the point of connection [9]. If $V_t$ is the voltage at the line and $V_i$ is the voltage at the inverter terminal, then

- $V_i > V_t$: The inverter generates $Q$ and the net effective is capacitive.

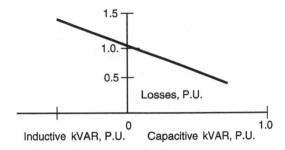

**Figure 14.16** Reactive power output versus losses in per unit.

- $V_i < V_t$: The inverter absorbs $Q$ and the net effective is inductive.
- $V_i = V_t$: No net exchange of $Q$ between the inverter and the line.

## 14.5 LOSSES AND HARMONICS IN THE SVC

The losses in an SVC are important to consider when selecting the rating. Since the thyristor controlled reactor operates at various firing angles depending on the circuit, it is difficult to specify the total losses at any given operating point. For a fixed capacitor and switched reactor configuration, at no-load operating point, the capacitive and inductive VARS cancel. That means that all the capacitive currents are circulated through the inductive circuit. Then the losses decrease with the increase in the capacitive VAR output, as shown in Figure 14.16 [7]. The overall expected losses in such a configuration are around 1% of the rating. In the switched capacitor and thyristor controlled reactor configuration, the losses are low at the no-load condition and increase with the increase in the capacitive VAR output as shown in Figure 14.17. The losses in the thyristor controlled reactor schemes will be higher than the estimated values due to the presence of the harmonics.

### 14.5.1 Harmonics

In thyristor controlled reactor schemes, harmonics are generated due to the phase control. The amount of harmonics

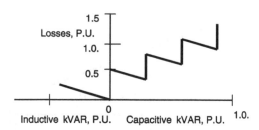

**Figure 14.17**  Losses in the thyristor controlled reactor and switched capacitor.

generated depends on the firing angle. There are advantages in controlling the harmonic voltage distortion in terms of decreased losses, decreased heating, less maintenance, reduced misoperation of relaying, etc. Some issues related to certain harmonics are discussed below.

### 14.5.1.1  Harmonics in Delta Connected Configurations

The triple harmonic currents produced by the thyristor operation circulate in the delta. Such harmonics will not be present in the system output.

### 14.5.1.2  Harmonics in the Wye Connected Configurations

Wye connected reactor and capacitor banks are commonly used and all types of harmonics are expected in the output. In order to control the harmonics, tuned filters are generally employed in the SVC installations. The filters can be 3rd, 5th, 7th, etc., depending on the harmonic magnitudes present at the given installation. In order to demonstrate this concept, the ratings of the switched reactor, capacitor, and the filter ratings of some SVC installations are presented in Table 14.1. It can be seen that SVC design is application specific and does not follow any general specific guidelines.

### Example 14.1

Consider a 17 mile distribution feeder supplied from a 69 kV/ 25 kV transformer (see Figure 14.18). A 600 hp, three-phase induction motor is supplied from the feeder through a 25 kV/

TABLE 14.1  Example Ratings of Some Practical SVC Installations

| Location | kV | Reactor MVAR | Cap. MVAR | Transformer kV | Filter 5th | Filter 7th | Notes |
|---|---|---|---|---|---|---|---|
| KG&E | 0.60 | – | 2.475 | 25/0.600 | – | – | Ref [2] |
| Chester | 18 | 163 | 363 | 138/18 | Two | Two | Filter 31 MVAR [4] |
| Beaver | 8.3 | 125 | 125 | 128/8.3 | One | One | Filter 8 MVAR [6] |

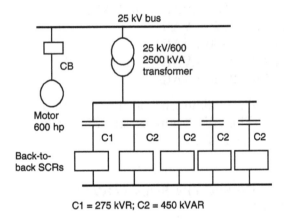

C1 = 275 kVR; C2 = 450 kVAR

**Figure 14.18**  One-line diagram of the SVC used in Example 14.1.

12.5 kV transformer. The motor, starting at the pumping station, was found to cause voltage dips of 12.5% on the 25 kV feeder. The voltage dip duration was up to 15 s. An SVC was considered for this application. Estimate the rating of the SVC. Assume a suitable starting current/rated current ratio.

Solution

Based on Reference [2] the SVC calculations are projected below.

Motor hp (assumed motor hp = kVA) = 600 kVA
Starting current/rated current = 4.125
Starting kVA of the motor (600 × 4.125) = 2,475 kVA

    Select an SVC voltage rating $= 600$ V
    Transformer voltage ratio $= 12.5/25$ kV/600 V
    Transformer kVA $= 2{,}500$
    SVC, use $5 \times 450$ kVAR banks $= 2{,}250$ kVAR
    Use one 225 kVAR bank $= 225$ kVAR
    Total $= 2{,}475$ kVAR

Eleven possible switching combinations are identified and listed below.

| | | |
|---|---|---|
| 225 kVAR | 1350 kVAR | 2475 kVAR |
| 450 kVAR | 1575 kVAR | |
| 675 kVAR | 1800 kVAR | |
| 900 kVAR | 2025 kVAR | |
| 1125 kVAR | 2250 kVAR | |

The one-line diagram of the system is shown in Figure 14.18. The SVC reduces the flicker levels during the motor starting to 1.55%.

## 14.6   EFFECT OF SVC ON BULK POWER SYSTEMS

The application of SVC on a power system contributes to damping power system oscillations, transient stability improvement, and voltage support to prevent voltage instability.

### 14.6.1   Power Oscillation Damping

The dynamic behavior of a system is described by the swing equation:

$$H\frac{d^2\delta}{dt^2} = P_m - P_e \qquad (14.10)$$

where $P_m$ = Mechanical power
    $P_e$ = Electrical power
    $H$ = System inertia
    $\delta$ = Rotor angle
    $P_m = P_e$ = Accelerating power

For small variations, assuming constant mechanical power and expressing the change in electrical power in terms of controllable voltage $V_m$ at the midpoint voltage:

$$H\frac{d^2\Delta\delta}{dt^2} + \frac{dP_e}{dV_m}\Delta V_m + \frac{dP_e}{d\delta}\Delta\delta = 0 \qquad (14.11)$$

In the above equation if $V_m = $ constant and $\Delta V_m = 0$, then the angle $\Delta\delta$ would oscillate with a frequency of:

$$\omega_0 = \sqrt{\frac{1}{H}\frac{\partial P_e}{\partial\delta}} \qquad (14.12)$$

To provide damping, the midpoint voltage must be varied as a function of $d(\partial\delta/\partial t)$:

$$\Delta V_m = K\frac{d(\Delta\delta)}{dt} \qquad (14.13)$$

where $K$ is a constant. If the midpoint voltage is increased by providing capacitive VARS, then $(d(\Delta\delta)/dt)$ must be positive in order to increase the transmitted electric power. It is decreased by absorbing inductive VARS when $(d(\Delta\delta)/dt)$ is negative. Power oscillation damping is achieved by alternating the maximum available VAR output.

### 14.6.2 Improvement of the Transient Stability

Suppose Equation (14.5) is applied to each half of the line. Then:

$$P = \frac{V^2}{(X/2)}\sin\frac{\delta}{2} \qquad (14.14)$$

The maximum power transfer obtained at $(\delta/2) = (\pi/2)$ is $(2V^2/X)$, twice the steady-state limit of the uncompensated case. In general, the transmission line reactance $X$ can be divided into equal sections with a perfect synchronous compensator. In such a case the power transfer is:

$$P = \frac{V^2}{(X/n)} \sin \frac{\delta}{n} \qquad (14.15)$$

which gives a maximum power transfer of $n(V^2/X)$, $n$ times the steady-state power limit of the uncompensated line. The improvement in transient stability will follow based on the increase in the steady-state maximum power transfer.

### 14.6.3 Voltage Support

The receiving end voltage of a transmission line is a function of the line impedance, the load, and the power factor. The magnitude of voltage at the receiving end of a typical line as a function of the receiving end power is given in Figure 14.19. In the case of a weak power system, load changes, switching of transmission lines, transformers, or large capacitor and reactor banks can cause significant voltage variation at the receiving end. In the extreme case, when the power demand of the load exceeds the transmittable power limit at the given load power factor, the receiving end voltage may collapse. A typical system configuration for potential voltage instability is a large load area supplied from two or more generator plants with independent transmission lines. The loss of one of the power sources could suddenly increase the load demand on

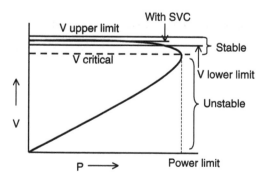

**Figure 14.19**   Variation of receiving end voltage as a function of $P$.

the remaining part of the system, causing the receiving end voltage to collapse. The maximum transmittable power over a given transmission line can be increased by increasing the reactive power at the receiving end using an SVC. This suggests that with a rapidly variable VAR source of appropriate rating connected to the receiving end terminal, voltage collapse could be prevented and constant terminal voltage can be maintained.

## 14.7 CONCLUSIONS

The basic concepts of shunt reactive power compensations are presented along with the effect on $P$ and $Q$. It is shown that by providing shunt compensation, the power transfer can be increased between two terminals. The following five SVCs are commonly used in the industry. Types of approximate reactor and capacitor ratings are presented for comparison purposes.

| Type | Capacitor (MVAR) | Reactor (MVAR) | Total (MVAR) |
|---|---|---|---|
| TCR and FC | 10 | 10 | 20 |
| TCR and TSC | 10 | 10 | 20 |
| TCR and MC | 10 | 10 | 20 |
| TSC | – | 10 | 10 |
| Microprocessor based [8] | 2,250 kVAR | 150 kVAR | 2,400 kVAR |

It can be seen that the thyristor switched capacitor scheme requires smaller bank size and the cost will be lower. The response offered by this scheme will be in steps and will require significant electronic control. Many of the schemes in operation are thyristor controlled, reactor supported, with fixed capacitors or switched capacitors. Such schemes require significant harmonic filtering. The effect of SVC on the bulk power system is to increase the power transfer limits, increase the stability limits, and improve the voltage profile.

SVS Specifications

| | |
|---|---|
| Nominal system voltage (line to line) | Nominal phase to ground voltage |
| Maximum voltage, P.U | Minimum voltage, P.U |
| Nominal frequency, Hz | Minimum frequency, Hz |
| Maximum frequency, Hz | Maximum frequency and duration |
| Minimum frequency and duration | Maximum short circuit current |
| Maximum system short circuit current, single-phase | |
| Minimum system short circuit current, three-phase | |
| Minimum system short circuit current, single-phase | |
| Rated surge arrester voltage, kV/phase | Rated basic insulation level, kV |
| Rated switching surge insulation level, kV peak | |
| Maximum allowable voltage distortion | Individual harmonic |
| Expected annual operating hours (−100 to +25%), (−25 to +25%), (+25 to +100%) | |
| Maximum operating temperature, °C | |
| Minimum operating temperature, °C | |
| Relative humidity, % | |
| Elevation above mean sea level, m or ft | |
| Maximum wind velocity, mph/h | |
| Isokereraunic level in G | |

## PROBLEMS

14.1. What are the different types of SVCs available? Which is the popular type for utility applications?

14.2. Name a few applications for the static VAR controller.

14.3. In a thyristor switched capacitor controller, if there are three capacitor banks per phase with ratings of 2, 4, and 8 MVAR, how many switching choices are available?

14.4. What is the role of a thyristor controlled reactor? What are the side effects of the controlled reactor?

14.5. What are the effects of SVC on a bulk power system?

14.6. There are a few SVCs installed in utility applications. What prevents the application of these devices on a large scale like shunt capacitors?

## REFERENCES

1. Gyugyi, L. (1988). Power electronics in electric utilities; Static var compensators, *Proceedings of the IEEE*, 76(4), 483–494.

2. McAvoy, J. (1989). Distribution static var compensator solves voltage fluctuation problem, *Transmission and Distribution*, 41(5), 60–68.

3. Hauth, R. L., Miske, S. A., Nozari, F. (1982). The role and benefits of static var systems in high voltage power system applications, *IEEE Transactions on Power Apparatus and Systems*, Vol. PAS–101(10), 3761–3770.

4. Dickmander. D., Thorvaldssn, B., Stromberg, G., Osborn, D. (1991). Control system, design and performance verification for the Chester, Maine Static Var Compensator, *IEEE/PES 1991 Summer Meeting*, Paper No. 91 SM 398–8 PWRD.

5. Bhargava, B. (1993). Arc furnace flicker measurements and control, *IEEE Transactions on Power Delivery*, 8(1), 400–410.

6. Torgerson, D. R. (1990). Static Var Compensators, planning, operating and maintenance experience, *IEEE Power Engineering Society Winter Meeting*, February 4–9, Reference No. 90$^{TH}$ 0320–2PWR.

7. Gyugyi, L., Matty, W. P. (1997). Static Var Compensator with minimum no-load losses for transmission line compensation, 1997 American Power Conference, 24 pp.

8. Kemerer, R. (1998). Distribution System Static Var Compensator; Field Experience, IEEE Transmission and Distribution Committee, Application of Static Voltage Conditioning Devices for Enhanced Power Quality, Tampa, Florida, February 4, website: http://grouper.ieee.org/groups/1409/9802_kemerer.pdf.

9. Sen, K. K. (1999). STATCOM – Static Synchronous Compensator: theory, modeling and applications, *IEEE Power Engineering Society Winter Meeting*, Paper No. 99 WM 706.

# 15

## PROTECTION OF SHUNT CAPACITORS

### 15.1 INTRODUCTION

Protective relaying is applied to power system components to separate the faulted equipment from the rest of the system so that the system can continue to function, to limit damage to faulted equipment, to minimize the possibility of fire, and to ensure personnel safety. The shunt capacitor banks are built using individual units connected in a series–parallel combination. Capacitor bank protection is provided against the following conditions [1–4]:

1. Overcurrent due to faults
2. Transient overvoltages
3. Overcurrents due to individual capacitor failure
4. Continuous capacitor unit overvoltage
5. Discharge of capacitor unit overvoltage
6. Inrush current during switching
7. Arc-over within the capacitor bank
8. Lightning

Table 15.1 identifies the types of faults in shunt capacitor banks and the protection methods. Capacitor overcurrent

**TABLE 15.1**  Shunt Capacitor Protection Methods

| Fault Condition | Protection Measures |
|---|---|
| Bus faults | Circuit breaker with overcurrent relay, fuses |
| System surge voltages | Surge arrester and spark gaps |
| Inrush currents | Series reactors or controlled switching |
| Discharge current from parallel banks | 1. Current limiting reactors<br>2. Proper bank selection |
| Overcurrent due to individual unit fuse operation | Current limiting or expulsion fuse |
| Continuous capacitor unit overvoltage | 1. Unbalance detection with current/voltage relays or double wye banks<br>2. Phase voltage relays |
| Rack faults | Unbalance detection, overcurrent relays |

protection, which is obtained through fuses, is outlined in Chapter 16. Overall fault protection is performed through the use of circuit breakers as discussed in Chapter 17. Surge protection, which is achieved through surge arresters, is presented in Chapter 18. Protective relaying arrangements are discussed in this chapter.

## 15.2  OVERCURRENT PROTECTION

Overcurrent condition may arise in individual capacitor units or in the capacitor bank within the capacitor phases and also due to system unbalances. In order to protect the capacitor banks from various failures, several protective measures are taken. The first line of overcurrent protection of the capacitor bank is the fuse. Protecting a capacitor bank against a single line-to-ground fault, line-to-line fault, or three-phase fault requires an overcurrent protection scheme. Consider power fuses or circuit breakers with associated relay circuits as shown in Figure 15.1. The primary and the secondary overcurrent relays are provided and are designated as:

1. 51, 51N for overcurrent protection.
2. 52, AC circuit breaker for tripping the capacitor bank.

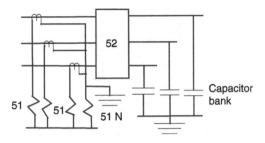

**Figure 15.1** Overcurrent protection for capacitor banks.

These relays are set to clear capacitor bank failures by sensing the overcurrent magnitudes. Usually the overcurrent magnitudes are sensed using the CT and the circuit breaker is tripped. Capacitor banks can operate at 135% of rated current according to the requirements of the standards, but with the switched capacitors, the current can vary depending on the number of capacitors in service, and setting the phase overcurrent relay may be difficult. Time overcurrent relays may be given normal settings without experiencing false operations during switching or inrush currents. Instantaneous relays must be set high to override these transient currents. The recommended instantaneous overcurrent relay setting is three times the capacitor current when no parallel banks are present.

## 15.3 PROTECTION AGAINST RACK FAILURES

Sometimes arc-over occurs in the capacitor units between the series or parallel units due to animal intrusion or contamination. A typical rack fault between two phases is shown in Figure 15.2 [5]. When allowed to burn, such faults may produce very little phase overcurrent. If allowed to continue, it may involve more and more series groups of the same phase until the instantaneous overcurrent relay trips or the fuses clear the fault. The time involved in such faults may be of the order of a few seconds. Such a failure may lead to case rupture and blown fuses. Therefore, instantaneous overcurrent relays and fuses may not be the best protection against rack failures.

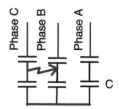

**Figure 15.2**   Phase-to-phase fault leading to a rack failure.

Unbalance protection schemes are used to protect against rack failures in the ungrounded capacitor banks. In the case of ungrounded schemes the midrack failures are not detectable using normal relaying. The most efficient protection for midrack phase-to-phase faults is the negative sequence current.

## 15.4   UNBALANCE PROTECTION

The purpose of an unbalance detection scheme is to remove a capacitor bank from the system in the event of a fuse operation. This will prevent damaging overvoltages across the remaining capacitor units in the group where the operation occurs, thereby protecting against a situation that can be immediately harmful to the capacitor units or associated equipment. Consider the capacitor connection shown in Figure 15.3. When all the four capacitors are in service, the voltage across each unit will be $V/2$. If one of the fuses is open, then the

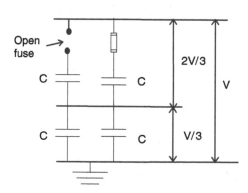

**Figure 15.3**   Open fuse and voltage distribution in a series group.

voltage across the upper branch is $\frac{2}{3}V$ and the lower branch is $\frac{1}{3}V$. Such a voltage increase in any capacitor unit is unacceptable. The unbalance in the voltage has to be detected and the unit must be isolated before significant damage occurs.

There are many methods available for detecting unbalances in capacitor banks, but there is no practical method that will provide protection under all possible conditions. All unbalance detection schemes are set up to signal an alarm upon an initial failure in a bank. Upon subsequent critical failures, where damaging overvoltages are produced, the bank would be tripped from the line. Typical detection schemes associated with grounded and ungrounded wye banks are discussed below. Since the delta connected banks are so seldom used and ungrounded wye banks serve the same purpose, delta configurations are not evaluated. Ten schemes are discussed in this chapter. The failure of one or more capacitor units in a bank causes voltage unbalance. Unbalance in the capacitor banks is identified based on the following considerations.

- The unbalance relay should provide an alarm on 5% or less overvoltage and trip the bank for overvoltages in excess of 10% of the rated voltage.
- The unbalance relay should have time delay to minimize the damage due to arcing fault between capacitor units. Also, the time delay should be short enough to avoid damage to sensors such as a voltage transformer or current transformer.
- The unbalance relay should have time delay to avoid false operations due to inrush, ground faults, lightning, and switching of equipment nearby. A 0.5 second delay should be adequate for most applications.

### 15.4.1 Scheme 1: Unbalance Relaying for Grounded Capacitor Banks

In Figure 15.4, a grounded capacitor arrangement is shown with neutral current relay. For a grounded wye bank or each wye of a grounded split-wye bank, the allowable number of

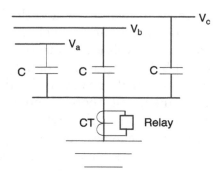

**Figure 15.4**  Neutral current sensing using a current transformer.

units that can be removed from one series group, given a maximum $\%V_R$ on the remaining units, can be calculated using the following formula [3]:

$$F = \left[\frac{NS}{S-1}\right] \quad \left[1 - \frac{V_{ph}}{SV_c} \times \frac{100}{\%V_R}\right] \tag{15.1}$$

If $F$ is fractional, use the next lower whole number. The relay is then set to signal the alarm upon failure of $F$ units. The neutral-to-ground current flow $I_N$ and relay setting upon loss of $F$ units for this scheme is determined by the following formula:

$$I_N = \left[\frac{I_c V_{ph}}{SV_c}\right] \times \left[\frac{NF}{S(N-F)+F}\right] \tag{15.2}$$

The relay would further be set to trip the bank upon loss of $F+1$ units. The neutral-to-ground current flow and relay setting can be determined using $F+1$ in place of $F$. The percent of overvoltage for any number of units removed from a series group can be determined by the following formula:

$$\%V_R = \left[\frac{V_{ph}}{SV_c}\right] \times \left[\frac{100SN}{S(N-F)+F}\right] \tag{15.3}$$

where $V_{ph}$ = Applied line-to-neutral voltage, kV

$V_c$ = Rated voltage of capacitor units, kV

$V_R$ = Voltage on remaining units in group, %

$F$ = Units removed from one series group

$I_N$ = Neutral-to-ground current flow, A

$I_U$ = Rated current of one unit, A

$S$ = Number of series groups per phase

$N$ = Number of parallel units in one series group

$F$ = Number of units removed from one series group

A typical unbalance protective scheme consists of a current transformer with a 5 A secondary using a burden of 10–25 Ω connected to a time-delayed voltage relay through suitable filters. The advantages of this scheme are:

1. The capacitor bank contains twice as many parallel units per series group compared to the double wye bank for a given kVAR size which reduces the over-voltage seen by the remaining units in a group in event of a fuse operation.
2. This bank may require less substation area and connections than a double wye bank.
3. Relatively inexpensive protection scheme.

The disadvantages of this scheme are:

1. Sensitive to system unbalance, which is a significant factor for large banks.
2. Sensitive to triple harmonics and will generally require a filter circuit.
3. Will not act when there is similar failure in all the phases.
4. It is not possible to identify the phase of the failed capacitor unit.

## 15.4.2 Scheme 2: Summation of Intermediate Tap-Point Voltage; Grounded Wye Capacitor Banks

Figure 15.5 shows an unbalance protection scheme for a grounded wye capacitor bank using capacitor tap point

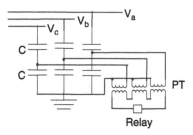

**Figure 15.5** Unbalance detection using summation of intermediate tap-point voltage in a grounded wye capacitor bank.

voltages. Any unbalance in the capacitor units will cause an unbalance in the voltages at the tap points. The resultant voltage in the open delta provides an indication of the unbalance. The changes in the neutral current magnitude and voltage are given by Equations (15.2) and (15.3), respectively.

### 15.4.3   Scheme 3: Neutral Current Differential Protection; Grounded Split-Wye Capacitor Banks

In this scheme shown in Figure 15.6, the neutrals of the two sections are grounded through separate current transformers. The CT secondaries are connected to an overcurrent relay, which makes it insensitive to any outside condition, which

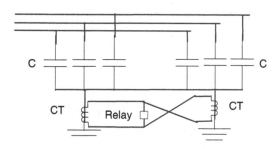

**Figure 15.6** Unbalance detection in a grounded split-wye capacitor bank using two CTs.

may affect both sections of the capacitor bank. The advantages of this scheme are:

1. The scheme is not sensitive to system unbalance and it is sensitive in detecting capacitor unit outages even on very large capacitor banks.
2. Harmonic currents do not affect this scheme.
3. For very large banks with more than one series group the amount of energy in the capacitors will decrease. This will lower the fuse interrupting duty and may reduce the cost of fuses.

By splitting the wye into two sections, the number of parallel units per series group is decreased, thereby increasing the over-voltages on the remaining units in the series group in the event of a fuse operation. A double-wye type of capacitor bank needs more substation area and connections. A balanced failure in each wye does not provide any indication in this scheme.

### 15.4.4 Scheme 4: Voltage Differential Protection Method for Grounded Wye Capacitor Banks

In this scheme, shown in Figure 15.7, two three-phase voltage transformer outputs are compared in a differential

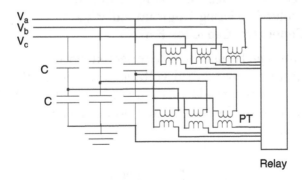

**Figure 15.7** Voltage difference prediction method for a grounded wye connected capacitor bank.

relay. Loss of capacitor unit in each phase can be detected independently. The zero sequence voltage is present during the unbalance in the shunt capacitor bank. The advantages of this scheme are:

1. The capacitor bank contains twice as many parallel units per series group compared to a split-wye bank. The overvoltages seen by the remaining units in a group in the event of a fuse operation will be less.
2. This capacitor bank may require less substation area.
3. This scheme is less sensitive to system unbalance. It is sensitive to failure detection in the series capacitors.

The main limitation of this scheme is that the number of PTs required is six and extensive connections are also required.

### 15.4.4.1  Unbalance Detection in Ungrounded Capacitor Banks

In order to detect the unbalance in ungrounded capacitor banks, the voltage transformer or current transformer sensors are used along with appropriate relays in the secondary circuit. Six different schemes for the detection of unbalance in the ungrounded capacitor circuits are presented.

### 15.4.5  Scheme 5: Neutral Voltage Unbalance Protection Using Ungrounded Wye Connected Capacitor Banks

Using a voltage transformer connected between the neutral and the ground, any neutral voltage shift due to the failure of a capacitor unit is sensed (see Figure 15.8) . The neutral voltage shift ($V_{NS}$) due to the loss of individual capacitor unit can be calculated as:

$$\% V_{NS} = \left[ \frac{100F}{3S(N - F) + 2F} \right] \tag{15.4}$$

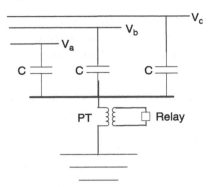

**Figure 15.8** Neutral voltage unbalance protection for ungrounded wye capacitor bank using a PT.

where $F$ is the number of units removed from one series group. The percentage overvoltage for any number of units removed from a series group is given by:

$$\% V_{\mathrm{R}} = \left[ \frac{V_{\mathrm{ph}}}{SV_{\mathrm{c}}} \right] \times \left[ \frac{300NS}{3S(N-F)+2F} \right] \tag{15.5}$$

The unbalance protective scheme consists of a time-delayed voltage relay with third harmonic filter connected across the secondary of the PT. The potential transformer may be a voltage transformer or a capacitive device. The voltage transformer for this application should be rated for full system voltage because the neutral voltage can be expected to rise above the rated voltage during certain switching operations. The advantages of this scheme are:

1. The capacitor bank contains twice as many parallel units per series group compared to a split-wye bank. The overvoltages seen by the remaining units in a group in the event of a fuse operation will be less.
2. This capacitor bank may require less substation area and connection in the power circuit.
3. This scheme is less sensitive to system unbalance.

### 15.4.6   Scheme 6: Neutral Voltage Unbalance Protection Method for Ungrounded Capacitor Banks Using Capacitive Voltage Divider

This scheme is similar to the PT scheme shown above (see Figure 15.9). A conventional inverse time voltage relay is connected across the grounded end capacitor. The grounded capacitor is a low voltage unit, 2400 V or less, sized to provide the desired unbalance voltage to the relay. In the event of one phase open, the voltage in the neutral relay exceeds the short time rating and a limiter has to be used. Scheme 6 has the same advantages and disadvantages as Scheme 5.

### 15.4.7   Scheme 7: Neutral Voltage Unbalance Detection Method for Ungrounded Wye Capacitor Banks Using Three PTs

This scheme is shown in Figure 15.10. This protection scheme uses three lines to neutral PTs with the secondary connected in the broken delta and an overvoltage relay. This scheme has advantages similar to Scheme 5. This scheme is sensitive to triple harmonics and it is expensive.

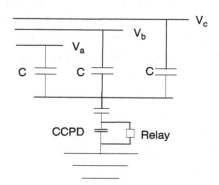

**Figure 15.9**  Neutral voltage unbalance protection for an ungrounded wye capacitor bank using a capacitor voltage divider.

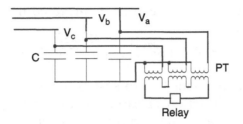

**Figure 15.10** Summation of line-to-neutral voltages with optional line-to-neutral overvoltage protection using three PTs.

### 15.4.8 Scheme 8: Neutral Current Unbalance Detection Method for Ungrounded Split-Wye Capacitor Banks

This scheme is shown in Figure 15.11. In this protection scheme, a current transformer is used in the neutral circuit to identify the unbalanced current. An overcurrent relay can be used to provide an alarm or trip signal. The neutral current due to the loss of individual capacitor units in a bank of two individual capacitors units, in a bank of two equal sections, can be determined [3]:

$$I_N = [I_c N]\left[\frac{V_{ph}}{SV_c}\right]\left[\frac{3F}{6S(N-F)+5F}\right] \qquad (15.6)$$

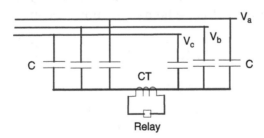

**Figure 15.11** Ungrounded split-wye connected capacitor bank; unbalance detection method using neutral current sensing.

The neutral voltage shift can be determined by:

$$\%V_R = \left[\frac{V_{ph}}{SV_c}\right]\left[\frac{600SN}{6S(N-F)+5F}\right] \tag{15.7}$$

The scheme is not sensitive to system unbalance. The scheme is sensitive to detection of capacitor unit outages and is not affected by the harmonic currents. This scheme contains only one CT and a relay. The disadvantages of this scheme are an increase in the overvoltages per unit because there are fewer parallel units per series group. The scheme requires more substation area compared to a wye connected capacitor bank.

### 15.4.9 Scheme 9: Neutral Voltage Protection Method for Ungrounded Split-Wye Connected Capacitor Banks

A schematic of this scheme is shown in Figure 15.12. This scheme is similar to Scheme 8. The sensor is a PT. This scheme is not sensitive to system unbalance, but it is sensitive to unit outage and is relatively inexpensive. The split-wye may require more substation area.

### 15.4.10 Scheme 10: Neutral Voltage Unbalance Protection Method for Ungrounded Split-Wye Connected Capacitor Banks

A schematic of this scheme is shown in Figure 15.13. The relay is 59N. This scheme is not sensitive to system

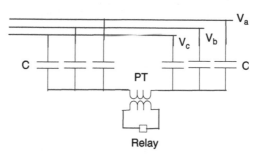

**Figure 15.12** Ungrounded split-wye connected capacitor bank; unbalance detection method using a PT.

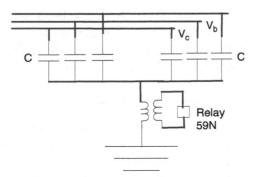

**Figure 15.13** Ungrounded split-wye connected capacitor bank; unbalance detection method using a neutral voltage sensing method.

unbalance, but it is sensitive to unit outage and is relatively inexpensive.

15.4.10.1 Overvoltage and
Undervoltage Protection

The relaying for the overvoltage and undervoltage are designated as:

59 for overvoltage protection
27 for undervoltage protection

These relays are normally set to coordinate with the system characteristics and with shunt capacitor banks on the system. Tripping for overvoltage typically occurs at 110% of the rated voltage. The low voltage tripping is set at 0.95 of the rated voltage. In certain circumstances, the undervoltage relays are used to trip the capacitor banks when the system is re-energized.

15.4.10.2 Voltage Differential Relays

The voltage differential relay is designated as 60 voltage or current unbalance relay that operates on a given difference. These relays compare the voltage across the total capacitor bank with the midpoint voltage of the bank for each phase. If one of the capacitor units is lost, then the ratio of the two voltages will change. The change in the voltage will be proportional to the change in the impedance in the capacitor bank. The voltage differential relays are set to alarm for greater than 0.7% but less than 1% change in the voltage ratio and will trip at greater than 2% change in the voltage ratio.

### 15.4.10.3 Voltage Detection Relays

The voltage detection relays use the midpoint voltage and are designated as 59–1/S and 59–2/S overvoltage relays. These relays are set to alarm for one capacitor unit out and will trip the circuit breaker for two capacitor units out.

### 15.4.10.4 Neutral Voltage Relays

The neutral voltage relays measure the voltages developed by the neutral current through the capacitor bank and are designated as 59–1/P and 59–2/P overvoltage relays. The neutral voltage relays need to filter the harmonics and only the voltage due to the fundamental frequency will be used to operate the relay. Loss of one capacitor unit is indicated by an alarm. Loss of two capacitor units indicates the capacitor bank was tripped.

**Example 15.1**

A 115 kV, 97 MVAR, three-phase, 60 Hz, grounded wye capacitor bank is used for power factor correction. The capacitor bank consists of 14 series capacitors per phase and 12 parallel capacitors per phase. Each capacitor is rated for 5 kV and 300 kVAR per unit. The allowable continuous overvoltage of these capacitor units is 110%. Calculate the relay settings for the alarm and trip signals.

Solution

V/phase $(115\,kV/1.732) = 66.397\,kV$
Voltage per unit $= 5\,kV$
kVAR per can $= 300$
Number of series units/phase, $S = 14$
Number of parallel units/phase, $N = 12$
Current per capacitor unit, $I_c = 60\,A$

The percentage of overvoltage when one capacitor unit fails $(F=1)$ is:

$$\%V_R = \left[\frac{V_{ph}}{SV_c}\right]\left[\frac{100SN}{S(N-F)+F}\right] = \frac{66.397\,(100\times14\times12)}{14\times5\,(14)(12-1)+1}$$
$$= 102.8\%$$

If the system voltage is 1.05 times the nominal voltage, the voltage will increase by 107.9%. This is within the acceptable limits for continuous operation of the capacitor. Loss of one unit should initiate an alarm. The neutral current of the transformer is given by:

$$I_N = \left[\frac{I_c/\text{unit}\times V_{ph}}{SV}\right]\left[\frac{NF}{S(N-F)+1}\right]$$
$$= \frac{60\times66.397}{14\times5}\frac{12\times1}{14(12-1)+1} = 4.4\,\text{A}$$

The percentage of overvoltage when a two-capacitor unit fails $(F=2)$ is:

$$\%V_R = \frac{66.397\,(100\times14\times12)}{14\times5\,(14)(12-2)+2} = 112.2\%$$

This voltage exceeds the voltage rating of the capacitor. Loss of two units should initiate a circuit breaker trip to protect the capacitor bank. The neutral current of the transformer is given by:

$$I_N = \frac{60\times66.397}{14\times5}\frac{12\times2}{14(12-2)+2} = 9.6\,\text{A}$$

The neutral currents calculated due to the loss of one and two units are based on a nominal voltage of 115 kV. If the system voltage reaches 1.05 P.U., then the neutral will increase directly with the system voltage.

## 15.5   CONCLUSIONS

The basic protection schemes for shunt capacitor banks are discussed, including overcurrent protection, rack fault failures, and unbalance protection. A total of 10 schemes are presented for grounded and ungrounded capacitor bank unbalance detection. A suitable unbalance detection scheme can be selected based on the capacitor bank connection.

## PROBLEMS

15.1.   What are the different fault conditions against which protection is required in a capacitor bank? Which is the most severe form of fault in a capacitor bank?

15.2.   What considerations must be taken into account to identify unbalance in capacitor banks?

15.3.   Identify an efficient unbalance detection scheme for a grounded wye capacitor bank.

15.4.   Identify an unbalance detection scheme suitable for an ungrounded wye capacitor bank. What are the advantages and disadvantages of this scheme?

15.5.   A 230 kV, three-phase, 60 Hz, grounded wye capacitor bank is used for power factor correction. The capacitor bank consists of 14 series capacitors per phase and seven parallel capacitors per phase. Each capacitor is rated for 10.9 kV and 300 kVAR per unit. The allowable continuous overvoltage of these capacitor units is 110%. Calculate the relay settings for the alarm and trip signals. What is the design MVAR of the capacitor bank? What is the delivered MVAR of this bank at rated operating voltage?

## REFERENCES

1. ANSI/IEEE Standard C37.99 (1990), IEEE Guide for Protection of Shunt Capacitor Banks.

2. Porter, G. A., McCall, J. C. (1990). Application and protection considerations in applying distribution capacitors, Conference

Proceedings of Pennsylvania Electrical Association, Hershey, September 25–26.

3. Bishop, M., Day, T., Chaudhary, A. (2001). A primer on capacitor bank protection, *IEEE Transactions on Industry Applications*, 37(4), 1174–1179.

4. Horton, T., Fender, T. W., Harry, S., Gross, C. A. (2002). Unbalance protection of fuseless split-wye grounded capacitor banks, *IEEE Transactions on Power Delivery*, 17(3), 698–701.

5. Brunello, G., Kasztenny, B., Wester, C. (2003). Shunt Capacitor bank fundamentals and protection, *Proceedings of the 2003 Conference for Protective Relay Engineers*, Texas A&M University, College Station, April 8–10, 17 pages. Website: http://www.geindustrial.com/products/applications/shunt.pdf.

# 16

## OVERCURRENT PROTECTION

### 16.1  INTRODUCTION

Capacitor units are protected from overcurrent by fuse links. A fuse is a protection device with a melting part that is heated by the passage of overcurrent through it. When a fuse opens, the device is protected [1–5].

### 16.2  TYPES OF FUSING

The capacitor banks are protected either by an individual fuse for each capacitor or by a group fuse.

#### 16.2.1  Individual Fusing

In individual fusing, each capacitor unit is provided with a separate fuse. The typical individual fusing arrangement for an ungrounded wye capacitor bank is shown in Figure 16.1. This type of fusing is used in outdoor substation capacitor banks. Individual fusing is not commonly used in the ungrounded wye capacitor banks due to the overvoltage stress on units adjacent to a unit isolated by a fuse operation.

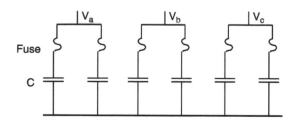

**Figure 16.1**   Example of individual fuse arrangement.

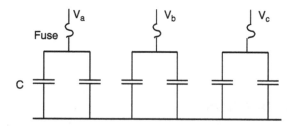

**Figure 16.2**   Example of group fuse arrangement.

### 16.2.2   Group Fusing

In group fusing, one fuse is used to protect several capacitor units. An example of a group fusing for an ungrounded wye capacitor bank is shown in Figure 16.2. The group fuse protection is commonly used for the pole-mounted capacitor banks. The fuses are mounted on cutouts and installed on the cross arm above the capacitor bank. The function of the group fuse is to detect the escalating failure of a single capacitor unit and remove the capacitor group from service to prevent case rupture and damage to other units. The group fuse should withstand the normal capacitor bank operating conditions without spurious fuse operations. The group fuse should have the following withstand capabilities for normal operation:

1. *Continuous current.* This includes consideration for a harmonic component, capacitance tolerance (maximum of 15%), and overvoltage ($+10\%$). The continuous

current capability of the fuse is a minimum of 125–135% of the capacitor normal current.

2. *Switching inrush current.* The minimum melting curve of the fuse is coordinated with the bank inrush current to minimize the possibility of nuisance tripping.

3. *Surge current.* The surge current due to lightning strike or a nearby arcing fault can be a concern in pole-mounted capacitor banks with fuse ratings of small value. In high lightning areas, slower fuse speeds with higher surge withstand capability (T-speed) are often used instead of high speed fuses (K-links).

4. *Rated fuse voltage.* The group fuse is rated for phase-to-phase voltage in ungrounded wye banks and in grounded wye banks. A higher voltage rating is needed for ungrounded wye banks due to the higher recovery voltage across the fuse when clearing a failed fuse unit.

To minimize the possibility of case rupture of the failed unit and damage to other units, the fuse should be selected to:

1. Interrupt the maximum 60 Hz fault current expected. In grounded wye and delta connected applications, the maximum current is the system available fault current at the capacitor location.

2. Coordinate with the capacitor case rupture curve for each capacitor unit. The maximum total clearing time of the fuse to be used must be on the left side of the case rupture curve.

3. Remove the failed unit without impressing the excessive overvoltages on good units. In ungrounded wye applications, the line-to-line voltage will be impressed on the good phases during shorting of a capacitor unit.

## 16.3 CAPACITOR FUSES

### 16.3.1 Expulsion Fuse

NEMA K-links usually are used in all the capacitor applications. NEMA K-link fuses are for faster applications.

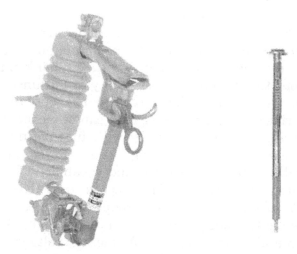

**Figure 16.3**   View of the cutout for an expulsion type of fuse and a fuse link. (Courtesy of S&C Electric Company, Chicago.)

The T-link fuses are used for slower operations. A typical cutout and a fuse link are shown in Figure 16.3. The typical minimum melting time–current characteristic curves of K-link fuses are shown in Figure E-1, Appendix E. The preferred ratings are 6, 10, 25, 40, 65, 100, 140, and 200 K. The intermediate ratings are 8, 12, 20, 30, 50, and 80 K. The corresponding melting time for K-link fuses are listed in Table E-1. The typical minimum melting time–current characteristic curves of T-link fuses are shown in Figure E-2. The preferred ratings of T-link fuses are the same as K-link fuses. The melting times of T-link fuses are shown in Table E-2.

A typical expulsion fuse has a short fusible element to sense the overcurrent and start the arcing required for inter- ruption. Attached to this fusible element is a large conductor (fuse leader), which then connects to the fuse hardware. During the fault, the fuse element will melt, producing an arc inside the fuse cartridge. When the arc is produced, it will rapidly create gases that will de-ionize and remove the

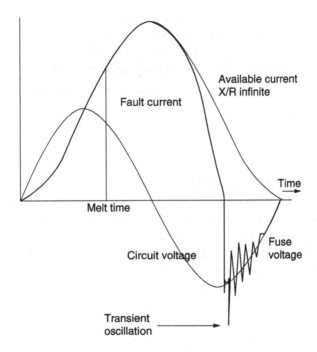

**Figure 16.4** Fault clearing with an expulsion type of fuse.

arc generated by the ionized gases and allow a rapid buildup of dielectric strength that can withstand the transient recovery voltage and steady-state power system voltage. The expulsion type of fuse link can be replaced with a new link. The difference between the K-link and the T-link fuses is the speed ratio. The K-links have speed ratios between 6 and 8.1. The T-links have speed ratios between 10.0 and 13.0 [1]. The fault clearing of an expulsion type of fuse is shown in Figure 16.4.

### 16.3.2  Current Limiting Fuse

The current limiting fuse limits the current to a substantially lower value than the peak value of the fault current. An outline of a current limiting fuse is shown in Figure 16.5. A current limiting fuse consists of a fusible element made of silver, packed with quartz filler, and sealed inside a strong ceramic case. During a fault, the fusible element vaporizes and the

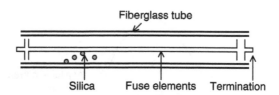

**Figure 16.5** The current limiting fuse.

resulting arcs melt the surrounding quartz. This absorbs the heat from the arcs, quickly extinguishing the arcs and clearing the fault. The molten quartz cools and solidifies, forming insulating plugs between the ends of the fuse links, thus preventing restriking. This type of current limiting fuse is a one-time, nonrenewable fuse that is discarded after operation and replaced by a completely new fuse. Current limiting fuses are available in two types, general purpose and backup. The fuses usually cover 150–200% of the rated capacitor current. The backup fuse is capable of interrupting all currents from the rated interrupting current to the rated minimum interrupting current. The backup fuse is used to clear high fault current [2].

Figure 16.6 shows the plot of the fault current against time in a typical current limiting fuse and the resulting ability of the fuse to limit the magnitude of the let-through current. The current through the fuse is asymmetrical, since the fuse acts fast within the first cycle of the prospective fault current. Heat is generated as the current rises rapidly, causing the fuse element to melt before the current can reach the instantaneous peak value. The area under the curve represents the total energy and is referred to as the $I^2t$ value.

## 16.4 SELECTION OF FUSE LINKS

The following factors are considered when selecting a fuse link for capacitor protection:

1. Continuous current
2. Voltage rating

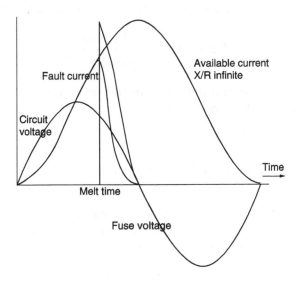

**Figure 16.6** Fault current interruption by the current limiting fuse.

3. Capacitor tolerance
4. Harmonic currents
5. Fault current magnitudes
6. Transient currents due to back-to-back switching
7. Case rupture characteristics
8. Voltage on good capacitor units
9. Energy discharge into failed units
10. Outrush current
11. Coordination with unbalanced detection schemes

Consideration of these factors in the selection of a suitable fuse link is discussed below.

### 16.4.1 Continuous Current

For capacitor applications, the fuse should withstand 135% of rated current. For a single-phase capacitor unit:

$$I\,(\text{Capacitor}) = \frac{\text{kVAR/unit}}{\text{kV/unit}} \qquad (16.1)$$

For a capacitor in a three-phase circuit, the capacitor current is:

$$I\,(\text{Capacitor}) = \frac{\text{kVAR/unit}}{\sqrt{3}\,\text{kV/unit}} \qquad (16.2)$$

Capacitors are manufactured in accordance with industry standards capable of operating at:

- 135% of rated kVAR
- 180% of rated current
- 110% of rated voltage continuously
- +15% tolerance on capacitor production tolerance

The rms current ($I_{\text{rms}}$) through the capacitor in a harmonic environment can be calculated, knowing the harmonic components and using Equation (16.3):

$$I_{\text{rms}} = \sqrt{I_1^2 + I_2^2 + I_3^2 \ldots I_n^2} \qquad (16.3)$$

The current through the individual fuse link required in a capacitor circuit is given by:

$$I\,(\text{link}) = \frac{\text{kVAR/unit}}{\text{kV/unit}} \times 1.35 \qquad (16.4)$$

In the case of NEMA K- and T-link fuses, which are 150% of the rated capacitor current, Equation (16.4) should be divided by 1.5 to get the desired fuse rating:

$$I\,(\text{link}) = \frac{\text{kVAR/unit}}{\text{kV/unit}} \times \frac{1.35}{1.5} \qquad (16.5)$$

The fuse current for a three-phase capacitor bank is given by:

$$I\,(\text{Group Fuse}) = \frac{\text{kVAR/unit}}{\sqrt{3}\,\text{kV/unit}} \times \frac{1.35}{1.5} \qquad (16.6)$$

### 16.4.2  Voltage Rating

In a power system, the acceptable tolerance on the voltage is ±5%. For example, the expected operating voltage range for a

2.4 kV system is 2.28–2.52 kV. The voltage rating of the fuse link has to be higher than the supply voltage. Based on Equation (16.5), the fuse links for individual capacitor units are shown in Table 16.1. Based on Equation (16.6), the fuses for group capacitors for a few selected ratings are listed in Table 16.2.

## Example 16.1

Select a fuse link for a 100 kVAR, 2.4 kV, single-phase capacitor unit.

Solution

$$kVAR = 100, \quad kV = 2.4$$

$$I \text{(link)} = \frac{100 \, kVAR}{2.4 \, kV} \times \frac{1.35}{1.5} = 37.4 \, A$$

Select a 40 K-link fuse with a voltage rating of 4.3 kV.

As indicated in the current limiting fuses, the capacitor banks are sometimes protected with fast acting current limiting fuses. The capacitor current is calculated using Equation (16.1) or (16.2). Then the fuse current is calculated as:

$$I \text{(Current limiting fuse)} = 1.5 \times \text{Capacitor current} \quad (16.7)$$

The factor 1.5 is a safety margin to account for overvoltage, harmonics, capacitor tolerances, and $I^2t$ melting characteristics. Typical current limiting fuses for various kVAR ratings based on Equation (16.7) are presented in Table 16.3.

## Example 16.2

Select a current limiting fuse for a 100 kVAR, 2.4 kV, single-phase capacitor unit.

TABLE 16.1 Expulsion Fuses for Individual Capacitors Based on Equation (16.5) for Grounded Wye Connection

| Cap Unit kV | Fuse Voltage kV | 50 kVAR Unit | | 100 kVAR Unit | | 150 kVAR Unit | | 200 kVAR Unit | | 300 kVAR Unit | |
|---|---|---|---|---|---|---|---|---|---|---|---|
| | | I (Fuse) A | Fuse Link | I (Fuse) A | Fuse Link | I (Fuse) A | Fuse Link | I (Fuse) A | Fuse Link | I (Fuse) A | Fuse Link |
| 2.4 | 4.3 | 18.8 | 25K | 37.5 | 40K | 56.3 | 65K | 75.0 | 80K | 112.5 | 140K |
| 2.77 | 4.3 | 16.2 | 25K | 32.5 | 40K | 48.7 | 65K | 65.0 | 80K | 97.5 | 140K |
| 4.16 | 4.3 | 10.8 | 25K | 21.6 | 40K | 32.5 | 40K | 43.3 | 65K | 64.9 | 80K |
| 4.8 | 5.5 | 9.4 | 10K | 18.8 | 25K | 28.1 | 40K | 37.5 | 40K | 56.3 | 65K |
| 7.2 | 15.5 | 6.3 | 10K | 12.5 | 25K | 18.8 | 25K | 25.0 | 40K | 37.5 | 40K |
| 7.62 | 15.5 | 5.9 | 10K | 11.8 | 25K | 17.7 | 25K | 23.6 | 40K | 35.4 | 40K |
| 7.96 | 15.5 | 5.7 | 10K | 11.3 | 25K | 17.0 | 25K | 22.6 | 40K | 33.9 | 40K |
| 8.32 | 15.5 | 5.4 | 8K | 10.8 | 8K | 16.2 | 25K | 21.6 | 40K | 32.5 | 40K |
| 12.47 | 15.5 | 3.6 | 8K | 7.2 | 10K | 10.8 | 12K | 14.4 | 25K | 21.7 | 40K |
| 13.28 | 35.0 | 3.4 | 8K | 6.8 | 10K | 10.2 | 12K | 13.6 | 25K | 20.3 | 40K |
| 13.80 | 35.0 | 3.3 | 8K | 6.5 | 10K | 9.8 | 12K | 13.0 | 25K | 19.6 | 40K |
| 14.40 | 35.0 | 3.1 | 8K | 6.3 | 10K | 9.4 | 12K | 12.5 | 25K | 18.8 | 40K |
| 19.92 | 35 | 2.3 | 8K | 4.5 | 10K | 6.8 | 10K | 9.0 | 10K | 13.6 | 20K |

TABLE 16.2 Expulsion Fuse for Group Capacitors Based on Equation (16.6) for Grounded Wye Connection (MVAR Given is Three Phase)

| System kV | Fuse kV | 150 kVAR | 300 kVAR | 450 kVAR | 600 kVAR | 900 kVAR | 1200 kVAR | 1800 kVAR | 2400 kVAR | 3000 kVAR |
|---|---|---|---|---|---|---|---|---|---|---|
| 2.4 | 4.3 | 40K | 80K | - | - | - | - | - | - | - |
| 2.77 | 4.3 | 40K | 80K | 100K | - | - | - | - | - | - |
| 4.16 | 4.3 | 30K | 65K | 65K | 100K | - | - | - | - | - |
| 4.8 | 5.5 | 25K | 40K | 65K | 80K | - | - | - | - | - |
| 6.64 | 15.5 | 20K | 40K | 50K | 65K | 80K | - | - | - | - |
| 7.2 | 15.5 | 20K | 30K | 40K | 65K | 80K | 100K | - | - | - |
| 7.96 | 15.5 | 20K | 30K | 40K | 65K | 80K | 100K | - | - | - |
| 8.32 | 15.5 | 20K | 30K | 40K | 50K | 65K | 100K | - | - | - |
| 9.96 | 15.5 | 10K | 25K | 30K | 50K | 65K | 80K | 140K | - | - |
| 12.47 | 15.5 | 10K | 25K | 30K | 40K | 50K | 65K | 100K | 120K | - |
| 13.8 | 35.0 | 10K | 25K | 25K | 40K | 50K | 65K | 80K | 120K | - |
| 14.4 | 35 | 10K | 25K | 25K | 30K | 40K | 65K | 80K | 100K | - |
| 19.2 | 35 | 6K | 10K | 20K | 30K | 40K | 40K | 65K | 80K | 100K |
| 21.6 | 35 | - | - | 20K | 25K | 40K | 40K | 65K | 80K | 100K |
| 23.0 | 35 | - | - | 12K | 25K | 30K | 40K | 65K | 80K | 80K |
| 24.8 | 35 | - | - | 12K | 12K | 30K | 30K | 65K | 65K | 80K |
| 34.5 | 35 | - | - | - | 12K | 20K | 30K | 40K | 50K | 65K |

**TABLE 16.3**   Current Limiting Fuses for Capacitor Applications Based on Equation (16.7)

| Capacitor kV | Fuse kV | 100 kVAR Unit | | | 200 kVAR Unit | | | 300 kVAR Unit | | | 400 kVAR Unit | | |
|---|---|---|---|---|---|---|---|---|---|---|---|---|---|
| | | I (Cap) | I (Fuse) | Fuse A | I (Cap) | I (Fuse) | Fuse A | I (Cap) | I (Fuse) | Fuse A | I (Cap) | I (Fuse) | Fuse A |
| 2.40 | 4.3 | 42 | 63 | 65 | 83 | 125 | 130 | 125 | 188 | 200 | | | |
| 2.77 | 4.3 | 36 | 54 | 65 | 72 | 108 | 130 | 108 | 162 | 200 | | | |
| 4.16 | 4.3 | 24 | 36 | 65 | 48 | 72 | 75 | 82 | 108 | 130 | 96 | 144 | 150 |
| 4.8 | 5.5 | 21 | 31 | 40 | 42 | 63 | 65 | 63 | 94 | 100 | 83 | 125 | 130 |
| 6.64 | 8.3 | 15 | 23 | 25 | 30 | 45 | 50 | 45 | 68 | 80 | 60 | 90 | 100 |
| 7.20 | 8.3 | 14 | 21 | 25 | 28 | 42 | 50 | 42 | 63 | 65 | 56 | 83 | 100 |
| 7.96 | 8.3 | 13 | 19 | 20 | 25 | 38 | 40 | 38 | 57 | 65 | 50 | 75 | 80 |
| 8.32 | 8.3 | 12 | 18 | 20 | 24 | 36 | 40 | 36 | 54 | 65 | 48 | 72 | 80 |
| 9.96 | 15.5 | 10 | 15 | 18 | 20 | 30 | 40 | 30 | 45 | 50 | 40 | 60 | 65 |
| 12.47 | 15.5 | 8 | 12 | 18 | 16 | 24 | 25 | 24 | 36 | 40 | 32 | 48 | 50 |
| 13.80 | 15.5 | 7 | 12 | 18 | 14 | 22 | 25 | 22 | 33 | 40 | 29 | 43 | 50 |
| 14.40 | 15.5 | 7 | 12 | 18 | 14 | 21 | 25 | 21 | 31 | 40 | 28 | 43 | 50 |
| 19.2 | 23 | 5 | 8 | 12 | 10 | 18 | 25 | 16 | 23 | 25 | 21 | 31 | 40 |
| 21.6 | 23 | 5 | 8 | 12 | 9 | 18 | 25 | 14 | 21 | 25 | 19 | 28 | 30 |

Solution

$$kVAR = 100, \quad kV = 2.4$$

$$I \text{ (Capacitor)} = \frac{100 \, kVAR}{2.4 \, kV} = 41.7 \, A$$

$$I \text{ (Current limiting fuse)} = 1.5 \times I \text{ (capacitor)}$$
$$= 1.5 \times 41.7 \, A = 62.5 \, A$$

Select a 65 A fuse and a 4.3 kV fuse link.

### 16.4.3 Capacitor Tolerance

The acceptable tolerance on the capacitor unit is +15%.

### 16.4.4 Harmonic Currents

When capacitor units are added to the power system for power factor correction, this will change the frequency response characteristics of the system. This is illustrated using the circuit in Figure 20.8 and the response without and with shunt capacitors in Figure 20.9 (Chapter 20). The effect of a tuned filter is studied using the circuit in Figure 20.10. The frequency response of the system with a tuned filter is shown in Figure 14.9. Results also show that the individual voltage and current harmonics are reduced by the use of tuned filters. The source of harmonic resonance and voltage magnification are discussed in Chapter 22, using Figure 22.22. During energization of high voltage capacitors, the low voltage capacitor circuit shows significant oscillations in the absence of suitable damping. Such a condition will increase the harmonic currents. The harmonic currents have to be accounted for in calculating the rms current through the capacitor using Equation (16.3).

### 16.4.5 Fault Current Magnitudes

The available current at the capacitor location should be considered in the selection of the fuse link. When there is a fault, the capacitor must be able to withstand the fault current until the fuse interrupts the fault current successfully.

The fault current duty will be very severe in the case of a bus fault [3], close to the capacitor location. An example circuit with such a fault is shown in Figure 16.7. During the bus fault, there is a small impedance between the capacitor and the fault location to control the fault current. A typical waveform of the current is shown in Figure 16.8. The frequency of the fault current is 5,000 Hz. For typical all-film capacitors, the available fault current should not exceed the limits given in Table 16.4 [4]. If these levels are exceeded, then current limiting fuses should be used. For specific capacitors, the fault current limit value should be obtained from the manufacturer.

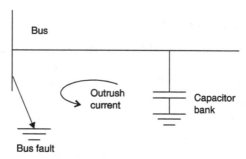

**Figure 16.7** Circuit diagram showing a bus fault with a shunt capacitor bank.

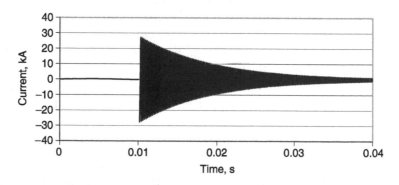

**Figure 16.8** The current during SLG fault close to the shunt capacitor location.

TABLE **16.4**  Fault Current Limits for All-Film Capacitors

| | Max. Sym. Fault Current rms A, When $X/R$ is | | | | Maximum Fuse Rating Used | |
|---|---|---|---|---|---|---|
| TV Level | 0 | 5 | 10 | 15 | K-Link | T-Link |
| 50 and 100 kVAR | | | | | | |
| Less than 5 kV | 4300 | 3000 | 2600 | 2500 | 50 | 25 |
| Above 30 kV | 3100 | 2200 | 1900 | 1800 | 30 | 20 |
| With 38 kV cutout | 2000 | 1700 | 1500 | 1400 | 30EK | 20ET |
| 150, 200, 300, and 400 kVAR | | | | | | |
| Less than 9 kV | 8400 | 5900 | 5100 | 4800 | 80 | 50 |
| Less than 9 kV | 7900 | 5600 | 4800 | 4600 | 100 | - |
| Greater than 9 kV | 6300 | 4400 | 3800 | 3600 | 65 | 40 |
| Greater than 9 kV | 5700 | 4000 | 3400 | 3300 | 80 | - |
| With 38 kV cutout | 4100 | 3400 | 3100 | 2900 | 65EK | 40ET |

## 16.4.6  Transient Currents due to Back-to-Back Switching

Consider two capacitor banks connected to the same bus as shown in Figure 16.9. The capacitor unit $C_1$ is in service and the capacitor unit $C_2$ is not in service. $C_1$ is charged and $C_2$ is not charged. Therefore, closing the switch to energize the capacitor bank produces a transient inrush current. A typical time domain plot is shown in Figure 16.10. This may not be the case in pole-mounted capacitors, but in many industrial and substation capacitors, such back-to-back switching can occur. The magnitude and frequency of transient inrush current depends on applied voltage, the point of closing, the capacitance MVAR rating of the banks, and damping due to closing resistance in the circuit breaker. The frequency of the transient current in the example is 1,300 Hz. The capacitor fuses are

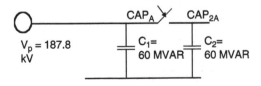

**Figure 16.9**  Back-to-back capacitor on a 230 kV system for phase A.

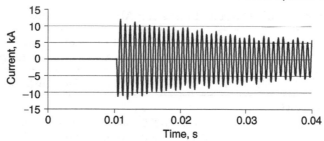

**Figure 16.10**   Current waveform during back-to-back switching in phase A.

also subject to high frequency transient currents due to lightning surges. To minimize spurious operations due to lightning surges, T-links are used in low current circuits below 25 A. The K-link fuses are used for circuits above 25 A. T-link fuses can withstand a higher surge current than K-link fuses.

### 16.4.7   Case Rupture Characteristics

The maximum clearing time characteristics curve for the fuse link must coordinate with the case rupture curve of the capacitor. Such coordination is necessary to ensure that the fuse will clear the circuit prior to case rupture when an internal fault occurs. The maximum clearing time of the fuse must fall to the left of the case rupture curve at and below the level of the available fault current. Figure 16.11 shows a typical case rupture curve for a capacitor. The safe and unsafe zones are

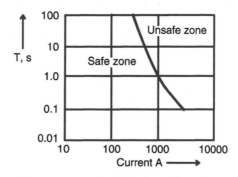

**Figure 16.11**   Typical capacitor tank rupture curve.

identified in this curve. The melting characteristics of the fuse have to be in the safe zone of this curve. In the event of a fault, the fuse will go first and the case rupture will not occur.

### 16.4.8 Voltage on Good Capacitor Units

In ungrounded wye connected capacitor banks, when a capacitor in one phase fails, the voltage on the unfaulted phases will be 1.732 times the rated voltage. If the failed unit is not removed quickly, then the other units will fail due to overvoltage, leading to phase-to-phase fault. Therefore, it is desirable to use a fault-clearing fuse to minimize the possibility of second unit failure. This criterion requires a fast acting fuse such as a K-link. The transient clearing requires a slow acting fuse such as a T-link. A smaller rating fuse can perform a faster clearing. Table 7.4, Chapter 7 presents the voltage on other series groups in the bank when a capacitor unit is shorted on phase A.

### 16.4.9 Energy Discharge into Failed Capacitor Units

Both the capacitor and the fuse should be capable of handling the energy available in the parallel units. When a capacitor failure occurs, all the stored energy of the parallel capacitor units can discharge through the failed capacitor and the fuse. Figure 16.12 shows a diagram illustrating this condition. The calculated stored energy should not exceed the energy capability of the capacitor unit and the fuse. Exceeding this rating may result in a fuse failure and rupture of the capacitor case. The calculated value of energy should not exceed 10,000 J (i.e., 3,100 kVAR in parallel) for paper or film capacitors and

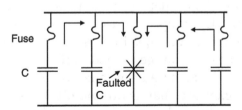

**Figure 16.12** Current flow into a failed capacitor unit; only one phase is shown.

15,000 J (i.e., 4,650 kVAR in parallel) for all-film capacitors. When the calculated value of the energy exceeds the limits, the following two options are available.

- Redesign the capacitor bank to reduce the amount of stored energy in the parallel units. This can be done using additional series groups or reconnecting the bank in a split-wye configuration.
- Use current limiting fuses. This involves more losses in the fuses and increased cost.

Therefore, it can be seen that fusing is not the limit in the design of large banks.

### 16.4.10 Outrush Currents

The fuses connected to the unfailed capacitor units must withstand high frequency $I^2t$ discharge of the failed capacitor. When a capacitor failure occurs, the remaining units will discharge energy into the failed unit. The fuses on the unfailed capacitor should be able to withstand the high frequency discharge in order to avoid multiple fuse operations. Proper consideration of the criteria requires an understanding of the $I^2t$ capabilities of the fuse and the $I^2t$ outrush capacity of the capacitor. The application requires the fastest fuse link to meet the outrush current.

### 16.4.11 Coordination with Unbalanced Detection Schemes

When a fuse operates in a capacitor bank, an increase in the voltage occurs on the remaining units in the series group. Usually, an unbalance detection scheme is used to monitor such conditions. Upon operation of the first fuse, if the overvoltage on the remaining capacitors is less than 110%, an alarm will sound. If the overvoltage exceeds 110%, the capacitor bank will be tripped. Load break switches or circuit breakers are used to switch capacitor banks on and off. Unbalance detection scheme settings should be coordinated with fuse TCC, so that the fuse will be allowed to clear a failed capacitor unit before the unbalance detection scheme trips the capacitor bank. If the capacitor bank is tripped before the fuse operates, there will be no visible indication of the cause for the removal of the bank. A proper time delay has to be included so that the fuse operates first and the unbalance detection operates at a delayed time.

## 16.5 EFFECT OF OPEN FUSES

In cable connected circuits, there is significant capacitance and the effect of one open phase has to be considered. If there is a potential transformer connected in the same circuit, then there will be interaction between the cable capacitance and magnetizing reactance of the potential transformer, causing ferroresonance. One such circuit used in ferroresonance simulation is shown in Figure 16.13. One open fuse will produce unbalanced voltage in all the 13.8 kV circuits, including the voltage transformer circuit. The fuse is modeled as a three-pole switch. The 75 ft cable between the fuse and the service transformer is represented by an impedance and $0.003\,\mu F$.

One phase open is a serious type of fault in unloaded lines because the voltage in the open phase can go up to 2 P.U. This is called neutral inversion. The motors can start running in the reverse direction with this type of supply condition. The issues related to open conductor and remedial approaches are discussed in IEEE Standard C57.105. Neutral inversion is a steady-state voltage condition and the corresponding unbalanced condition is difficult to control. If the transformer is not protected properly, it will fail due to overload (thermally). The lightning arrester in the circuit will fail because

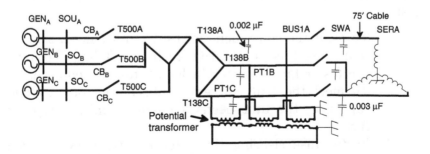

**Figure 16.13** Circuit diagram to simulate one fuse open.

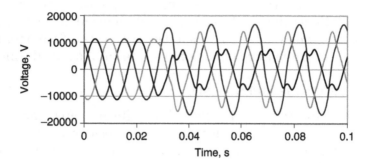

**Figure 16.14**  Voltage waveforms in a 13.8 kV circuit with one fuse open.

the MCOV voltage rating is exceeded for a prolonged duration. Unbalanced voltage is produced in the broken delta. The voltages for the three phases are presented in Figure 16.14. The voltages in the phases are 17, 14, and 7 kV, respectively. The corresponding per unit values are 1.5, 1.24, and 0.62 P.U., respectively. The voltage in the open circuit is 1.5 P.U., which will cause damage to sensitive equipment. Any capacitor bank in such a circuit has to be tripped in order to avoid equipment damage.

## 16.6  CONCLUSIONS

The following conclusions are made regarding capacitor fusing.

1. To meet the continuous current requirement, a minimum fuse link is required, such as a T-link. To meet the faster fault clearing, a K-link fuse is required.
2. To meet the fault current to discharge into a failed capacitor, the unit needs a current limiting type of fuse.
3. Faster fuse clearing time characteristics are needed to coordinate with the capacitor case rupture curve.

A faster fuse clearing also minimizes the overvoltages on good capacitor units during a unit failure.

4. Slow fuse melting characteristics are needed to withstand the transient currents due to back-to-back switching and outrush current due to unit failure.

It is not always possible to meet all the criteria. A fuse link is chosen based on the steady-state current and the desired safety factors.

## PROBLEMS

16.1. What type of fusing arrangement can be provided for capacitors?

16.2. What fuses are used in capacitor protection?

16.3. List the considerations when selecting a fuse for a capacitor circuit.

16.4. Consider a 300 kVAR, 13.8 kV, 60 Hz, single-phase capacitor unit. Select a suitable fuse link for this capacitor application. If this capacitor is applied in a 6.6 kV circuit, then what is the delivered kVAR? What rating of fuse link is to be used for proper overcurrent protection?

16.5. It was identified that the transient inrush current during a back-to-back capacitor switching contains high frequency components. What is the source of this high frequency?

16.6. Consider a three-phase, 480 V, 60 Hz supply. The circuit is provided with grounded wye capacitors for power factor correction. The fuse in one phase is open. The fuses in the other two phases are in good condition. Can you estimate the voltage in the other phases? Use phasor, symmetrical components, or any other technique. Explain why such voltage magnitudes are observed.

16.7. Select a current limiting fuse for a single-phase, 400 kVAR, 13.8 kV capacitor unit.

## REFERENCES

1. Burke, J. J. (1994). *Power Distribution Engineering*, Marcel Dekker, New York.

2. ANSI Standard 37.46 (1981), Specifications for Distribution Cutouts and Fuse Links.

3. Sobot, A., Morin, C., Guillaume, C., Pans, A., Taisne, J. P., Pizzo, G. L., Morf, H. L. (1993). A unique multipurpose damping circuit for shunt capacitor bank switching, *IEEE Transactions on Power Delivery,* 8(3), 1173–1183.

4. Mendis, S. R., Bishop, M. T., McCall, J. C., Hurst, W. M. (1993). Overcurrent protection of capacitors applied on industrial distribution systems, *IEEE Transactions on Industry Applications*, 29(3), 541–547.

5. IEEE Standard 1036 (1992), IEEE Guide for Application of Shunt Power Capacitors.

# 17

## CIRCUIT BREAKERS

### 17.1 INTRODUCTION

Circuit breakers are electromechanical devices used in the power system to connect or disconnect the power supply at the generator, substation, or load location. The circuit breaker consists of current-carrying contacts called electrodes. When the power is on, these electrodes are normally in contact and they will be apart when the power is disconnected. During the disconnection of the power supply, the contacts are separated and an arc is struck between them. The arc plays an important role in the interruption process by providing for the gradual transition from current-carrying to voltage-withstanding state of the contacts. Most of the technical problems in the circuit breaker are related to arc quenching and transient overvoltages across the circuit breaker blades in the open condition. The opening and closing of the circuit breakers in a capacitive circuit poses additional technical problems due to the energy stored in the capacitor elements.

### 17.2 TYPES OF CIRCUIT BREAKERS

The circuit breaker is a mechanical device capable of making, carrying, and breaking currents under normal conditions.

The circuit breaker also should have the capability of making, carrying, and breaking currents under specific abnormal circuit conditions. The circuit breakers are classified into various categories based on voltage levels, switching duty, installation, and interrupting medium [1–5].

### 17.2.1 Circuit Breakers Based on Voltage Levels

Industry standards identify circuit breakers as (a) below 72.5 kV and (b) above 121 kV. Such classification is applicable to all circuit breakers including gas insulated switchgear [4]. The list of circuit breakers based on the above classification is presented in Appendix C.

### 17.2.2 Circuit Breakers Based on Switching Duty

Circuit breakers are used for switching resistive, inductive, and capacitive load currents. The resistive and inductive current handling can be performed by circuit breakers designed for normal switching operations. The circuit breakers intended for capacitive current switching operations are special purpose devices. The classification based on the switching duty is:

- Normal duty circuit breakers for inductive currents.
- Special circuit breakers for capacitive current applications.

Such classification of circuit breakers is identified in Appendix C for voltages below 72.5 kV and above 121 kV levels.

### 17.2.3 Circuit Breakers Based on Installations

Circuit breakers can be used in either indoor or outdoor installations. Indoor circuit breakers are designed for use only inside buildings or weather-resistant enclosures. For example, the medium voltage circuit breakers are designed for use inside a metal-clad switchgear enclosure. The circuit breakers for outdoor applications are equipped with enclosures resistant to weather changes. The internal parts in both the indoor and outdoor circuit breakers are the same.

The outdoor circuit breakers can be classified as dead tank or live tank. A dead tank circuit breaker is a device in which the tank of the circuit breaker is at ground potential. The interrupters are surrounded by an insulating medium inside the tank. A live tank breaker is a switching device in which the tank housing the interrupters and insulating medium is above the ground potential. The IEC standard on the live tank circuit breakers claims the following advantages [6]:

- Low cost and smaller size.
- Less installation space requirements.
- Use of lesser amount of insulating medium.

However, the live circuit breakers are likely to cause problems during an earthquake. The dead tank circuit breakers are commonly used in U.S. utilities. The ANSI standard claims the following advantages:

- Better seismic withstand capability.
- Factory assembled and contains safe enclosure.
- The separate input and output bushing helps to install current transformers for measurement.

The dead tank circuit breakers are more robust and expensive compared to the live tank circuit breakers.

### 17.2.4 Circuit Breakers Based on Interrupting Medium

In an air circuit breaker, the main components are the interrupting mechanisms, the interrupting medium, and the tank. The methods that are used to achieve interaction between the interrupting medium and the electric arc are different in various circuit breakers. Some of the interrupting media used in circuit breaker applications are air, oil, compressed air, vacuum, and $SF_6$. A description of the circuit breakers using these media is presented below.

### 17.2.4.1  Air Circuit Breakers

This type of circuit breaker consists of three single-pole units linked together by an insulating crossbar. The pole assembly is mounted on a panel of insulating material with arc resisting barriers on both sides of each pole. The arc interruption process of air circuit breakers is based on the de-ionization of the gases by a cooling action. In the simplest type of air circuit breaker, the contacts are made in the form of two horns. The arc initially strikes across the shortest distance between the horns. Then the arc extends upwards across the horns and slowly decays as the distance between the electrodes increases.

In the magnetic blowout type of air circuit breaker, the arc extinction is performed by means of a magnetic blast. In this case, the arc is subjected to the action of a magnetic field setup by the coils connected in series with the circuit being interrupted. The arc is blown magnetically into the vertical arc chutes, where the arc is lengthened, cooled, and extinguished. The arc chutes help break the arc and prevent the spreading of the arc to adjacent metal parts. This type of circuit breaker is used at voltage levels up to 11 kV.

Another air circuit breaker design is based on the arc splitter type. In this design, the blowouts consist of steel inserts in the arc chutes. The arc chutes divide the moving arc in the vertical direction. When the arc comes into contact with long steel plates, it cools rapidly. Such construction is useful for heavy duty applications. Air circuit breakers are suitable for the control of power plant auxiliary systems and industrial plants. These circuit breakers are installed on the ground and provide a high degree of safety.

### 17.2.4.2  Oil Circuit Breakers

In this type of circuit breaker, the interrupting contacts are immersed in oil. The arc produced during the separation of electrodes evaporates the surrounding oil. The arcing in the oil produces the hydrogen gas. The presence of oil helps extinguish the arc rapidly and provides much needed insulation for the live exposed metal parts. Usually, the oil is flammable and may cause fire hazards if a failure occurs. Carbon is formed during the arcing process and the oil is polluted, decreasing the dielectric strength over a period of time. Therefore, periodic maintenance and replacement of oil are necessary. There are different versions of these breakers such as plain oil circuit breakers, bulk oil circuit breakers, and minimum oil circuit breakers. The oil breakers are commonly used up to 110 kV levels.

### 17.2.4.3 Air Blast Circuit Breaker

In the high voltage systems, high speed operations are achieved by using compressed air as a circuit breaking medium. Compressed air, nitrogen, carbon dioxide, hydrogen, and freon type of gases are used in the circuit breaker applications. The compressed air is passed through the parting electrodes and the arc is extinguished rapidly. The three types of air blast arrangements used are axial blast, radial blast, and cross blast. This type of circuit breaker construction is used for currents of 2,000–4,000 A.

### 17.2.4.4 Vacuum Circuit Breakers

Vacuum has the highest insulating strength. When the electrodes of a circuit breaker are opened, the interruption occurs at the first current zero. Therefore, the dielectric strength across the electrodes will build up faster than that obtained in a conventional circuit breaker. The vacuum circuit breaker is a very simple device compared to the oil or air blast circuit breakers. Two contacts are mounted inside an insulating vacuum-sealed container. One of the electrodes is fixed and the other may be moveable through a short distance. A metallic shield surrounds the contacts and protects the insulating container. One end of the fixed contact is brought out of the chamber, to which external connection can be made. The moving contact is firmly joined to the operating rod of the mechanism. The lower end is fixed to a spring-operated mechanism, so that the metallic bellows inside the chamber are moved upward or downward during closing and opening operations, respectively. These operations are performed without bouncing. These features enable the vacuum circuit breaker to operate for a large number of opening and closing operations without significant maintenance. Since there is no bouncing during the operation, vaccum circuit breakers can interrupt the capacitive currents without restrike.

### 17.2.4.5 Sulphur Hexafluoride ($SF_6$) Circuit Breakers

$SF_6$ is about five times heavier than air and is used as the medium of insulation and arc interruption in this type of circuit breaker. The gas is claimed to be chemically stable, odorless, inert, nonflammable, and nontoxic. The gas has high dielectric strength and outstanding arc quenching properties. The main parts of an $SF_6$ circuit breaker are a tank, the interrupter units, the operating mechanisms, the bushings, and the gas system. Since the by-products of the decomposed $SF_6$ are unhealthy, the circuit breaker requires a sealed construction, which is expensive. The overall performance of the $SF_6$ circuit breaker is superior to similar equipment operating in the presence of other media. An example of a definite purpose, 120 kV, 40 kA, three-phase, $SF_6$ circuit breaker is shown in Figure 17.1.

This type of circuit breaker is suitable for shunt capacitor switching and has a significant effect on the magnitude of overvoltages produced. For many years, oil circuit breakers were commonly used to switch shunt capacitor banks. The interrupters in these circuit breakers require several cycles of arcing prior to successful interruption of the fault current. Such a delayed interruption results in restriking and high transient overvoltages. Although oil circuit breakers are still used by utilities for capacitor switching, these circuit breakers are known to produce overvoltage problems.

The gas circuit breakers such as air blast and $SF_6$ can interrupt the capacitive currents without restrike during de-energization and are extensively used in high voltage systems. It becomes expensive to install closing resistors in gas breakers at higher voltage levels in order to handle the energization transients.

Vacuum circuit breakers are known to produce significant overvoltages when switching shunt capacitor banks. It is difficult to fit vacuum interrupters with closing or opening resistors. Vacuum interrupters can interrupt high

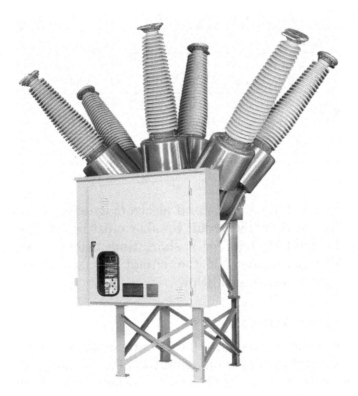

**Figure 17.1** Definite purpose, 120 kV, three-phase, 40 kA dead tank SF$_6$ circuit breaker. (Courtesy of Mitsubishi Electric Products Inc., Warrendale, PA.)

frequency currents successfully, and they can produce transients on closing due to multiple prestrikes. If a restrike occurs during de-energization of a vacuum interrupter, then that will produce transients higher than any other types of circuit breakers can produce.

## 17.3 CIRCUIT BREAKER RATINGS

A circuit breaker should have the capability of carrying the rated current at rated voltage conditions. During a short circuit, the circuit breaker must be capable of opening upon occurrence of the fault and performing fault clearing. Also, the

TABLE 17.1  Preferred Ratings of ANSI and IEC for Voltages

(a) Below 72.5 kV

| ANSI | 4.76 | 8.25 | 15.0 | 15.5 | 25.8 | 38.0 | 48.3 | 72.5 |
|------|------|------|------|------|------|------|------|------|
| IEC | 3.6 | 7.2 | 12.0 | 17.5 | 24.0 | 36.0 | 52.0 | 72.5 |

(b) Above 72.5 kV

| ANSI | – | 121 | 145 | 169 | 242 | – | 362 | – | 550 | 800 |
|------|-----|-----|-----|-----|-----|-----|-----|-----|-----|-----|
| IEC | 100 | 123 | 145 | 170 | 245 | 300 | 362 | 420 | 525 | 800 |

circuit breaker must withstand electromagnetic forces during the faults. Further, the circuit breaker must be capable of carrying the fault current for a short time before clearing the fault. The parameters of the circuit breaker covering all these aspects are defined below.

### 17.3.1  Rated Voltage

A circuit breaker is designed for a nominal voltage for a specific system condition. The same circuit breaker is assigned with a maximum operating voltage, which should not be exceeded. The standard voltage rating of a circuit breaker is presented in terms of the three-phase, line-to-line voltage. The specified ratings are based on operation at altitudes of 3,300 ft or less. At higher altitudes the rating should be de-rated. The preferred voltage ratings recommended by ANSI [4] and IEC [6] are listed in Table 17.1(a) and (b).

### Example 17.1

Circuit breakers are needed for a substation project with operating voltages of 345 kV and 110 kV. Based on the voltage rating, select suitable circuit breaker ratings from the ANSI and IEC preferred ratings.

Solution

Rated voltage = 345 kV
Maximum voltage $(345 \times 1.05) = 362$ kV

A 362 kV circuit breaker is adequate to accommodate the maximum system voltage requirement.

Rated voltage $= 110\,\mathrm{kV}$
Maximum voltage $(110 \times 1.05) = 115.5\,\mathrm{kV}$

The 121 kV circuit breaker from the ANSI rating and the 123 kV circuit breaker from the IEC rating are suitable for this application.

### 17.3.2 Frequency

The standard power system circuit breakers are rated at 60 or 50 Hz. Service at other frequencies must be given special consideration.

### 17.3.3 Rated Continuous Current

The nominal current rating of a circuit breaker is the rms current, which the circuit breaker shall be able to carry continuously at the rated voltage and frequency. The ambient temperature outside the enclosure is 40°C.

### 17.3.4 Rated Short Circuit Current

This is the highest value of symmetrical rms short-circuit current (see Figure 17.2).

### 17.3.5 Symmetrical Interrupting Current

At any operating voltage, the symmetrical interrupting current is given by:

$$I(\mathrm{int}) = I_{\mathrm{sc}}(\mathrm{rated})\left(\frac{V_{\mathrm{max}}}{V_{\mathrm{operating}}}\right) \qquad (17.1)$$

### 17.3.6 Latching Current

This is the current rating of the circuit breaker that will latch when it closes on a fault. The latching current is measured during the maximum cycle and the interrupting current is measured at the time of contact. This rating is equal to 1.6 times the rated short-circuit current (see Figure 17.3). If expressed in peak value, it is equal to 2.7 times the rated short-circuit current. Sometimes, this current is called momentary current and is measured one half cycle after the fault occurs.

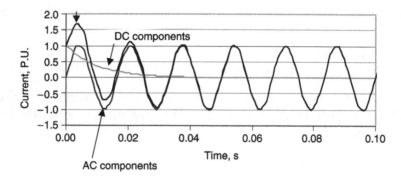

**Figure 17.2**  Components of the fault current.

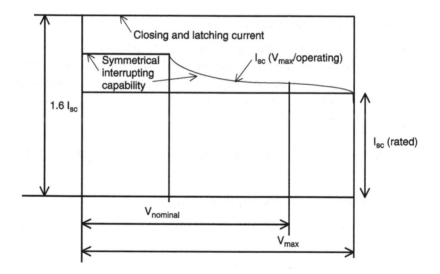

**Figure 17.3**  Relation of short-circuit current to closing and latching current.

### 17.3.7  Rated Interrupting Time

This is the maximum permissible interval between energizing the trip circuit at rated control voltage and interrupting the main circuit on all the poles when interrupting a current. The response time of the relaying is 0.5–1 cycle. The typical normal fault clearing time is 6 cycles, as shown in Figure 17.4.

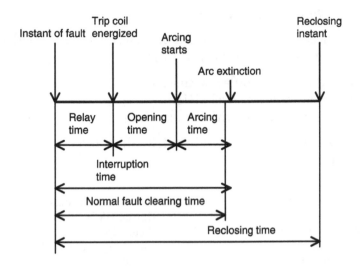

**Figure 17.4** Components of the circuit breaker operating time.

### 17.3.8 The Reclosing Time

In order to avoid power interruption followed by a circuit breaker tripping, the circuit breaker is closed after a small time delay. See Figure 17.4. Typical reclosing time in a high voltage circuit breaker is of the order of 16 cycles.

### 17.3.9 The Symmetrical Breaking Current

This is the rms value of the AC component in the pole at the instant of separation:

$$I_{\text{sym}} = \frac{I_{\text{AC}}}{\sqrt{2}} \tag{17.2}$$

### 17.3.10 The Asymmetrical Breaking Current

This is the rms value of the total current comprising the AC and DC components of the current at the pole at the instant of contact separation:

$$I_{\text{asym}} = \sqrt{\left[\left(\frac{I_{\text{AC}}}{\sqrt{2}}\right)^2 + I_{\text{DC}}^2\right]} \tag{17.3}$$

### 17.3.11  Rated Dielectric Strength

The required dielectric strength of a circuit breaker is defined in terms of the following three ratings:

- Lightning impulse withstand
- Basic impulse level (BIL)
- Chopped wave withstand voltage

Relations between the ANSI and IEC short-circuit currents [7] are given below.

| ANSI Current Ratings | IEC Current Ratings |
|---|---|
| First cycle | Initial (Ik") |
| Closing and latching | Peak (Ip) |
| Interrupting | Breaking current (Ib) |
| Time delayed | Steady-state current |

| ANSI Current Ratings | IEC Current Ratings |
|---|---|
| Momentary (1) | Steady state × 1.6 |
| Interrupting | Interrupting or breaking (Ib) |

Note (1): Momentary rating = Symmetrical short-circuit current × 1.6.

### Example 17.2

A circuit breaker has a specification of 25 kV, three-phase, 60 Hz, 1,200 A, 2000 MVA. What are the rated current, latching current, and breaking current? If the circuit breaker handles a capacitor bank with a steady-state current of 850 A/phase, what is the available safety margin?

Solution

Rated current $= 1,200\,\text{A/phase}$

$$\text{Symmetrical breaking current} = \frac{2,000\,\text{MVA}}{\sqrt{3}(25\,\text{kV})} = 46.2\,\text{kA}$$

Latching current $(46.2\,\text{kA} \times 1.6) = 73.9\,\text{A}$
Steady-state current $= 850\,\text{A/phase}$
Safety margin to handle tolerances, etc. $= 1.3$
Required steady-state current $(850 \times 1.3) = 1,105\,\text{A/phase}$

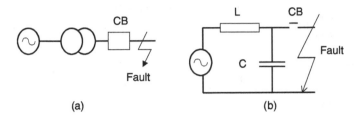

**Figure 17.5** Power system (a) and the equivalent circuit (b).

There is adequate safety margin available in the steady-state current rating.

## 17.4 CIRCUIT BREAKER PERFORMANCE

The performance of the circuit breakers is measured in terms of their behavior in the arc interruption, transient recovery voltage (TRV), rate of rise of recovery voltage (RRRV), and the performance towards the capacitive high-frequency currents during transient conditions. Figure 17.5 shows a simple power system and the equivalent circuit. Both the circuit inductance $L$ and the system capacitance are responsible for storing energy during normal operation of the system. Switching and faults cause overvoltages. The transients resulting from switching are caused by stored energy in both $L$ and $C$ when it undergoes a change from one state to another due to circuit interruption. The capacitor voltage during the opening of the circuit breaker is shown in Figure 17.6. The source voltage is displayed as a reference voltage.

### 17.4.1 Arc Phenomena and Arc Voltage

The arc is produced between the electrodes during the opening of the circuit breaker. This is due to the column of ionized gases produced during the circuit breaker opening. The arc voltage is mainly resistive and is in phase with the arc current. The magnitude of the arc voltage increases in each successive current loop because the contacts of the circuit breaker are

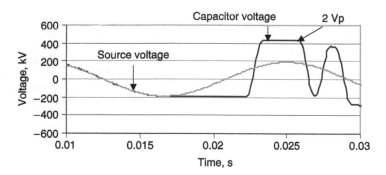

**Figure 17.6**  Capacitor voltage during de-energization.

assumed to be separating, thereby increasing the arc length. The voltage appearing across the opening contacts is the arc voltage.

### 17.4.1.1  Arc Interruption

As the contacts are separated, the effective resistance between the contacts increases with time. The corresponding time domain current decreases to a value insufficient to maintain the arc. With an alternating current, the total resistance of the circuit tends to bring the current in phase with the voltage at the current zero and the arc extinguishes.

### 17.4.2  Transient Recovery Voltage (TRV)

The TRV is defined as the normal frequency voltage appearing between the circuit breaker poles after the final arc extinction. Sometimes, after the extinction of the arc, the voltage across the circuit breaker blades can re-ignite the arc. Such a transient voltage is called restriking voltage. The arc voltage across the contacts at the instant of arc extinction is very low, whereas the power frequency voltage in the circuit is at the peak value. Considering Figure 17.7, the voltage across the opened contacts of the circuit breaker is given by:

$$V_c = \frac{(V_{max}/p)}{pL + 1/pC} = \frac{V_{max}}{L}\left[\frac{1}{p(p^2 + \omega^2)}\right] \qquad (17.4)$$

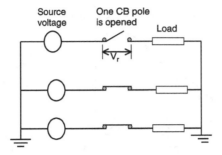

**Figure 17.7** Diagram showing the TRV across the circuit breaker blades ($V_r$).

where $\omega = (1/\sqrt{LC})$ and $V_{max}$ is the peak value of the recovery voltage. Transforming Equation (17.4) into time domain function:

$$V_c = V_{max}(1 - \cos \omega t) \qquad (17.5)$$

The maximum value of the recovery voltage is $2V_{max}$ and occurs at $t = \pi/\omega$. The oscillatory transient voltage has a frequency of $(1/2\pi\sqrt{LC})$ Hz. One of the important factors affecting the restriking voltage of the circuit breaker performance is the rate of rise of recovery voltage (RRRV).

### 17.4.3 Rate of Rise of Recovery Voltage (RRRV)

With a single frequency transient:

$$\text{RRRV} = \frac{\text{Peak restriking voltage}}{\text{Time between voltage zero and peak voltage}} \qquad (17.6)$$

The RRRV concept is shown in Figure 17.8. According to the terminology in Figure 17.8, $\text{RRRV} = V_{max}/T_p$. The expression for RRRV can be derived from the restriking voltage, given

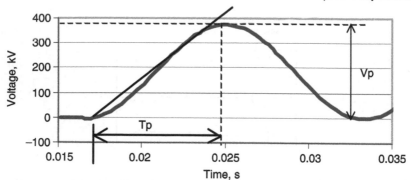

**Figure 17.8** Diagram to illustrate the definition of RRRV.

by Equation (17.5):

$$\text{RRRV} = \frac{dV_c}{dt} = V_{\max}\,\omega\,\sin\,\omega t \qquad (17.7)$$

The maximum value of RRRV occurs when $\omega t = \pi/2$, $t = \pi/2\omega$, or $t = \sqrt{LC}(\pi/2)$. The maximum RRRV value is $V_{\max}\,\omega$.

## 17.5 CIRCUIT BREAKERS FOR CAPACITOR SWITCHING

The circuit breakers for capacitor switching require additional withstand capabilities compared to a normal circuit breaker. The following items need careful consideration.

### 17.5.1 Voltage Rating

A service factor has to be applied when calculating the capacitance current. This factor can be 1.1 because the capacitors can be operators continuously up to 10% above the nameplate rating.

### 17.5.2 Capacitor Tolerance

A multiplier in the range of 1.05–1.15 should be used to adjust the nominal current to account for the manufacturing tolerance.

### 17.5.3 Harmonic Current

Capacitor banks provide a low impedance path for the flow of harmonic currents. When capacitor banks are ungrounded, no path is provided for zero sequence (third, sixth, etc.) currents. A multiplier of 1.1 is used for grounded neutral banks and a multiplier of 1.05 is used for ungrounded banks. In the absence of specific information on the capacitor bank, a total multiplier of 1.25 is used for ungrounded systems and 1.35 is used for grounded capacitor banks.

### 17.5.4 Isolated Capacitor Banks

A capacitor bank is considered to be isolated when the inrush current on energization is limited by the inductance of the source and capacitance of the bank. A capacitor bank is also considered to be isolated if the maximum rate of change, with respect to time, of the transient inrush current on energizing an uncharged capacitor bank does not exceed the maximum rate of change of symmetrical interrupting current of the circuit breaker. The limiting value is:

$$\left(\frac{di}{dt}\right)_{max} = \omega\sqrt{2}\left[\frac{\text{Rated maximum voltage}}{\text{Operating voltage}}\right]I \text{ A/s} \qquad (17.8)$$

where $I$ is the rated short-circuit current in amperes.

### 17.5.5 Back-to-Back Capacitor Banks

The inrush current of a single capacitor bank will be increased when other capacitor banks are connected to the same bus. Provisions should be made to control the inrush current due to back-to-back capacitor switching currents.

### 17.5.6 Inrush Currents

The energization of a capacitor bank by closing a circuit breaker produces significant transient inrush currents. The magnitude and frequency of the inrush current is a function of the applied voltage, capacitance of the circuit, inductance of the circuit, charge on the capacitance, instant of closing, and damping, if any, in the circuit. The circuit breaker should have sufficient momentary current rating to meet the transient inrush current. The same logic is applicable to the back-to-back energization. Table 17.2 summarizes the inrush current and frequency for switching capacitor banks. The following notations are used in Table 17.2:

$f_s$ = Supply frequency
$L_{eq}$ = Equivalent inductance, mH
$I_1$ = Steady-state current of bank 1 with a multiplier of 1.15
$I_2$ = Steady-state current of bank with a multiplier of 1.15
$V_{ll}$ = Line-to-line voltage, kV
$I_{sc}$ = symmetrical rms short-circuit current, A

**TABLE 17.2**  Inrush Current and Frequency

| Condition | $I_{peak}$ and Frequency |
|---|---|
| Isolated bank | $I_{max,peak} = 1.41 \sqrt{I_{sc} I_1}$ <br><br> $f = f_s \sqrt{\dfrac{I_{sc}}{I_1}}$ |
| Energize a capacitor bank with another of the same size | $I_{max,peak} = 1.747 \sqrt{\dfrac{(V_{ll})(I_1 I_2)}{L_{eq}(I_1 I_2)}}$ <br><br> $f(\text{kHz}) = 9.5 \sqrt{\dfrac{f_s(V_{ll})(I_1 I_2)}{L_{eq}(I_1 I_2)}}$ |
| Energize a capacitor bank with an equal bank energized on the same bus | $I_{max,peak} = 1.747 \sqrt{\dfrac{(V_{ll})(I_1)}{L_{eq}}}$ <br><br> $f(\text{kHz}) = 13.5 \sqrt{\dfrac{f_s(V_{ll})}{L_{eq}(I_1)}}$ |

**Example 17.3**

In a 230 kV, three-phase, 60 Hz system, a short circuit occurs very close to a circuit breaker. The maximum restriking voltage is 2 P.U. The natural frequency of the recovery voltage is 15,000 Hz. Estimate the RRRV.

Solution

$$V \text{ (line-to-line)} = 230 \, \text{kV}, \quad V \text{ (phase)} = 132.8 \, \text{kV}$$
$$\text{Maximum restriking voltage} = 2 \times 132.8 \times 1.414 = 376 \, \text{kV}$$

$$t = \frac{1}{2f_n} = \frac{1}{2 \times 15{,}000} = 3.333 \times 10^{-5} \, \text{s}$$

$$\text{RRRV} = \frac{\text{Peak restriking voltage}}{t}$$

$$= \frac{376 \, \text{kV}}{3.333 \times 10^{-5}} = 11.28 \, \text{kV}/\mu\text{s}$$

### 17.5.7 Effect of Transients on the Circuit Breaker

The capacitor switching operations include energization, de-energization, fault clearing, reclosing, restriking, pre-strike, back-to-back switching, and similar events as discussed in Chapter 22 for a 230 kV system. The circuit breaker is expected to withstand each of these switching transients. The overvoltage magnitudes and the TRVs in each of the selected cases are presented in Table 17.3. For the practical cases listed in Table 17.3, the corresponding maximum overvoltages, TRV, and frequency of oscillation are identified. In some cases, the maximum overvoltage is excessive, TRV is high, or the frequency of oscillation is excessive. The acceptable TRV is 2.4 P.U. and the frequency of oscillation is 4,250 Hz. Therefore, it can be seen that the capacitor switching requires additional measures to control the overvoltages, TRV, and frequency of oscillation.

### 17.5.8 Additional Considerations in Transient Control during Capacitor Switching

In capacitor switching applications, the selection of a specific circuit breaker may not meet all the system requirements.

**TABLE 17.3**   Overvoltages, Oscillation Frequencies, and TRV

| Description | $V_{max}$, P.U. | $f$, HZ | TRV, P.U. | Remarks |
|---|---|---|---|---|
| Energization | 2.1 | 588 | – | Acceptable |
| De-energization | 1.0 | DC | 2.0 | Acceptable |
| Fault clearing | 2.3 | 5,000 | 2.3 | Frequency of oscillation is unacceptable |
| Reclosing | 2.5 | 400 | – | $V_{max}$ is unacceptable |
| Restriking | 3.2 | 400 | 6.4 | $V_{max}$ and TRV are high |
| Prestrike | 3.28 | – | – | $V_{max}$ is unacceptable |
| Back-to-back | 1.48 | 400 | – | Acceptable |
| Voltage magnification | 2.1, 1.6 | | – | $V_{max}$ is high |

In order to limit the short-circuit currents and transient oscillatory voltages, other control approaches are used, including series reactors, high impedance transformers, and high resistance grounding. The series reactor can be used in the generator circuits, bus bars, feeders, and the shunt capacitance circuits. There are advantages and limitations to these approaches. With the application of shunt capacitor banks for power factor correction, there is always the inrush current issue during energization. The outrush current from the capacitor banks is a concern when a line circuit breaker closes into a nearby fault. In order to limit both the inrush and outrush currents, series reactors are used. Three schemes of series reactors for shunt capacitor application are discussed.

*Scheme 1: Series reactor with each capacitor bank.* Such a scheme is shown in Figure 17.9. In order to satisfy the criteria $(I_{ph}f)$ to less than 2.0E + 7, there will be two reactors with two capacitor banks.

*Scheme 2: Series capacitors in the bus and capacitor circuits.* The required scheme is shown in Figure 17.9. The reactor size for each capacitor bank will be small to limit the inrush current. A third reactor will be used to limit the outrush current.

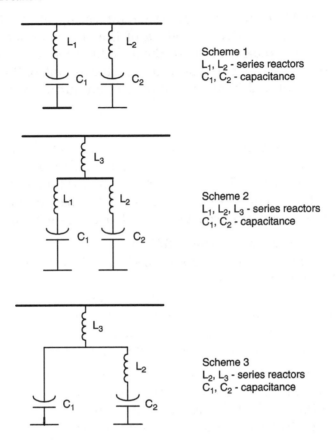

**Figure 17.9** Series reactor schemes for current limiting.

*Scheme 3: Reactor to limit the outrush current and breaker to limit the inrush current.* The inrush current can be controlled by using a circuit breaker with controlled switching or by using a closing resistor/inductor. The outrush current can be controlled by using a series reactor. Such a scheme is shown in Figure 17.9.

**Example 17.4**

Select a circuit breaker for switching a three-phase, 345 kV, 60 Hz shunt capacitor bank. The capacitor bank consists of two 50 MVAR banks. The short-circuit rating of the 345 kV system is 40 kA. Verify the rating of the circuit breaker with the industry standards. In case the switching requirements are not satisfied, select a suitable series reactor to control the short-circuit current and frequency of oscillation during switching.

Solution

The circuit breaker is intended to switch 50 MVAR shunt capacitor banks and should meet the performance criteria described in ANSI C37.06. The desired performance specifications of the circuit breaker to meet the capacitor switching application (definite purpose) of the 345 kV systems are:

> Nominal voltage rating $= 345$ kV
> Maximum voltage rating $(345$ kV $\times 1.05) = 362$ kV
> Rated current $= 2,000$ A
> Three-phase, short-circuit rating $= 40$ kA
> Lightning impulse withstand voltage $= 1,175$ kV
> Power frequency withstand voltage $= 450$ kV
> Switching impulse withstand voltage $= 950$ kV

When a system fault occurs near the capacitor bank location, the electrical energy stored in the capacitor bank discharges through the low fault impedance with considerable magnitude and at high frequency. Such a system is shown in Figure 17.10. The size of the 345 kV capacitor bank is 50 MVAR, three-phase. The expected outrush current magnitude and frequency, for a single 50 MVAR, 345 kV capacitor bank is given by:

$$C = \frac{\text{MVAR}}{2\,\pi \times 60 \times \text{kV}^2} = \frac{50}{2\,\pi \times 60 \times 345^2} = 1.11 \text{ MFD}$$

$$I_{pk} = \frac{V_{pk}}{\sqrt{L_f/C}} = \frac{345 \text{ kV} \times \left(\sqrt{2}/\sqrt{3}\right)}{\sqrt{36\,\mu\text{H}/1.11\,\mu\text{F}}} = 49.5 \text{ kA}$$

$$f = \frac{1}{2\pi\sqrt{L_f \times C}} = \frac{1}{2\pi\sqrt{36\,\mu\text{H} \times 1.11\,\mu\text{F}}} = 25.2 \text{ kHz}$$

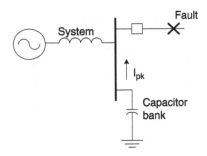

**Figure 17.10** Fault outside the circuit breaker without series reactor.

An inductance of $10\,\mu H$ for the bank and $0.26\,\mu H/ft$ with a 100 ft cable length is used.

The calculated outrush current magnitude of 49.5 kA is much higher than the allowable IEEE Standard C37.06 value of 20 kA. The frequency of the outrush current is also higher than the allowed value of 4,250 Hz. The outrush current from the capacitor bank needs to be controlled using current limiting series reactors. The minimum series reactor needed to limit the $(I_{pk}f)$ product to less than $2.0 \times 10^7$ is given by:

$$L_{min} = \frac{V_{pk}}{2\pi \times 2 \times 10^7} \qquad (17.9)$$

*With current limiting reactor*

The equivalent circuit with a series reactor in the shunt capacitor circuit is shown in Figure 17.11. For the proposed 345 kV, 50 MVAR bank the minimum reactor needed is:

$$L_{min} = \frac{345\,kV \times \left(\sqrt{2}/\sqrt{3}\right)}{2\pi(2 \times 10^7)} = 2.24\,mH$$

For the high outrush current to occur, a breaker must close into a fault very close to the 345 kV substation. A series inductor of 3 mH is selected for the 345 kV circuit and the corresponding

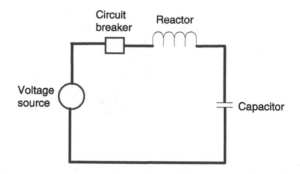

**Figure 17.11**  Capacitor circuit with series reactor.

$I_{pk}$ and the frequency of oscillation are given by

$$I_{pk} = \frac{V_{pk}}{\sqrt{L_f/C}} = \frac{345\,\text{kV} \times \left(\sqrt{2}/\sqrt{3}\right)}{\sqrt{3{,}000\,\mu\text{H}/1.1\,\mu\text{F}}} = 5.42\,\text{kA}$$

$$f = \frac{1}{2\pi\sqrt{L_f \times C}} = \frac{1}{2\pi\sqrt{3{,}000\,\mu\text{H} \times 1.1\,\mu\text{F}}} = 2.76\,\text{kHz}$$

This peak current and the frequency of oscillation are below the ANSI C37.06 values. Therefore, the circuit breaker is acceptable for the energization of the 345 kV, 50 MVAR shunt capacitor bank. From this example, it is clear that series reactors can be useful in shunt capacitor circuits to limit the fault current magnitudes to protect circuit breakers.

## 17.6  CONCLUSIONS

The types of circuit breakers used in the switching applications are classified based on voltage levels, switching duty, installation, and interrupting medium. The existing capacitor installations use circuit breakers with air, oil, air blast, vacuum, and SF$_6$ medium. For the new projects, circuit breakers using vacuum and SF$_6$ media are selected. The circuit breakers used in capacitor switching applications have to satisfy the following:

- Higher TRV magnitudes
- Restriking due to the stored energy in the capacitors
- Prestrike in the case of vacuum circuit breakers
- Overvoltages in the remote substation or transformer
- High inrush currents
- Overvoltages due to inrush or outrush currents
- Higher outrush currents due to fault or back-to-back switching
- Voltage magnifications in the low voltage capacitors due to the switching of high voltage capacitor banks
- High-frequency transients in the circuit breaker transients

Therefore, in many cases, just selecting a circuit breaker may not be enough to meet all the required technical requirements. In all the capacitor switching applications, suitable measures to limit the transients are to be implemented. Some of the approaches to controlling system transients due to the capacitor switching include:

- Series reactor (see Example 17.4)
- Closing resistor to control the inrush related transients (see Section 21.3)
- Controlled closing
- Staggered closing of the circuit breaker contacts
- Surge arresters to control the overvoltages
- Suitable capacitor bank connection based on experience

Usually, series inductors are very effective in controlling the high-frequency transients. An example calculation is shown in Example 17.4. The usefulness of closing resistors in the control of overvoltages is discussed in Chapter 21. The circuit breaker rating and typical performance features are also presented in this chapter.

## PROBLEMS

17.1. Circuit breakers are needed for a substation project with operating voltages of 500 kV and 138 kV.

Based on the voltage ratings, select a suitable circuit breaker rating from the ANSI and IEC preferred ratings.

17.2. Define TRV. Derive an expression for TRV.

17.3. Define RRRV. Derive an expression for RRRV.

17.4. What is restriking?

17.5. A three-phase, 100 MVA, 13.8 kV generator is connected to a bus through a circuit breaker and tie line. The reactance of the cable is 3 Ω/phase. The capacitance of the cable is 0.015 µF/phase. Calculate the frequency of restriking, TRV, and RRRV when the circuit breaker is used to de-energize the circuit.

17.6. Select a circuit breaker for switching a three-phase, 110 kV, 60 Hz shunt capacitor bank. The capacitor bank consists of two 30 MVAR banks. The short-circuit rating of the 110 kV system is 30 kA. Verify the rating of the circuit breaker with the industry standards. If the switching requirements are not satisfied, select a suitable series reactor to control the short-circuit current and frequency of oscillation during switching.

17.7. Classify the circuit breakers based on the medium of operation. Which of these circuit breakers is suitable for capacitor switching?

## REFERENCES

1. ANSI Standard C37.10 (1979), American National Standard Requirements for Transformers 230,000 Volts and Below.

2. ANSI Standard 242 (1986), IEEE Recommended Practice for Protection and Coordination of Industrial and Commercial Power Systems.

3. ANSI/IEEE Standard C37.010 (1989), IEEE Application Guide for AC High Voltage Circuit Breakers Rated on a Symmetrical Current Basis.

4. ANSI Standard C37.06 (2000), AC High Voltage Circuit Breakers Rated on a Symmetrical Current Basis.

5. ANSI Standard C37.012 (1998), Application Guide for Capacitance Current Switching for AC High Voltage Circuit Breaker Rated on a Symmetrical Current Basis.

6. IEC Standard 909 (1988), International Electrotechnical Commission, Geneva.

7. Rodolakis, A. (1993). A Comparison of North American (ANSI) and European (IEC) Fault Calculation Guidelines, *IEEE Transactions on Industry Applications*, 29(3), 915–921.

# 18

## SURGE PROTECTION

### 18.1  INTRODUCTION

Shunt capacitors are installed on pole-mounted structures and in substation locations for power factor improvement. Like any other transmission or distribution equipment such as transformers, circuit breakers, or transmission lines, shunt capacitors are exposed to both switching and lightning surges. The overvoltages from the above sources can be limited to acceptable levels by using surge arresters. The source of overvoltages, types of surge arresters available, and the selection and application of surge arresters for capacitors are discussed in this chapter [1–4].

### 18.2  LIGHTNING SURGES

Rain clouds are formed from the collection of water vapor from the Earth's surface. A charge of a specific polarity occurs within the cloud. An equivalent and opposite charge is produced on the earth under the cloud as shown in Figure 18.1. When the potential difference between the cloud and earth reaches a critical value, a flashover occurs between the cloud and the ground point. This type of flashover can be initiated

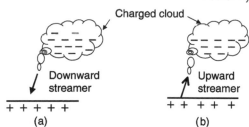

**Figure 18.1** The concept of lightning: (a) downward streamer; (b) upward streamer.

from the cloud to the ground as a downward streamer as shown in Figure 18.1(a). Sometimes such a stepped leader can be initiated from the ground to the cloud. Such a surge is an upward streamer as shown in Figure 18.1(b). During the flashover the air column breaks and a loud noise, called thunder, is produced. The flashover produces a visible light called lightning. A large current flow occurs during the flashover and passes through the object on the ground and produces overvoltage magnitudes on electrical equipment. Significant overvoltages can cause failure in electrical insulation, leading to equipment outage. A typical photograph of a direct lightning stroke from cloud to ground is shown in Figure 18.2. The downward streamer, the branching streamer, and the visible light are observable in this photograph. Several other strokes also visible in this photograph. The arc is larger at the cloud and smaller near the ground and is therefore a downward stroke equivalent to Figure 18.1(a) from cloud to ground.

## 18.3   TYPES OF LIGHTNING SURGES

The lightning surges interacting with the power system are identified as direct lightning strokes, back flashover strokes, and multiple strokes.

### 18.3.1   Direct Lightning Strokes

A photograph of direct lightning stroke from cloud to ground is shown in Figure 18.2. Consider a charged cloud with lightning

**Figure 18.2** Photograph of two direct strokes from cloud to ground. Bright light = downward leader. Small lines = branching streamers. (Courtesy of National Oceanic and Atmospheric Administration, U.S. Department of Commerce.)

stroke to ground as shown in Figure 18.3. In this photograph, the downward stroke is traveling from the cloud to the ground. At the same instant, the upward stroke is traveling from the ground to cloud as identified by a larger current magnitude at the ground level. Both the downward and upward strokes meet at a middle location as shown in Figure 18.3. Lightning strokes with magnitudes of a few tens of kA can bypass the overhead shield wire and can strike directly on the phase conductor. From the geometry of the tower, the maximum lightning current that can strike the phase conductors of an overhead transmission line can be estimated. A typical lightning surge can be represented by a wave as shown in Figure 18.4. The wave is characterized by [1]:

- Crest or peak value; usually 25% of the strokes are below 10 kA, 60% of the strokes are below 50 kA, 11% of the strokes are 50–100 kA, and so on.

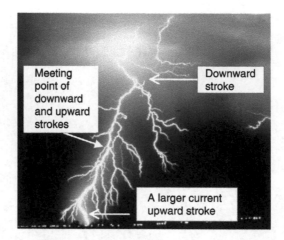

**Figure 18.3** A lightning stroke from ground to cloud, edited version. (Courtesy of National Oceanic and Atmospheric Administration, U.S. Department of Commerce.)

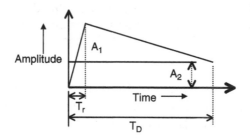

**Figure 18.4** A typical model of a lightning surge.

- The rise time varies from 1–10 μs.
- The duration of the wave TD varies from 10–100 μs.

If a direct stroke hits an overhead line, the expected overvoltage is given by:

$$V = (I_S)\,(Z/2) \qquad\qquad (18.1)$$

where $I_S$ is the magnitude of the lightning current and $Z$ is the surge impedance of the line. With a 50 kA surge current,

$Z = 500\,\Omega$, and the voltage rise is $25,000\,\text{kV}$. The corona effects associated with lightning current will reduce the overvoltages significantly. Direct strokes are infrequent.

### 18.3.2 Back Flashover

Lightning strokes are of very high potential with the capacity to discharge hundreds of kA with low rise time. The surges can strike overhead neutral wires, towers, or phase conductors, and they may produce overvoltages sufficient to cause sparkover across the insulators. Since most of the stroke current flows into the ground during the back flashover, the tower footing resistance has a major impact on the overvoltages generated. The back flashover causes a line-to-ground fault that will be cleared by a circuit breaker. A line outage will result until the circuit breaker is reclosed. Typical ranges of lightning surge characteristics causing the back flashover are [1]:

Peak current = 5–10 kA
Rate of rise = 5–30 kA/μs
Rise time = 0.5–30 μs
Tail time = 20–200 μs

The surge current from the tower or neutral conductor to the phase conductor is characterized by a sharp rise time and much smaller magnitude of the order of 10–20 kA. As a conservative approach, both the direct strokes and flashover caused by a stroke to the tower (back flashover) are modeled with a standard 1.2/50 μs current wave, with a peak of appropriate magnitude.

### 18.3.3 Multiple Strokes

A photograph of multiple lightning strokes is shown in Figure 18.5. The cloud has a very large charge center and the lightning strike to ground occurs at many locations. Many lightning strikes are actually multiple strokes and have serious consequences for the electrical system. Multiple strokes are caused by quickly charging the cloud area after the first stroke occurs. Sometimes there may be a time delay between strokes. Equipment such as a surge arrester is designed to

**Figure 18.5**  Photograph of multiple lightning strokes from a large cloud to ground. (Courtesy of National Oceanic and Atmospheric Administration, U.S. Department of Commerce.)

withstand one lightning stroke successfully. Subsequent lightning strokes will damage the surge arrester, and hence the importance of multiple strokes cannot be underestimated.

## 18.4  SURGE ARRESTERS

Before the 1970s, most of the surge arresters used in the utility industry were the gapped type silicon carbide. These devices are equipped with silicon carbide valve elements and a series of gaps that act as insulators. These gaps provide protection from continuous power frequency voltage. These series gaps will break down during high transient overvoltage conditions. The voltage and current zeros occur simultaneously, permitting the gap to clear the circuit established through the surge arrester [2,3].

In a metal oxide varistor (MOV) type of surge arrester, the disks are mounted one over the other and there are no gaps. The metal oxide disks insulate the arrester electrically from the ground. The disk is made up of the zinc oxide type of semiconductor material. The surge arrester starts conduction at a

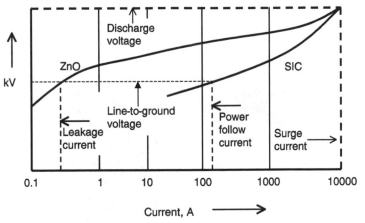

**Figure 18.6** Volt-ampere curves of silicon carbide and MOV surge arrester.

specific voltage and stops conduction when the voltage falls below a threshold voltage. The main differences between the silicon carbide and MOV type of surge arresters are:

1. There are no gaps in MOV surge arresters.
2. Disks are used in the zinc oxide arresters and draw a very small current through the arrester at normal operating conditions. The volt-ampere characteristics of the silicon carbide and the MOV surge arresters are shown in Figure 18.6. The cross-sectional view of typical MOV surge arresters is shown in Figure 18.7.

### 18.4.1 Classes of Surge Arresters

There are three classes of surge arresters available: distribution, intermediate, and station. Distribution arresters are used in distribution systems and intermediate and station class arresters are used in substation applications. Intermediate class arresters may be used on smaller substations, subtransmission lines, and cable terminal poles. Station class arresters are used on large substations and equipment to be protected under partly shielded environment. The energy levels of these arresters differ from one another. The typical data for

**Figure 18.7** A typical cut view of a MOV surge arrester. (Courtesy of Ohio Brass (Hubbell Power Systems), Aiken, SC.)

the distribution, intermediate, and station class surge arresters are presented in Appendix D. The proper application of surge arresters requires knowledge of the maximum operating voltage, magnitude, and duration of the temporary overvoltages (TOV) during abnormal operating conditions. Such data must be compared with the MCOV rating of the surge arrester. The gas-insulated substations and the gas-insulated transformers are provided with the gas-insulated type of surge arresters. For the traction power system, the surge arresters are constructed with a robust casing to withstand the vibrations caused by continuous movement. Typical surge arresters are shown in Figure 18.8 [6]. The preferred voltage ratings of surge arresters are listed in Table 18.1.

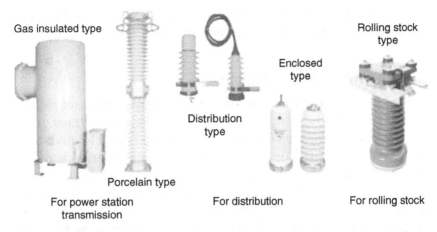

Gas insulated type

Rolling stock type

Enclosed type

Distribution type

Porcelain type

For power station transmission

For distribution

For rolling stock

**Figure 18.8** Types of surge arresters; the station class, distribution. (Reproduced from Toshiba Power Systems & Services, Sydney, Australia.)

TABLE **18.1** Voltage Ratings of Surge Arresters

| Distribution | 3 | 6 | 9 | 10 | 12 | 15 | 18 | 21 | 24 | 27 | 30 | 36 |
|---|---|---|---|---|---|---|---|---|---|---|---|---|
| Intermediate | 3 | 6 | 9 | 10 | 12 | 15 | 18 | 21 | 24 | 27 | 30 | 36 |
| Station Class | 3 | 6 | 9 | 10 | 12 | 15 | 18 | 21 | 24 | 27 | 30 | 36 |
| Intermediate | 39 | 45 | 48 | 54 | 60 | 72 | 90 | 96 | 108 | 120 | | |
| Station Class | 39 | 45 | 48 | 54 | 60 | 72 | 90 | 96 | 108 | 120 | | |
| Station Class | 132 | 144 | 168 | 172 | 180 | 192 | 228 | | | | | |

The voltage ratings mentioned above are typical; surge arresters with other voltage ratings are available. In the station class, additional ratings available beyond 228 kV are 240, 258, 264, 276, 288, 294, 300, 312, 336, 360, 396, 420, 444, and 588 kV.

## 18.4.2 The Gas Insulated Surge Arresters

The protection of GIS equipment against overvoltages is performed by the use of metal-enclosed surge arresters in which the ZnO blocks are contained in an earthed vessel filled with $SF_6$ gas under pressure. The GIS switchgear is known to produce fast front transients and the surge arrester is specially tested for current wave shapes of 8/20 and 30/60 µs having a front time shorter than 1 μs. Such equipment is larger than conventional surge arresters and it is expensive.

### 18.4.3 Polymer Surge Arresters

The high voltage surge arresters are equipped with porcelain housing. Recently, surge arresters housed in polymer have become available, and several advantages are claimed over the porcelain-housed units. The polymer type of surge arrester has a better mechanical behavior and a higher safety level in the case of a failure. The polymer arresters are much lighter in weight and are claimed to be superior in the polluted environment. Typical polymer types of surge arresters are shown in Figure 18.9.

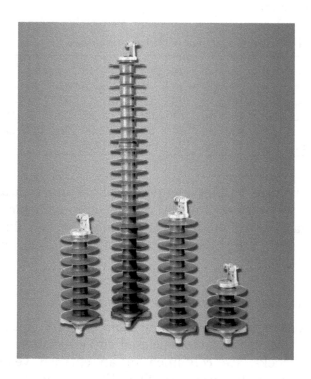

**Figure 18.9**  Typical polymer types of surge arresters. (Courtesy of Ohio Brass (Hubbell Power Systems), Aiken, SC.)

## 18.5 SURGE ARRESTER CHARACTERISTICS

The metal oxide varistor type of surge arrester has nonlinear volt-ampere characteristics. It performs as a blocking device at lower voltages and as a good conducting device at higher voltages. Typical volt-ampere characteristics of a MOV arrester are shown in Figure 18.10. The important operating features of such a surge arrester can be identified as:

- Very small leakage current at the operating voltage.
- The MCOV level between the operating voltage and rated voltage.
- The TOV voltage and a current of around 10 A.
- The switching surge capability with a current of up to 1 kA.
- The lightning surge conduction capability of about 10 kA.

The allowable time durations for various operating points are different and are defined accordingly.

### 18.5.1 Rated Voltage

The rated voltage of a surge arrester is presented in line-to-line voltage. This voltage rating must be higher than the system voltage in order to withstand the voltage rise during the system fault conditions. The voltage rating of the arrester is equal to the line-to-line voltage × the coefficient of earthing to account for the system voltage change.

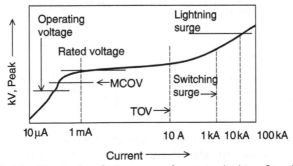

**Figure 18.10** Typical volt-ampere characteristics of an MOV type of surge arrester.

### 18.5.2 Temporary Overvoltage (TOV)

The common source of temporary overvoltage is the rise in unfaulted phases during the line-to-ground fault. Such voltage shift depends on the type of grounding. On a delta connected system, a single line-to-ground (SLG) fault produces line-to-line voltage on the unfaulted phases. Such a condition is shown in Figure 18.11. If the surge arrester is connected phase-to-ground, such arresters will experience 1.732 times the phase voltage. In a solidly grounded wye connected system, during an SLG fault, there will be phase voltages on faulted phases as shown in Figure 18.12. There will be certain increases in the voltages on the unfaulted phases due to line charging. Such voltages can be calculated, and the results are available in the form of a graph as shown in Figure 18.13 [4]. If the $X_0/X_1$ and $R_0/X_1$ ratios are known, the multiplying factor can be identified.

### 18.5.3 Maximum Continuous Operating Voltage (MCOV)

The MCOV rating of a surge arrester is the maximum rms value of the power frequency voltage that may be applied continuously between the terminals of the surge arrester. The MCOV of the surge arrester is approximately 84% of the arrester rating. For example, if the arrester rating is 10 kV,

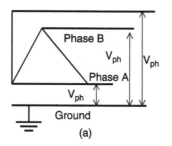 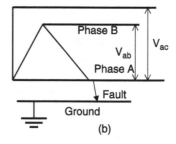

**Figure 18.11** The concept of voltage rise in unfaulted phases during SLG fault in the delta system. (a) Normal operation. (b) SLG fault, phase voltages in unfaulted phases are equal to line-to-line voltages.

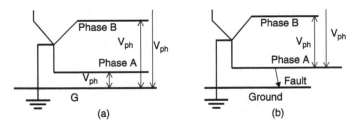

**Figure 18.12** The concept of voltage rise in unfaulted phases during SLG fault in a grounded wye system. (a) Normal operation. (b) SLG fault, phase voltages in unfaulted phases are equal to $V_{\text{ph}}$.

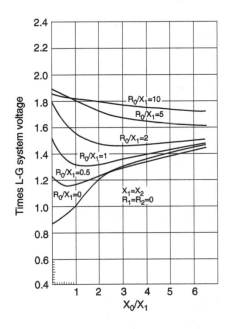

**Figure 18.13** The multiplying factor based on $X_0/X_1$ and $R_0/X_1$.

the MCOV of the device is 8.4 kV. The minimum required MCOV for a solidly grounded system is the maximum phase-to-ground voltage divided by $\sqrt{3}$. In the case of phase-to-phase arresters, the required MCOV is the maximum phase-to-phase operating voltage. Commonly used surge arresters in the utility system are listed in Table 18.2.

**TABLE 18.2**   Commonly Used Surge Arresters [4]

| System Voltage, Line-to-Line, kV rms | | Surge Arrester Rating kV, rms | |
| --- | --- | --- | --- |
| Nominal | Maximum | Solidly Grounded | Ungrounded |
| 2.40 | 2.52 | | 3 |
| 4.16 | 4.37 | 3–4.5 | 6 |
| 4.80 | 5.04 | | 6 |
| 6.90 | 7.25 | 4.5–5.1 | 9 |
| 12.47 | 13.09 | 9–10 | |
| 13.20 | 13.86 | 10 | 15–18 |
| 13.80 | 14.49 | 10–12 | 15–18 |
| 20.70 | 21.74 | 15 | |
| 23.00 | 24.15 | | 24–27 |
| 24.90 | 26.15 | 18–21 | |
| 27.60 | 28.98 | 21–24 | 27–30 |
| 34.50 | 36.23 | 27–30 | 36–39 |
| 46.00 | 48.30 | | 48 |
| 69.00 | 72.45 | 54–60 | 66–72 |
| 115.00 | 120.75 | 90–96 | 108–120 |
| 138.00 | 144.90 | 108–120 | 132–144 |
| 161.00 | 169.05 | 120–144 | 144–168 |
| 230.00 | 241.50 | 172–192 | 228–240 |
| 345.00 | 362.25 | 258–312 | – |
| 500.00 | 525.00 | 396–444 | – |
| 765.00 | 803.25 | 588 | – |

For EHV systems 230 to 765 kV, the maximum voltages are 1.1 P.U.

## Example 18.1

A surge arrester is needed for a 13.8 kV, 60 Hz, ungrounded system. Select a surge arrester and find out the MCOV of this device. If the system can be grounded, then what is the rating of the surge arrester? Use a safety of factor of 1.2 due to temporary overvoltages during single line-to-ground fault.

Solution

In an ungrounded system, the expected phase voltage during a single line-to-ground fault is the rated line voltage. The metal oxide surge arrester has to withstand the rated line-to-line voltage of 13.8 kV. A station class surge arrester is selected.

Temporary overvoltage $(13.8\,\text{kV} \times 1.2) = 16.6\,\text{kV}$
Select a surge arrester from Table D-3, Appendix D with
    MCOV above 16.6 kV.
Surge arrester voltage rating $= 21\,\text{kV}$
MCOV of the surge arrester $= 17\,\text{kV}$
MCOV of the arrester $(21\,\text{kV} \times 0.84) = 17.64\,\text{kV}$ (verified)

For a grounded system, the expected maximum phase voltage during a single line-to-ground fault is the rated phase voltage. The metal oxide surge arrester has to withstand the rated phase voltage $(13.8\,\text{kV}/1.732) = 7.97\,\text{kV}$. A station class surge arrester is selected.

Temporary overvoltage $(7.97\,\text{kV} \times 1.2) = 9.56\,\text{kV}$
Select a surge arrester from Table D-3, Appendix D with
    MCOV above 9.56 kV.
Surge arrester voltage rating $= 12\,\text{kV}$
MCOV of the surge arrester $= 10.2\,\text{kV}$
MCOV of the arrester $(12\,\text{kV} \times 0.84) = 10.08\,\text{kV}$ (verified)

### 18.5.4 Nominal Discharge Current

A discharge current having a specified crest value shape is used because of its protective characteristics. Usually 1.5 kA, 3.0 kA, 5.0 kA, 10 kA, 15 kA, 20 kA, and 40 kA ratings are specified using a $8 \times 20$ µs current wave. The recommended use of these ratings is:

1. *10 kA current and above.* For the protection of major power stations and substations with frequent lightning and system voltages above 66 kV.
2. *5 kA current rating.* For protection of substations on systems with nominal system voltages not exceeding 66 kV.
3. *3 kA current rating.* For protection of small substations where diverters of higher current rating are

not economically justified. System voltage can be of the order of 22 kV.

4. *1.5 kA current rating.* For protection of rural distribution systems with system voltages not exceeding 22 kV.

### 18.5.5    Front of Wave Protective Level (FOW)

This is the discharge voltage for a faster ($0.6 \times 1.5$ μs), 10 kA impulse current (15 kA for ratings 396–444 kV, 20 kA for 588 kV), which results in a voltage wave cresting in 0.5 μs. The resultant crest wave is specified and is used to calculate the percentage of margin.

### 18.5.6    Switching Surge Protective Level

The switching surge discharge voltage of a surge arrester increases with increasing current. A switching surge coordination current of 3 kA ($45 \times 90$ μs) is used for ratings 54–588 kV and 500 A is used for ratings 2.7–48 kV.

### 18.5.7    Impulse Discharge Voltage Using
###               $8 \times 20$ μs Wave

The resultant voltage across the surge arrester due to the forced current is usually presented for various current levels.

### 18.5.8    Magnitude of Discharge Currents

The discharge current through a surge arrester depends on the flash overvoltage of the insulation, surge impedance of the incoming line, and the type of grounding. For effectively shielded installations, the discharge current will vary from 1 to 20 kA, depending on the system voltage. A conservative approximation of the maximum discharge current $I_A$ for a specific application can be calculated using the following equation:

$$I_A = \frac{2V_0 - V_A}{Z} \qquad (18.2)$$

TABLE **18.3** Energy Capability of Surge Arresters

| kV Range | Energy, kJ/kV |
|---|---|
| 2.7–4.8 | 4 |
| 54–360 | 7.2 |
| 396–588 | 13.1 |

where

$V_0 = 1.2 \times$ line insulation level, critical flash overvoltage, kV

$Z =$ Surge impedance, $\Omega$

$V_A =$ Arrester discharge voltage, kV

### 18.5.9 Energy Absorption Capability

Any discharge of a capacitor bank through a surge arrester results in significant currents. Thus energy absorption capability is a critical factor to consider when selecting a surge arrester for capacitor bank applications. If $V_0$ and $V_a$ are the voltages without and with surge arrester, then the energy due to the presence of a capacitor can be expressed as:

$$E = \frac{C}{2}\left(V_0^2 - V_a^2\right) \text{ J} \qquad (18.3)$$

Therefore, the available energy capability of a surge arrester is to be examined carefully for shunt capacitor application. The typical energy capability of the surge arresters available for utility applications is listed in Table 18.3.

### Example 18.2

A surge arrester is to be used in a 230 kV, three-phase, 60 Hz system for the surge protection of a shunt capacitor bank. What is the energy capability of the surge arrester? Assume a surge arrester with a 180 kV rating.

Solution

Voltage rating of the 230 kV surge arrester $= 180$ kV
Energy of the surge arrester $= 7.2$ kJ/kV

Rated energy capability $(7.2\,\text{kJ/kV} \times 180\,\text{kV}) = 1{,}296$ J
Energy at the operating voltage $(7.2\,\text{kJ/kV} \times 132.8\,\text{kV}) = 956.2$ J

## 18.6   NATURE OF TEMPORARY OVERVOLTAGES

The power system equipment is exposed to sustained over-voltages, temporary overvoltages (TOV), switching surges, and lightning surges. The nature of the lightning surge is discussed in Section 18.1. The important switching surges are discussed below for the switching operations.

### 18.6.1   Energization

The energization of a capacitor bank produces oscillatory voltage and current waveforms. An example voltage waveform due to energization of a capacitor bank is shown in Figure 22.2, Chapter 22. The maximum overvoltage magnitude is 2.1 P.U. and the frequency of oscillation is 588 Hz.

### 18.6.2   De-Energization

The de-energization of a capacitor bank leaves a DC voltage in the opened line and the capacitor circuit. An example voltage waveform due to de-energization of a capacitor bank is shown in Figure 22.4, Chapter 22. The corresponding TRV waveform is shown in Figure 22.5. The maximum TRV voltage in the example case is 2.0 P.U.

### 18.6.3   Fault Clearing

Most power system faults are single line-to-ground faults. Opening the circuit breaker on both ends of the affected circuit performs the fault clearing. Such fault clearing produces over-voltages in the unfaulted phases and oscillatory currents in the faulted phase. An example voltage waveform due to fault clearing is shown in Figure 22.7, Chapter 22. The corresponding TRV waveform is shown in Figure 22.8. The maximum TRV voltage in the example case is 2.3 P.U.

### 18.6.4 Backup Fault Clearing

During normal fault clearing, the circuit breaker blades sometimes get stuck and fail to open. Then the backup fault clearing is performed using the circuit breaker in the upstream of the failed breaker. Such a fault clearing procedure is similar to the normal fault clearing and produces larger overvoltage magnitudes because of the delayed clearing.

### 18.6.5 Reclosing

In order to improve the power system reliability, the circuit breakers sometimes are closed after a fault clearing with a small time delay. Such reclosing operation is performed with trapped charges on the line from the previous opening. If no discharge mechanisms are available on the lines or capacitors, such reclosing produces significant overvoltages. An example voltage waveform during the reclosing is shown in Figure 22.9, Chapter 22. The maximum overvoltage is 2.5 P.U. and the frequency of oscillation is 400 Hz.

### 18.6.6 Restriking

During the de-energization of a capacitor bank, if the transient recovery voltage across the circuit breaker exceeds the circuit breaker capability, then the arc across the interrupting contacts is re-established. Such a phenomenon, called restriking, produces significant overvoltages on the system. A typical voltage waveform during a restrike is shown in Figure 22.10, Chapter 22. The overvoltage magnitude is 2.4 P.U. and the TRV is 6.4 P.U.

### 18.6.7 Prestrike

During energization of the capacitor banks, an arc can establish even before the physical circuit breaker blades are closed. This phenomenon is called prestrike. During the prestrike, the normal high frequency current flow occurs. When the circuit breaker blades actually establish the physical contacts, the transients are much higher than the ordinary energization transients. The typical capacitor voltage and the capacitor current are shown in Figures 22.13 and 22.14, respectively, in Chapter 22. The peak transient voltage in the capacitor circuit has increased from 2.0 P.U. to 3.28 P.U., which will result in a surge arrester operation.

### 18.6.8   Current Chopping

During de-energization of an unloaded transformer, circuit breakers are allowed to interrupt the magnetizing currents. When such currents are chopped, voltages many times the value of the phase voltages are produced.

### 18.6.9   Sustained Overvoltages

Overvoltages may be produced due to circuit conditions and will prevail for long durations. Every attempt should be made to avoid such voltages in order to prevent damage to the equipment connected to the power system. Some of the sources of sustained overvoltages are load rejection in a shunt compensated line, ferroresonance, and ground fault conditions.

### 18.6.10   Ferroresonance

The overvoltages produced due to the interaction of the transformers and capacitive elements in the power system are classified in this category. For example, a ferroresonance can occur due to the opening of one phase of a three-phase system equipped with a three-phase potential transformer for instrumentation. The resulting overvoltage in the open-phase is significantly higher than the normal operating voltage. The one-phase condition may occur due to a blown fuse in a circuit, failure of circuit breaker poles to close properly, or a broken conductor. The voltage waveforms during a one-phase open are shown in Figure 16.14, Chapter 16. The overvoltage magnitudes observed in this condition are 1.6 P.U.

## 18.6.11    Ground Faults

When there is a single line-to-ground fault, there will be an increase in the voltages on the unfaulty phases in the effectively grounded power systems. The highest overvoltage on an unfaulted phase during faulted condition is of the order of 1.4 times the rated value.

## 18.7   PROTECTIVE COORDINATION

Once the system specifications, including the voltage rating, number of phases, and ratios $X_0/X_1$ and $R_0/X_1$ are known, then the surge arrester can be selected following these six steps [5]:

1. Determine the system phase-to-neutral voltage.
2. Select the multiplying factor from Figure 18.13, knowing $X_0/X_1$ and $R_0/X_1$.
3. The TOV rating = (Phase voltage × multiplying factor). Select a surge arrester.
4. Compare the MCOV/TOV ratio; this has to be greater than 1.0.
5. Calculate the margins.
6. Calculate the current through the surge arrester during conduction.

Once a surge arrester is selected, then the arrester characteristics such as the MCOV, front of wave (FOW) sparkover, lightning protective level (LPL), and the switching surge protective (SSP) level voltages are known. Then the system temporary overvoltage (TOV) is to be compared with the MCOV. The equipment insulation withstand voltages are to be identified. These include the chopped wave withstand CWW, BIL, and BSL. These voltages are discussed below.

### 18.7.1   Basic Impulse Insulation Level (BIL)

A basic insulation level is expressed in terms of the crest value of a standard switching impulse. The BIL values, which are related to the system voltage, are listed in Table 18.4.

**TABLE 18.4**  Insulation Withstand for Various Equipment [4]

| Test | Withstand Voltage | Equipment |
|------|-------------------|-----------|
| Front of wave 0.5 μs | (1.3–1.5) BIL | Transformers, reactors |
| Chopped wave 2 μs | 1.29 BIL | Circuit breakers 15.5 kV and above |
| Chopped wave 3 μs | (1.1–1.15) BIL<br>1.15 BIL | Transformers and reactors<br>Circuit breakers 15.5 kV and above |
| Switching surge 100 × 200 μs | 0.83 BIL | Transformers and reactors |
| Switching surge 250 × 2500 μs | (0.63–0.69) BIL | Circuit breakers (362 × 765) kV |

### 18.7.2  Chopped Wave Withstand (CWW)

The chopped wave withstand voltage is a crest value of $1.2 \times 50$ μs impulse wave chopped by the action of a rod gap placed in parallel with the insulation.

### 18.7.3  Basic Switching Impulse Insulation Level (BSL)

A specific insulation level is expressed in terms of the crest value of a standard switching impulse. The typical insulation withstands for various equipment are shown in Table 18.4.

The process of correlating the insulation withstand levels of electrical equipment with expected overvoltages and with the characteristics of the surge arrester is called insulation coordination. The degree of coordination is measured by the protection margin (PM). The fundamental definition is given as:

Protective margin (PM)

$$= \left( \frac{\text{Insulation withstand level}}{\text{Voltage at protected equipment}} - 1 \right) \times 100 \qquad (18.4)$$

In the three-point method of insulation withstand-to-arrester protection levels are verified for the front of wave, full wave, and switching surge protective levels. The following three protective margins are used:

$$PM(1) = \left[\frac{CWW}{FOW} - 1\right]100 \tag{18.5}$$

$$PM(2) = \left[\frac{BIL}{LPL} - 1\right]100 \tag{18.6}$$

$$PM(3) = \left[\frac{BSL}{SSP} - 1\right]100 \tag{18.7}$$

where CWW, BIL, and BSL are equipment insulation capabilities as defined above. The parameters FOW, LPL, and SSP are related to the surge arrester. The acceptable margins are:

$$PM(1) \geq 20\%, PM(2) \geq 20\% \text{ and } PM(3) \geq 15\% \tag{18.8}$$

The results can be presented in graphical format as shown in Figure 18.14. The current through the surge arrester ($I_A$) is calculated using Equation (18.2).

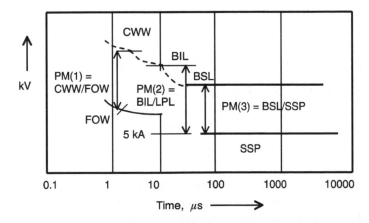

**Figure 18.14** The coordination curve for surge arrester selection.

## Example 18.3

A 60 MVAR, three-phase, 115 kV, 60 Hz capacitor bank needs surge protection. The BIL rating of the bank is 550 kV. The three-phase, short-circuit rating of the 115 kV system is 22,760 A. The single line-to-ground fault current is 21,340 A. The flash overvoltage of the insulator is 610 kV. The surge impedance of the line is 375 $\Omega$. The system is effectively grounded. Select a suitable surge arrester for this application.

Solution

Assume a 100 MVA base.

Base current = 100,000 kVA/(1.732 × 115 kV) = 502 A
Three-phase, short-circuit current = 22,760 A/502 A = 45.33 P.U.
SLG short-circuit current = 21,340/502 = 42.51 P.U.
$X_1$ on 100 MVA base = 1/45.33 = 0.0221 P.U.

$$I(\text{SLG}) = \frac{3E}{X_0 + 2X_1}; \quad 42.51 = \frac{3(1)}{X_0 + (2 \times 0.0221)}$$

Solving the above equation, $X_0 = 0.0264$ P.U.

$$\frac{X_0}{X_1} = \frac{0.0264}{0.0221} = 1.2$$

Assume $X/R = 30$ for the solidly grounded system.

$R_0 = 0.0264/30 = 0.0009$ P.U.

$$\frac{R_0}{X_1} = \frac{0.0009}{0.0221} = 0.04$$

The K constant from Figure 18.13 = 1.18
The system voltage (line-to-line) = 115 kV
Phase voltage (115 kV/1.732) = 66.4 kV
Maximum temporary overvoltage (66.4 × 1.18) = 78.35 kV

Select a surge arrester of 108 kV with MCOV of 84 kV. For the surge arrester of 108 kV:

FOW = 279 kV, LPL = 254 kV, SSP = 202 kV

For the 115 kV capacitor bank, BIL = 550 kV, CWW = 605 kV, BSL = 465 kV

$$\frac{\text{MCOV}}{\text{Maximum System Voltage}} = \frac{84 \text{ kV}}{78.35 \text{ kV}} = 1.07$$

This ratio is above unity and hence the arrester rating is acceptable for the steady-state voltage.

*Calculation of protective margins*:

$$\text{PM}(1) = \frac{605}{279} - 1 = 1.16 \text{ P.U.}$$

$$\text{PM}(2) = \frac{550}{254} - 1 = 1.16 \text{ P.U.}$$

$$\text{PM}(3) = \frac{465}{202} - 1 = 1.30 \text{ P.U.}$$

Protective margins PM(1) and PM(2) are above 0.2 P.U. Protective margin PM(3) is above 0.15 P.U. and is adequate.

$$V_A = 202 \text{ kV}, \quad Z_s = 375 \ \Omega, \quad \text{FLV} = 610 \ \Omega$$

The current through the surge arrester:

$$I_A = \frac{(2 \times 1.2 \times 610) - 202}{375}$$

$$= 3.36 \text{ kA}$$

A 10 kA surge arrester is suitable for this application.

## 18.7.4 Location of Surge Arrester

It is always good practice to connect the surge arresters very close to the equipment to be protected. In the case of effectively shielded substations, where the chance of direct stroke to the equipment is minimal, the surge impedance limits the discharge current. Therefore, sometimes a surge arrester can protect more than one piece of equipment. In the noneffectively shielded substations, the surge arresters should be installed at the terminals of the equipment to be protected.

### 18.7.4.1 Separation Effect

Sometimes it may not be practical to install the surge arrester at the equipment terminals. In such cases a cable of definite length is used to connect the equipment to the surge arrester. A typical connection of a surge arrester with separation distance $d$ is shown in Figure 18.15. A traveling wave coming into the substation is limited in magnitude at the surge arrester location to the value of discharge voltage. When a surge arrester is separated from the protected equipment by leads of significant length, oscillations occur which result in a voltage higher than the arrester voltage at the equipment. The voltage at the equipment $V_t$ at a distance of $d$ from the surge arrester is given by:

$$V_t = V_a + 2\frac{de}{dt} \cdot \frac{d}{1,000}; \quad \text{upon the maximum of } 2V_a \quad (18.8)$$

where   $V_a$ = Discharge voltage of the surge arrester, kV
$de/dt$ = Rate of rise of the wave front, kV/μs
$d$ = Separation distance, ft

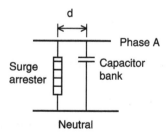

**Figure 18.15**   Diagram showing the separation effect.

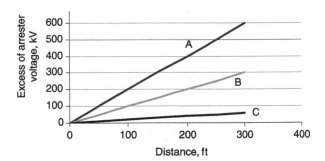

**Figure 18.16** Effect of separation with different incoming waves. A = 1000 kV/μs; B = 500 kV/μs; C = 100 kV/μs.

Using Equation (18.8), the effect of separation distance on the voltage at the equipment is shown in Figure 18.16 for different incoming waves. In addition to the reflected wave, it is always possible that still higher peak voltages can occur at the equipment due to the oscillations caused by the inductance of the cable between the surge arrester and the protected equipment.

### Example 18.4

A surge arrester is located at a distance of $d$ feet from a 40 MVAR shunt capacitor bank. The BIL rating of the capacitor bank is 550 kV. A traveling wave is arriving at the surge arrester with a rate of rise of 600 kV/μs. The discharge voltage of the surge arrester is 280 kV. Determine whether the surge arrester can protect the capacitor banks adequately in the following two cases: (a) $d = 20$ ft; (b) $d = 300$ ft.

Solution

$$V_a = 280 \text{ kV}, \quad \frac{de}{dt} = 600 \text{ kV/μs}$$

(a) Using Equation (18.8):

$$V_t = 280 + (2)(600 \text{ kV/μs}) \times 20 \text{ ft}/1{,}000 = 304 \text{ kV}$$

The BIL rating of the capacitor bank is 550 kV. The magnitude of the incoming wave is 304 kV. The capacitor bank will be protected.

(b) $V_t = 280 + (2)(600 \text{ kV}/\mu\text{s}) \times 300 \text{ feet}/1{,}000 = 640 \text{ kV}$

The BIL rating of the capacitor bank is 550 kV. The magnitude of the incoming wave is 640 kV. The capacitor bank is not protected suitably since the incoming surge voltage is much higher than the BIL of the equipment.

### 18.7.5   Effect of Transients on the Surge Arrester

It is necessary to understand the effect of transient overvoltages due to various switchings on the surge arrester. The typical switching surges due to various switching operations are discussed in Chapter 22 and summarized in Table 18.5. For the practical cases listed in Table 18.5, the corresponding maximum overvoltages, TRV, and frequency of oscillation are identified. In some cases, the maximum overvoltage is excessive, TRV is high, or the frequency of oscillation is excessive. The acceptable TRV is 2.4 P.U. and the frequency of oscillation is 4,250 Hz. Therefore, it can be seen that the capacitor switching produces overvoltages in some cases and the surge

TABLE 18.5   Overvoltages, Oscillation Frequencies, and TRV

| Description | $V_{\max}$, P.U. | $f$, Hz | TRV, P.U. | Remarks |
|---|---|---|---|---|
| Energization | 2.1 | 588 | – | Arrester will conduct |
| De-energization | 1.0 | DC | 2.0 | Acceptable |
| Fault clearing | 2.3 | 5,000 | 2.3 | Frequency of oscillation is unacceptable |
| Reclosing | 2.5 | 400 | – | $V_{\max}$ is unacceptable |
| Restriking | 3.2 | 400 | 6.4 | $V_{\max}$ and TRV are high |
| Prestrike | 3.28 | – | – | $V_{\max}$ is unacceptable |
| Back-to-back | 1.48 | 400 | – | Acceptable |
| Voltage magnification | 2.1, 1.6 | | – | Arrester will conduct |

arrester will conduct. In some cases, however, the overvoltages are excessive, and the effectiveness of the surge arrester has to be evaluated suitably. It should be noted that if there is continuous conduction, the surge arrester may fail.

## 18.8 CONCLUSIONS

The surge arrester is used in the power system for surge protection. Lightning surges and switching operations produce overvoltages. The important characteristics of the metal oxide varistor type of surge arrester are presented. The approach for surge arrester selection for capacitor application is shown with the help of an example. The separation effect is discussed. The effects of various switching operations on the surge arrester conduction are identified.

## PROBLEMS

18.1.   What are the different types of lightning surges that affect a power system?

18.2.   What are the types of surge arresters available for protection of shunt capacitor banks?

18.3.   Identify the typical specification of a surge arrester.

18.4.   Define MCOV. How is it important in the MOV arresters? In this respect, identify the differences between MOV and silicon carbide surge arresters.

18.5.   A 30 MVAR, three-phase, 13.8 kV, 60 Hz capacitor bank is to be provided with surge arresters. The BIL rating of the bank is 95 kV. The ratios are $X_0/X_1 = 2$ and $R_0/X_1 = 0.35$ at the substation location. Evaluate 10 kV and 12 kV surge arresters for this application. Which arrester is suitable for this application and why?

18.6.   What are the different types of overvoltages expected in a capacitor bank installation? Which one is the most severe?

18.7.   A surge arrester is needed for the protection of a 30 MVAR shunt capacitor bank. The BIL rating of the capacitor bank is 550 kV. The expected traveling wave at the surge arrester has a rate of rise of 800 kV/μs. The discharge voltage of the surge arrester is 280 kV. Since there is no space for installation of the surge arrester at the terminal of the capacitor bank, it has to be installed at a distance. What is the recommended separation distance?

## REFERENCES

1. Ali, A. F., Durbak, D. W., Elahi, H., Kolluri, S., Lux, A., Mader, D., McDermott, T. E., Morched, A., Mousa, A. M., Natarajan, R., Rugeles, L., Tarasiewicz, E. (1996). Modeling guidelines for fast front transients, *IEEE Transactions on Power Delivery*, 11(1), 493–506.

2. IEEE PES Surge Protective Devices Committee (1991). Surge protection of high voltage shunt capacitor banks on AC power systems – Survey results and application considerations, *IEEE Transactions on Power Delivery*, 6(3), 1065–1072.

3. IEEE PES Surge Protective Devices Committee (1996). Impact of capacitor banks on substations surge environment and surge arrester applications, *IEEE Transactions on Power Delivery*, 11(4), 1798–1809.

4. Surge Protection in Power Systems (1978), *IEEE Tutorial Course*, 79EHO144-6-PWR.

5. IEEE Standard C62.22 (1991), IEEE Guide for the Application of Metal Oxide Surge Arresters for Alternating-Current Systems.

6. Website of Toshiba Power Systems & Services, Sydney, Australia, www.tic.toshiba.com.au.

# 19

## MAINTENANCE AND TROUBLESHOOTING

### 19.1  INTRODUCTION

For proper operation of capacitor banks, it is necessary to maintain balanced phase voltages and current flows through the three-phase power system. Any unbalance in the phase voltages or currents is an indication of a system problem. The actual harmonic voltages and harmonic currents at any given operating point may be measured using oscillograph records or harmonic analyzers.

### 19.2  MAINTENANCE

Capacitor banks require very little maintenance because they are static equipment. However, regular inspection of the capacitor units should include a check of ventilation, fuses, ambient temperature, phase voltages, line currents, and cleaning of the surfaces for removal of dust. The recommended inspection and maintenance including initial inspection, maintenance before energization, periodic inspection, and required measurements are identified below [1].

### 19.2.1  Inspection and Maintenance

Standard safety practices should be followed during installation, inspection, and maintenance of capacitors. In addition, there are procedures that are unique to capacitor equipment that should be followed to protect operating staff and equipment in accordance with the National Electrical Safety Code. Some of the items of importance are listed below.

#### 19.2.1.1  Clearance and Grounding

After a capacitor bank is de-energized, there will be residual charges in the units. Therefore, wait at least 5 min before approaching it to allow sufficient time for the internal discharge resistors in each capacitor unit to dissipate the stored energy. These resistors are designed to reduce the voltage across the individual capacitor units to less than 50 V within 5 min. However, the grounding leads should be applied to all three phases to short out and ground the capacitor bank. On larger substations, permanent grounding switches may be used to achieve this function. Even after grounding, it is recommended that individual capacitor units be shorted and grounded before personnel come into contact with them to ensure that no stored energy is present.

#### 19.2.1.2  Bulged Capacitor Units

One of the failure modes of capacitor units is bulging. Excessively bulged units indicate excessive internal pressure caused by overheating and generation of gases due to probable arcing condition. These units should be handled carefully. The manufacturer should be consulted regarding the handling of bulged units.

#### 19.2.1.3  Leaking from Capacitor Units

Another mode of failure in the capacitor bank is leaking due to the failure of the cans. When handling the leaking fluid, avoid contact with the skin and take measures to prevent entry into sensitive areas such as eyes. Handling and disposal of capacitor insulating fluid should comply with state, federal, and local regulations. Some capacitor units contain combustible liquid and require careful handling.

### 19.2.1.4 Re-Energization of the Capacitor Banks

When returning to service, verify that all ground connections that were installed for maintenance purpose are removed. Allow a minimum of 5 min between de-energization of the capacitor bank and re-energization of the capacitor bank to allow enough time for the stored energy to dissipate.

### 19.2.1.5 Initial Inspection Measurements and Energization Procedures

During the initial inspection before energization of the capacitor banks the following measures should be taken:

1. Verify proper mechanical assembly of the capacitor units, clearances as per the electrical code, and soundness of the structure of all capacitor banks.
2. It may be useful to measure the capacitance of the banks and keep the measurements as benchmark data for future comparison.
3. Check the electrical connections for proper installation and good electrical contact. Verify that all terminal connections are tightened properly. Check the individual fuse connections to ensure that they are tight and make good contact.
4. Clean all insulators, fuses, and bushings to prevent the possibility of dirty porcelain creating a flashover danger. Inspect all porcelain insulators for cracks or breaks.
5. Test the operation of all controls and load break, disconnect, and grounding switches prior to energizing the capacitor banks.
6. Prior to energizing the bank, verify the capacitance values of each phase and compare them with the coordination values used in the relay settings. The capacitance unbalance should not result in a voltage of more than 110% of the rated voltage on any one unit.

7. Immediately after energization, verify the voltage increase on the terminals and compare it with the calculated values. Also, measure and verify if the supply voltage, phase currents, and the kVAR of the capacitor bank are within the allowed limits. Approximately 8 h after energization, conduct a visual inspection of the bank for blown fuses, bulged units, and proper balance in the currents.

### 19.2.1.6 Periodic Inspection, Measurement, and Maintenance

The substation and distribution capacitor banks should be inspected and electrical measurements be made periodically. The frequency of the inspection should be determined by local conditions such as environmental factors and type of controller used to switch the capacitors on and off.

### 19.2.1.7 Visual Inspections

Visual inspection of the capacitor bank must be conducted for blown capacitor fuses, capacitor unit leaks, bulged cases, discolored cases, and ruptured cases. During such inspection, check the ground for spilled dielectric fluid, dirty insulating surface on the bushings, signs of overheated electrical joints, open switches, and tripped protective devices. An infrared camera is very useful for inspecting the substation equipment for overheated joints and surfaces, and records can be maintained for future reference. Pole-mounted capacitor units should be examined for exterior corrosion.

### 19.2.1.8 Physical Inspection and Measurements

Physical inspection and measurements should include loose connections, overheated lead wires, and faulty fuse tubes. Fuses should be inspected for evidence of overheating or other such damage. The protection devices should be inspected for proper settings including the position of the current transformer and the potential transformer. The capacitance of the bank should be measured and compared with the previous measurements.

## 19.2.2 Difficulties in Field Measurements

Capacitor bank installations present inherent difficulties with respect to the measurement process. These include the following [2]:

1. *Accessibility of equipment.* The capacitor banks are usually installed in rack mountings in the substation environment. Gaining access to the capacitors for measurement requires a boom truck or extended ladder. Space is limited in capacitor banks.
2. *Terminal connections.* In order to make measurements on a capacitor bank, the terminals have to be disconnected. Such practices are not advisable in an existing bank. If the connections were not made properly after the measurements, there may be additional problems.
3. *Measuring environment.* In a substation, there will be induced voltages due to the electrostatic and electromagnetic couplings on other equipment such as bus bars for various lines and circuit breakers. Therefore, the measurements made at the substation location may be inaccurate.
4. *Impact due to the number of units.* In capacitor banks, there are several series and parallel units, often hundreds in number. Troubleshooting by measuring unit by unit can be tedious and time consuming.
5. *Historical data.* Capacitance measurements often are made in the factory and the data are retained in an office location. Such historical data may not be easily retrievable by maintenance personnel for comparison purposes.

## 19.2.3 Measuring Techniques

Many electrical devices are available for measuring capacitance. Most of these devices are delicate and are operated indoors. The following categories can be identified:

1. *Laboratory type meter.* This type of equipment is delicate and produces accurate readings. Portability is limited due to its sensitivity and sometimes requires accessory equipment. Such equipment will be installed in a manufacturing plant or testing laboratory. Some modified equipment is available, but it is heavy and is used for measuring the capacitance in a restricted laboratory.

2. *Handheld digital capacitance meter.* The capacitance meter is a small, battery-operated, low voltage device suitable for capacitance measurement in different environments. Some of these devices are accurate enough to measure the loss of one capacitor unit, with a precision of 0.5%. The digital capacitance meter can be used to measure the capacitance of the units in the field. A typical digital capacitance meter is shown in Figure 19.1. This hand-held digital meter looks like a multimeter and can be used to measure resistance, inductance, and capacitance. The range of measurement in the capacitance is 20 nF–200 mF. This range is suitable for measuring capacitors used in power factor correction applications.

3. *Substation capacitance devices.* These are devices suitable for direct measurement of capacitor units in a rack assembly without disconnecting the connections. A low voltage is applied to the group of units to be measured. A clamp-on meter is placed around the bushing of the unit to be measured. A capacitance reading is obtained from the current measured by the clamp. Even though there is improvement in the measurement process, it is difficult to obtain repetition of the measurement. The degree of precision (2–5%) is not enough to accommodate the internally fused and fuseless capacitor technologies [2].

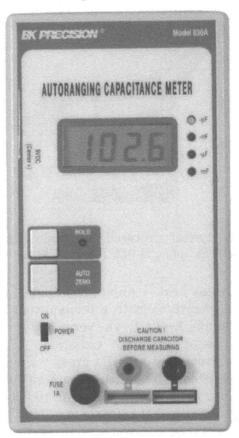

**Figure 19.1** A digital capacitance meter, type 830A. (Courtesy of B&K Precision, Melrose, MA.)

4. *Infrared camera for temperature measurement.* An infrared camera is a very useful tool for measuring the temperature of joints and other critical parts from the substation yard. The infrared camera can be used to "see" and "measure" thermal energy emitted from an object. Thermal or infrared energy is light that is not visible because its wavelength is too long to be detected by the human eye. It is the part of the electromagnetic spectrum that we

**Figure 19.2** A typical infrared camera, ThermaCAM P60. (Courtesy of FLIR Systems, North Billerica, MA.)

perceive as heat. Unlike visible light, in the infrared world, everything with a temperature above absolute zero emits heat. Even very cold objects, like ice cubes, emit infrared radiation. The higher the object's temperature, the greater the infrared radiation emitted. The infrared camera allows us to see what our eyes cannot. The infrared cameras produce images of invisible infrared or "heat" radiation and provide precise noncontact temperature measurement capabilities. Nearly everything gets hot before it fails, making infrared cameras extremely cost-effective and valuable diagnostic tools in many diverse applications. As industry strives to improve manufacturing efficiency, improve product quality, and enhance worker safety, new applications for infrared cameras continually emerge. An example of an infrared camera is shown in Figure 19.2. This camera can produce images suitable for jpeg format.

## 19.3 TROUBLESHOOTING

Some of the failure problems associated with capacitor banks are well known. A few of the failures are traceable to the

**Figure 19.3**  A failed capacitor bank due to fire. (Courtesy of Duke Energy, Charlotte, NC.)

original source and sometimes that may be difficult to do. In many instances, the final result of a failure may be catastrophic explosion of the capacitor into pieces or fire. A capacitor bank failure caused by a fire in a utility application is shown in Figure 19.3. Although the capacitor unit does not seem damaged, the damage to the doors is visible. There is debris on the floor. The typical causes for capacitor failure are discussed below.

1.  *Capacitor failure due to inadequate voltage rating.* In the filter banks, the capacitor units are connected in series with inductors. Sometimes the voltage across the capacitor units exceeds the design values. In such circumstances, the capacitor units

fail catastrophically due to inadequate voltage rating. This is demonstrated using an example.

## Example 19.1

Consider a second harmonic filter, a part of a multiple filter bank located at a 13.8 kV, 60 Hz, three-phase system. The reactance of the inductor and the reactance of the capacitor units are 4.5 Ω and 17.7 Ω, respectively, at 60 Hz. The capacitor bank consists of 400 kVAR, 9.96 kV, single-phase units, 21 units in parallel per phase. The filter bank is wye connected and ungrounded. Analyze the adequacy of the filter bank.

Solution

kVAR/unit = 400
kV/unit = 9.96 kV
System line-to-line voltage = 13.8 kV
Phase voltage (13.8 kV/1.732) = 7.97 kV
The capacitor voltage = 9.96 kV (There seems to be adequate margin)
Reactance of the inductor = 4.5 Ω
Reactance of the capacitor units = 17.7 Ω

$$\text{Expected MVAR delivered} = \frac{kV^2}{(X_c - X_l)} = \frac{13.8^2}{(17.7 - 4.5)}$$

$$= 14.42 \text{ MVAR}$$

$$\text{Circuit current } I = \frac{kVAR}{\sqrt{3}(kV)} = \frac{14{,}420}{1.732 \times 13.8\,kV} = 603\,A$$

Voltage across the inductor $= (I \times X_l) = (603\,A \times 4.5) = 2{,}713\,V$
Voltage across the capacitor units $= (I \times X_c) = (603\,A \times 17.7) = 10{,}673\,V$
The rating of the capacitor unit = 9.96 kV
Max. continuous capacitor voltage $= 9.96 \times 1.1 = 10{,}956\,V$

The nominal system voltage is 13.8 kV and the allowed limit on the high side is 1.05 P.U. The corresponding voltage across the capacitor unit is (1.05 × 10.673), i.e., 11.2 kV. The capacitor voltage is higher than the allowed maximum voltage of 10.956 kV. The capacitor unit will fail catastrophically due to excessive continuous voltage if the system is operated at 1.05 times the rated setting. Although it looks as if the selection of the capacitor voltage is acceptable at the beginning, careful consideration of the capacitor voltage shows that it is not adequate.

2. *Fuse blowing.* The blowing of a fuse may be due to short circuit in a capacitor unit, overcurrent due to an overvoltage, or harmonics. A short-circuited capacitor unit can be determined by inspecting the capacitor can for bulging or case rupture. Sometimes the fuse rating can be lower than the necessary rating. High contact resistance of the fuse unit can cause excessive temperatures and failure of the fuse. Failure of a fuse means blowing of the fuse under circumstances other than when performing its designed function. Thus a fuse that blows after a capacitor short is not classified as a fuse failure. Fuse failure may occur due to fatigue, incorrect application, and improper branch protection.

**Example 19.2**

In Example 19.1, the fuse selected for the application is 40 T, 9.96 kV. Perform a suitability analysis.

Solution

kVAR per unit = 400
kV = 9.96
Capacitor current (400 kVAR/9.96 kV) = 40 A
Desired fuse current (40 A × 1.35/1.5) = 36 A

A 40 T fuse is applicable. The fuse voltage is 9.96 kV. As per Example 19.1, the maximum capacitor voltage will be

11.2 kV. The fuse will also fail due to inadequate voltage rating if the system voltage is at 1.05 times the nominal rating. Select the next nearest fuse voltage, i.e., 15 kV.

3. *Thermal failure.* Capacitors operated at extreme hot conditions can fail due to excessive temperature. The excessive heat can be due to high ambient temperature, radiated heat from adjacent equipment, or extra losses.

4. *Ferroresonance.* The capacitor banks tend to interact with the source or transformer inductance and produce ferroresonance. This can produce undamped oscillations in the current or voltage, depending on the type of resonance. If the system is not adequately damped, then there is a possibility of capacitance or transformer failure. This is due to the system condition and cannot be identified by visual inspection.

5. *Harmonics.* Any nonlinear load in the system such as an arc furnace or converter equipment produces harmonics. Filters are used to control the harmonics. If the tuning of the filters is not sharp enough, then there may be excessive harmonic currents through the capacitor bank. Harmonics cause overheating and failure of the capacitor units.

6. *Open circuited capacitor units.* Field measurement can determine if the fuse is blown and if there is an open circuit due to capacitor failure.

7. *Dielectric failures.* When an internal series group of a capacitor unit fails, the voltage on the remaining internal series groups in the string increases. It is desirable to remove the bank from service when the voltage applied to the remaining internal series groups exceeds 110% of their rated voltage. The sensitivity necessary to detect individual internal series group failures requires an unbalance detection scheme that is not sensitive to system unbalance and can be adjusted to null out the capacitor bank's

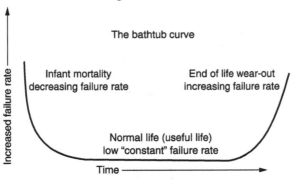

**Figure 19.4** A typical bathtub curve.

inherent unbalance. It is important to note that as
the internal design of the capacitor units used in
fuseless banks becomes an integral part of the
relay scheme, replacement of faulty units in a bank
should be made only with units sharing the same
internal electrical configuration.

8. *Rack faults and insulation failures.* The two failure
modes of greatest concern in a fuseless bank are
the failure of a unit's major insulation and the flashover of a unit's bushing. If a unit in an externally
fused bank experiences either of these two failure
modes, the external fuse will operate and remove
the unit from service; however, fuseless banks have
no external fuses.

9. *Manufacturing defects.* These defects are to be identified during the testing of capacitor units in the
factory. The typical bathtub curve of failure of any
electrical component is shown in Figure 19.4. The
bathtub curve consists of three periods: an infant
mortality period with a decreasing failure rate;
then a normal life period, also known as useful life
with a low, relatively constant failure rate; and
finally, a wear-out period that exhibits an increasing failure rate. If there are any manufacturing defects in the capacitor units, the failure will

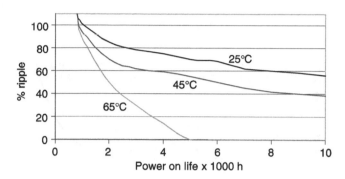

**Figure 19.5** Typical variation of capacitor life with ripple current at various ambient temperatures.

occur in the factory during testing. During useful life, any capacitor failure may be due to stress-related causes.

10. *Failures due to internal stress in the capacitor units.* The application of ripple currents, surge voltages, and high frequency oscillatory currents can cause internal stress in the capacitor units and premature failure. Figure 19.5 shows how the ripple current affects the life of capacitors. Considering a typical case at 45°C ambient temperature, a capacitor with a predicted life of 10,000 h at 39% ripple current has a life of 600 h at 100% ripple current. In addition to ripple current and overvoltages, the type of waveforms and duration of the current pulses also affect the life of the capacitor units.

11. *Failures due to external stress.* Sometimes the capacitor banks are exposed to extreme operating conditions, including excessive ambient temperatures, humidity, temperature cycling, vibrations, shock, and lack of ventilation. Such conditions can occur in substation capacitor installations. In certain applications, forced air cooling is used. In many applications, the capacitor banks are exposed without

any forced air cooling. Extra cooling of these units is expected to improve the life expectancy.

12. *Human error.* Sometimes human error is responsible for capacitor bank failure. If the protection coordination of the fuse selection is not performed correctly, fuse or capacitor failure may occur. For energization of the capacitor banks, a circuit switcher equipped with closing resistor is used. When a capacitor bank is tripped due to a fault, the circuit breaker is open. The circuit switcher is still in the closed position. Now, if the circuit breaker is used to energize the capacitor bank, there is no closing resistor in the circuit and the capacitor bank may fail due to excessive energization transients. To perform this operation correctly, the circuit switcher is opened. Then the circuit breaker is closed. The energization of the capacitor bank is performed using the circuit switcher with closing resistance in the circuit. Such human error is common but very difficult to prove, unless the operator admits or records such error.

13. *Surge arrester connection.* The surge arrester is connected across the filter bank for protecting from switching and lightning surges. Sometimes the surge arrester is connected across the reactor as shown in Figure 19.6(a). In such a connection, during overvoltage, the surge arrester conducts and forces large currents through the capacitors. Such large currents can cause capacitor failure momentarily. At the same time, there is no surge protection for the capacitors. The recommended connection of the surge arrester in a filter bank is shown in Figure 19.6(b). Applying a surge arrester in an ungrounded wye connection has other performance issues as outlined in IEEE Standard C62.22. A solidly grounded filter circuit with surge arrester across the filter circuit is recommended.

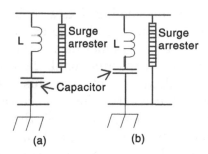

**Figure 19.6** Surge arrester application for the filter bank per phase. (a) Surge arrester across the inductor. (b) Surge arrester across the filter.

14. *Converter harmonic production and related issues.* If the converter produces more than acceptable harmonics, the filter components such as capacitor and inductor will be overloaded, leading to premature failure. If firing timing is not identical for a parallel set of thyristors, uneven loading can result, producing individual semiconductor failures and gravitating toward a subsequent cascade of failures through the system as fewer devices carry more and more of the load current. Even harmonics can be created in rectifier systems by firing irregularities. The various modes of timing irregularities in a converter occur for the following reasons [3]:

Pulse deviation related: Consider a six-pulse recti-fier, in which one of the six pulses does not occur in the correct time. This results in an across-the-board increase in harmonic currents, with poor cancellation of odd harmonics and production of even harmonic currents due to half-wave dissymmetry about zero.

Phase unbalance related: Phase unbalance does not produce even harmonics. It acts like a single-phase rectifier and produces the full spectrum of odd harmonics with modulation components of two of the normal harmonic frequencies.

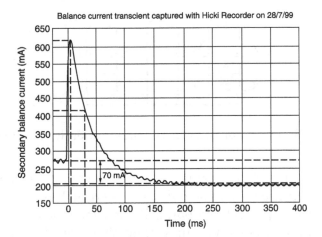

**Figure 19.7** Secondary current of a current sensor during a capacitor failure. (From Reference [4].)

> Group unbalance related: Sometimes pulses 1, 3, and 5 are displaced by an equal amount from 2, 4, and 6. This results in the generation of even harmonics, that is, multiples of $3n \pm 1$. In such cases, the second, fourth, and other such even harmonics can be observed. In Reference [3], the fourth harmonic resonance was observed.

15. *Failure problems associated with the H-configuration capacitor banks.* Consider an H-capacitor bank as shown in Figure 19.8. Each quadrant of the bank consists of several series–parallel capacitor units. The H-configuration is used in order to identify the failure in the groups using the resultant current through the bridge. When one element fails, it produces a temporary short circuit. The measured current on the secondary of the current transformer is shown in Figure 19.7. It can be seen that the peaky short-circuit current remains for a duration of 50 ms. Then the adjacent elements discharge into the failed element and the fuse blows under excessive current. The unbalance current decays and settles to a lower

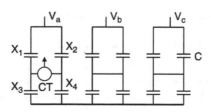

**Figure 19.8**  Capacitor bank in H-configuration for Example 19.2.

value. If the value is large enough to produce an alarm or a trip signal, then there will be some activity to protect the bank. If the capacitor elements fail in adjacent quadrants, then there will be a cancellation effect on the unbalance current level. Therefore, if elements are failing in the adjacent quadrant, the alarm or trip settings for unbalance protection may not be effective [4].

## Example 19.3

Consider an H-configuration of a capacitor bank as shown in Figure 19.8. The current transformer ratio used in the bridge circuit is 50:5. The reactance of $X_1 = X_2 = X_3 = X_4 = 10\,\Omega$. Due to the failure of one internal capacitor, the value of $X_1$ changes to $9.5\,\Omega$. Assume a capacitor bank current of $500\,A$. Solve for the following.

(a) Write an equation to find the secondary current in the bridge circuit for relaying applications. What is the secondary current in the CT circuit?
(b) Assume that due to the failure of some capacitor elements in the first quadrant, $X_1 = 9.5\,\Omega$. What is the secondary current in the CT circuit?
(c) Assume that $X_1 = X_2 = 9.5\,\Omega$ due to failures. What is the secondary current in the CT circuit?
(d) Assume that $X_1 = X_2 = X_3 = X_4 = 9.5\,\Omega$. What is the secondary current in the CT circuit? State your conclusions.

Solution

$I_{bank} = 500\,A$

(a) $X_1 = X_2 = X_3 = X_4 = 10\,\Omega$
   Current transformer ratio $= 500\,A/5\,A$

$$I_b = \frac{(X_1)(X_2) - (X_2)(X_4)}{(X_1 + X_2)(X_3 + X_4)} \times I_{bank} \times \frac{N_2}{N_1}$$

where $X_1 =$ Reactance of quadrant 1
   $X_2 =$ Reactance of quadrant 2
   $X_3 =$ Reactance of quadrant 3
   $X_4 =$ Reactance of quadrant 4
$N_1 =$ Primary current rating of the CT
$N_2 =$ Secondary current rating of the CT

$$I_b = \frac{(10 \times 10) - (10 \times 10)}{(10 + 10)(10 + 10)} \times 500\,A \times \frac{5}{50} = 0\,A$$

When the reactances of all the four branches of the H-configuration are equal, there will be no resultant current in the bridge circuit.

(b) $X_1 = 9.5\,\Omega;\ X_2 = X_3 = X_4 = 10\,\Omega$

$$I_b = \frac{(9.5 \times 10) - (10 \times 10)}{(9.5 + 10)(10 + 10)} \times 500\,A \times \frac{5}{500} = 0.0128\,A$$

When there is a failure in one quadrant of the H-configuration, such a failure can be identified.

(c) $X_1 = X_2 = 9.5\,\Omega;\ X_3 = X_4 = 10\,\Omega$

$$I_b = \frac{(9.5 \times 10) - (9.5 \times 10)}{(9.5 + 9.5)(10 + 10)} \times 500\,A \times \frac{5}{500} = 0\,A$$

When there are an equal number of failures in one and two quadrants of the H-configuration, such a failure cannot be identified.

(d) $X_1 = X_2 = X_3 = X_4 = 9.5 \, \Omega$

$$I_b = \frac{(9.5 \times 9.5) - (9.5 \times 9.5)}{(9.5 + 9.5)(9.5 + 9.5)} \times 500 \text{ A} \times \frac{5}{500} = 0 \text{ A}$$

When there is an equal number of failures on all four branches of the H-configuration, such a failure cannot be identified.

The symptoms, possible cause, and remedial approach for various types of capacitor failures are identified in Table 19.1. It can be seen from the table that the possible symptoms are capacitor failure, fuse failure, false tripping of the control or relay equipment, failure of surge arresters, reactor failure, overheating of transformer or machines, and telephone interference. The possible causes are resonance, excessive harmonics, and insufficient filtering. The possible remedy in many causes is to tune the existing filter correctly. Although it is not mentioned specifically to include a resistance to modify the filter as a high-pass filter, it is a very effective solution in the case of oscillatory cases. In some cases, it may be necessary to have multiple filters to control several dominant harmonics. In certain light loaded systems, it is possible to control the harmonic distortion by switching off the filter or capacitor bank. These issues are all application specific and require careful examination of the given system condition.

## PROBLEMS

19.1. What are the different types of maintenance requirements for capacitor banks?

19.2. Why is field-testing required for the capacitor banks? How can thermal photography be used to evaluate the soundness of capacitor units?

19.3. An open circuit is one of the suspected faulted conditions in a capacitor bank. How can this be detected?

19.4. There are three filter banks in an arc furnace installation. The filter banks are tuned to third,

**TABLE 19.1** Capacitor Unit Failures: Symptoms, Causes, and Remedies

| Symptom | Possible Cause | Remedy |
|---|---|---|
| Capacitor failure near a harmonic load | Harmonic resonance | Detune the filter bank; change the capacitor size |
| Misoperation of electronic controllers | Distorted voltages; see the waveforms | Use a harmonic filter |
| Capacitor fuse blowing | Resonance | Detune the filter |
| Power transformer overheats below rated load | Excessive harmonic currents | Use a filter bank; detune the existing filter |
| Induction motors overheat at part load | Excessive harmonic currents | Use a filter bank; detune the existing filter |
| Capacitor failure | Inadequate voltage rating | Check the voltage rating and use adequate voltage rating |
| Surge arrester failure | Harmonic resonance; arrester rating | Tune the filter; check the arrester rating |
| Filter reactor failure | Harmonic resonance | Tune the filter |
| False relay trips | Harmonic resonance | Perform system study and determine what to do |
| Telephone interference | Harmonic currents in the path of telephone circuit due to inductive coupling | Reduce system harmonics; examine ungrounded filter bank or change filter location |
| Voltage exceeds limit | Too much harmonics | Apply necessary filtering |
| Harmonic distortion at light loads | The harmonics exist only at light loads | Avoid switching capacitor banks at light loads |
| Harmonic distortion at peak loads | Inadequate filtering | Detune the existing filters; apply filters at the load |

fifth, and seventh harmonics. If the capacitor units in the fifth harmonic filter alone fail on a regular basis, what items should be examined in order to identify the cause of the failure? The other banks are operating normally.

19.5.   What are the possible human errors that lead to the failure of capacitor units?

19.6.   List the various causes of capacitor failure.

19.7.   Consider a fifth harmonic filter, a part of a multiple filter bank located at a 13.8 kV, 60 Hz, three-phase system. The capacitor bank consists of 400 kVAR, 9.96 kV units, 24 in parallel. The rating of the reactor is 3.13 mH/phase and 13.8 kV. The filter bank is wye connected and ungrounded. Analyze the adequacy of the filter bank. The system operating voltage is 1.05 P.U.

19.8.   If the fuse used in the fifth harmonic filter (Problem 19.7) is 40 T, is this rating adequate for continuous operation? Assume that the system voltage setting is changed to 1.05 P.U. for several hours during the day.

## REFERENCES

1.  IEEE Standard 1036 (1992), IEEE Guide for Application of Shunt Capacitors.

2.  Sévigny, R., Ménard, S., Rajotte, C., McVey, M. (2000). Capacitor measurement in the substation environment: A new approach, *Proceedings of the IEEE 9th International Conference on Transmission and Distribution*, 299–305.

3.  Buddingh, P. C. (2003). Even harmonic resonance—An unusual problem, *IEEE Transactions on Industry Applications*, 39(4), 1181–1186.

4.  Illing, K. J. (2003). Capacitor fuse fail detector, B.E. Thesis, The University of Queensland, Australia.

# 20

## HARMONIC FILTERING

### 20.1 INTRODUCTION

Until the 1960s, the main harmonic sources in the power system were the arc furnace and a few converter loads. In the 1970s, with thyristors and static power supplies, many variable speed drives were introduced in all industries. With the increase in the converter load in the power system, several new problems became noticeable, including:

- Flow of harmonic currents from the converter to the AC system
- Poor power factor on the AC side
- Poor voltage regulation on the AC side due to low power factor
- Interference in the telecommunication systems due to induced voltages
- Distortions of AC supply voltages that affect the performance of computer equipment
- Error in the metering
- Continuous neutral currents in the neutral conductors of four-wire systems

Therefore, it is necessary to understand the behavior of industrial power systems with converter/inverter equipment. With the introduction of the new filtering devices and the need to improve the power factor and control the harmonics, utilities can encounter new system problems. In this chapter, the sources of harmonics, the system response, modeling of the system for harmonic analysis, acceptable harmonic limits, and the approach for harmonic analysis are presented. In order to show the effect of harmonics on the system, consider a fundamental and third harmonic, both in phase, as shown in Figure 20.1. The total amplitude is less than the maximum fundamental. Now consider the fundamental and out-of-phase third harmonic along with the total magnitude in Figure 20.2. It can be seen that the total magnitude is higher

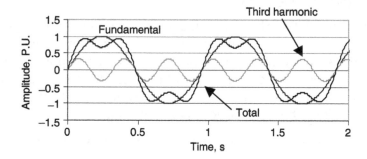

**Figure 20.1** Fundamental and third harmonic in phase with the total amplitude.

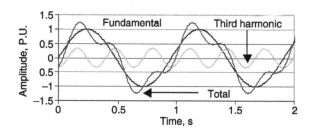

**Figure 20.2** Fundamental and third harmonic out of phase with the total amplitude.

than the fundamental. This illustrates that, depending on the phase angle of the harmonics, the total amplitude of the current can exceed the normal rating. The overloaded current will produce excessive heating. Therefore, the effect of harmonics requires careful consideration.

## 20.2 HARMONIC SOURCES

There are several harmonic sources in the distribution systems. These are loads with nonlinear characteristics. The AC–DC converters, pulse width modulated converters, cycloconverters, arc furnaces, static VAR compensators, and switched mode power supplies are typical nonlinear loads producing harmonics. The typical harmonic frequencies and the corresponding magnitudes produced by various harmonic producing equipment are listed below.

### 20.2.1 Converters

A six-pulse type of AC to DC converter is used in variable speed drives, battery charging applications, and HVDC circuits. In the converter circuit, each pair of thyristors is triggered and conducts until it is reverse biased. If a thyristor is triggered at zero firing angle, it acts exactly as a diode. With different firing angles, there will be increased harmonics. An analysis of the converter harmonics will yield the following harmonic contents:

$$h = kq \pm 1 \tag{20.1}$$

$$I_h = I_1/h \tag{20.2}$$

where $h$ = Harmonic order
$\quad k$ = Any integer (0, 1, 2 . . .)
$\quad q$ = Pulse number of the circuit
$\quad I_h$ = Magnitude of the harmonic component
$\quad I_1$ = Magnitude of the fundamental component

The harmonic contents of a six-pulse converter are listed in Table 20.1 [1].

**TABLE 20.1**   Harmonic Contents of a Six-Pulse Converter

| $h$ Order | Magnitude | Angle |
|-----------|-----------|-------|
| 1  | 100  | −75  |
| 5  | 33.6 | −156 |
| 7  | 1.6  | −151 |
| 11 | 8.7  | −131 |
| 13 | 1.2  | 54   |
| 17 | 4.5  | −57  |
| 18 | 1.3  | −226 |
| 23 | 2.7  | 17   |
| 25 | 1.2  | 149  |

## 20.2.2 Pulse Width Modulated (PWM) Converters

PWM converters use power electronic devices that can be turned on and off. The input power is usually obtained from a converter source and the output voltage is shaped according to requirement using thyristor switching. The output pulse widths are varied to obtain a three-phase voltage wave at the load. The load is usually an AC motor used as a variable speed drive. The harmonic contents due to a typical PWM drive at various load conditions are listed in Table 20.2 [1].

## 20.2.3 Cycloconverters

These are devices that convert AC power at one frequency to AC power at a lower frequency. These converters are used to drive large AC motors at lower speeds. These types of converters produce a significant amount of harmonics. The harmonic components due to the operation of a cycloconverter are given by:

$$f_h = f_1(kq \pm 1) \pm 6\,n\,f_0 \qquad (20.3)$$

where $f_h$ = Harmonic frequency imposed on AC system
   $k, n$ = Integers
     $f_0$ = Output frequency of the cycloconverter
     $f_1$ = Fundamental frequency of the AC system

TABLE 20.2  Harmonic Contents of a PWM Drive

| | 100% Load | | 75% Load | | 50% Load | |
|---|---|---|---|---|---|---|
| h Order | Magnitude | Angle | Magnitude | Angle | Magnitude | Angle |
| 1 | 100.00 | 0 | 100.00 | 0 | 100.00 | 0 |
| 3 | 0.35 | −159 | 0.59 | −44 | 0.54 | −96 |
| 5 | 60.82 | −175 | 69.75 | −174 | 75.09 | −174 |
| 7 | 33.42 | −172 | 47.03 | −171 | 54.61 | −171 |
| 9 | 0.50 | 158 | 0.32 | −96 | 0.24 | −102 |
| 11 | 3.84 | 166 | 6.86 | 17 | 14.65 | 16 |
| 13 | 7.74 | −177 | 4.52 | −178 | 1.95 | 71 |
| 15 | 0.41 | 135 | 0.37 | −124 | 0.32 | 28 |
| 17 | 1.27 | 32 | 7.56 | 9 | 9.61 | 10 |
| 19 | 1.54 | 179 | 3.81 | 9 | 7.66 | 16 |
| 21 | 0.32 | 110 | 0.43 | −163 | 0.43 | 95 |
| 23 | 1.08 | 38 | 2.59 | 11 | 0.94 | −8 |
| 25 | 0.16 | 49 | 3.70 | 10 | 3.78 | 7 |

TABLE 20.3  Harmonic Contents of the Arc Furnace Current

| | Harmonic Order in % | | | | | |
|---|---|---|---|---|---|---|
| Furnace Condition | 1 | 2 | 3 | 4 | 5 | 7 |
| Initial melting | 100 | 7.7 | 5.8 | 2.5 | 4.2 | 3.1 |
| Refining | 100 | – | 2.0 | – | 2.1 | – |

### 20.2.4  Arc Furnace

The harmonic produced by an electric arc furnace is very difficult to predict due to the variation of the arc impedance on a cycle-by-cycle basis. Therefore, the arc current is nonperiodic and the analysis shows both integer and noninteger harmonic. The harmonic content is different both for melting and refining periods. Table 20.3 presents the harmonic contents of the arc furnace operation [1].

### 20.2.5  Static VAR Compensator (SVC)

The thyristor controlled reactor with fixed capacitors has been used to control the power factor of the electric arc furnaces and

**TABLE 20.4**   Typical Harmonic Amplitudes due to an SVC

| Harmonic Order | Amplitude, % | Harmonic Order | Amplitude, % |
| --- | --- | --- | --- |
| 1 | 100 | 3 | 13.78 |
| 5 | 5.05 | 7 | 2.59 |
| 9 | 1.57 | 11 | 1.05 |
| 13 | 0.75 | 15 | 0.57 |
| 17 | 0.44 | 19 | 0.35 |
| 21 | 0.29 | 23 | 0.24 |
| 25 | 0.20 | | |

**TABLE 20.5**   Harmonic Contents of a Switched Mode Power Supply

| Harmonic Order | Amplitude, % | Harmonic Order | Amplitude, % |
| --- | --- | --- | --- |
| 1 | 100.0 | 3 | 81.00 |
| 5 | 60.6 | 7 | 37.00 |
| 9 | 15.7 | 11 | 2.40 |
| 13 | 6.30 | 15 | 7.90 |

similar distribution loads to reduce the voltage flicker. Harmonics are produced because the thyristor controlled reactor current is adjusted to correct the power factor. Typical harmonic components produced due to the operation of an SVC are listed in Table 20.4 [1].

### 20.2.6   Switched Mode Power Supplies

In all personal computers, switched mode power supplies are used. These are very economical designs in which energy is stored in a capacitor and discharged in order to get a DC voltage through an electronic circuit. Since the load seen by the AC side is capacitive, the current flow is not continuous. The typical harmonic components due to a switched mode power supply are shown in Table 20.5 [1].

### 20.3   SYSTEM RESPONSE TO HARMONICS

The effect of harmonics on the power system depends on the frequency response characteristics of the system. Some of the important contributing factors are discussed below.

## 20.3.1  System Short-Circuit Rating

A system with a large short-circuit capacity will produce a low voltage distortion. A system with a lower short-circuit rating will produce a large voltage distortion. The system short-circuit rating depends on the amount of generation, transmission voltage level, number of parallel lines, and other system characteristics.

## 20.3.2  Load Characteristics

The resistive component of the load produces damping in the circuit and hence reduces voltage magnification. The reactive component of the load can shift the point at which resonance occurs, hence a reactive load can amplify the voltage magnification. A lightly loaded system is likely to have less damping and hence a higher voltage distortion. A heavily loaded system is likely to offer better damping.

## 20.3.3  Parallel Resonance

Parallel resonance occurs when the system inductive reactance and the capacitive reactance are equal at some frequency. If the resonant frequency happens to coincide with a harmonic frequency of a nonlinear load, then an oscillatory current flow will occur between the inductive source and the capacitance. A typical parallel resonant circuit is shown in Figure 20.3, with a resonant frequency of $f_0$.

## 20.3.4  Series Resonance

Series resonance is the result of the series combination of the capacitor banks and the transformer inductance as shown in Figure 20.4. A series resonant circuit offers a low impedance path for the harmonic currents and traps any harmonic current at the tuned frequency. The series resonant circuit will cause voltage distortion. The resonant frequency ($f_0$) for both parallel and series resonance is given by Equation (20.4):

$$f_0 = \frac{1}{2\pi\sqrt{LC}} \tag{20.4}$$

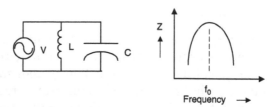

**Figure 20.3**   Parallel resonant circuit and frequency response.

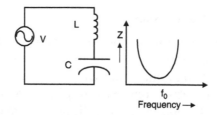

**Figure 20.4**   Series resonant circuit and frequency response.

where $L$ is the circuit inductance and $C$ is the capacitance. Both the parallel and series resonance can cause oscillatory currents in the power system.

## 20.4   ACCEPTABLE PERFORMANCE

### 20.4.1   Power Factor

When installing a filter bank for controlling the harmonic currents, the capacitor banks improve the power factor of the system. Most utilities like the customer to operate the load at a power factor of 95%. Sometimes a better power factor is prescribed.

### 20.4.2   Total Harmonic Distortion (THD) of the Voltage

This is a commonly used index for measuring the harmonic distortion in the voltage waveform. It is defined as:

$$\text{THD} = \frac{\sqrt{V_2^2 + V_3^2 + V_4^2 + \dots V_n^2}}{V_1} \tag{20.5}$$

**TABLE 20.6** Maximum Voltage Distortion in Accordance with IEEE Standard 519

| Maximum Distortion % | System Voltage | | |
| --- | --- | --- | --- |
| | Below 69 kV | 69–161 kV | > 161 kV |
| Individual harmonic | 3.0 | 1.5 | 1.0 |
| Total harmonic | 5.0 | 2.5 | 1.5 |
| For periods less than 1 h/day increase limit by 50%. | | | |

where $V_2$, $V_3$, $V_4$, ... $V_n$ are individual rms harmonic voltage components and $V_1$ is the fundamental frequency rms voltage. This is the ratio of the total rms value of the harmonic voltages to the rms value of the fundamental component. The voltage THD is important in industrial power systems because the voltage distortion affects the other loads in parallel with the harmonic producing load. The intent of IEEE Standard 519 is to show that the power supplier is responsible for maintaining the quality of the voltage of the power system. The acceptable voltage distortion limits for different system voltage levels are presented in Table 20.6.

### 20.4.3 Total Demand Distortion (TDD)

The total demand distortion is the total harmonic current distortion. It is defined as:

$$\text{TDD} = \frac{\sqrt{I_2^2 + I_3^2 + I_4^2 + \ldots I_n^2}}{I_{\text{Load}}} \tag{20.6}$$

where $I_2$, $I_3$, $I_4$, ... $I_n$ are the individual rms harmonic current components and $I_{\text{Load}}$ is the maximum load current at the point of common coupling. As per IEEE Standard 519, the customer is responsible for keeping the current harmonic components within acceptable limits. The current harmonic is defined in terms of the total demand distortion based on the customer load demand. Faced with a proliferation of harmonic producing loads, utilities attempt to use IEEE 519 to limit harmonics from individual customers or even individual loads. However,

**TABLE 20.7**   Harmonic Current Limits in %, IEEE Standard 519

| $I_{SC}/I_{LOAD}$ | Harmonic Order | | | | | |
| | < 11 | 11–16 | 17–22 | 23–24 | > 35 | TDD |
|---|---|---|---|---|---|---|
| < 20 | 4.0 | 2.0 | 1.5 | 0.6 | 0.3 | 5.0 |
| 20–50 | 7.0 | 3.5 | 2.5 | 1.0 | 0.5 | 8.0 |
| 50–100 | 10.0 | 4.5 | 4.0 | 1.5 | 0.7 | 12.0 |
| 100–1000 | 12.0 | 5.5 | 5.0 | 2.0 | 1.0 | 15.0 |
| > 1000 | 15.0 | 7.0 | 6.0 | 2.5 | 1.4 | 20.0 |

where $I_{SC}$ = Maximum short circuit current at point of common coupling.
$I_{Load}$ = Maximum demand load current at fundamental frequency.
TDD = Total demand distortion in % of maximum demand.
For conditions lasting more than 1 h/day. For shorter periods increase the limit by 50%.

this approach has limitations because the voltage distortion on the utility system is also a function of the system frequency response characteristic and harmonic sources from all customers. Large customers face stricter limits because they have more impact on voltage distortion. The acceptable TDD is listed in Table 20.7.

### 20.4.4   Frequency Domain Analysis

Filters are added to the power system to improve both the power factor and the harmonic performance. The addition of shunt capacitors introduces resonance peaks in the system. The resonant harmonic number $h$ can be calculated using the equation:

$$h = \sqrt{\frac{MVA_S}{MVAR_C}} \tag{20.7}$$

where $MVA_S$ is the short-circuit rating of the system and $MVAR_C$ is the rating of the shunt capacitor. In a system with many components, the resonant peaks can be predicted using the frequency scanning approach. By this method, one ampere of current is injected at the bus where the harmonic source is connected. The frequency domain characteristics of the system are typically plotted up to 3,000 Hz. If the impedance value at some harmonic $h$ is less than 1.0, then the filters are attenuating currents at that harmonic. If the impedance value is greater than 1.0, then the filters are amplifying the harmonic. A near-zero value on an amplification curve indicates a series resonance (see Figure 20.4). This is the value where a filter branch is tuned to provide maximum attenuation. A sharp maximum amplification curve indicates a parallel resonance (see Figure 20.3). This occurs at a harmonic where the net resonance of a filter branch is capacitive and equal in magnitude to the system or transformer reactance.

### 20.4.5 Acceptable Capacitor Performance

The allowable rms current loading of a capacitor is 180%. Knowing the various current components, $I_{\text{rms}}$ can be calculated as:

$$I_{\text{rms}} = \sqrt{I_1^2 + I_3^2 + I_5^2 + \ldots I_n^2} \qquad (20.8)$$

The rms voltage magnitude $V_{\text{rms}}$ can be calculated knowing the various currents $I_n$ and the reactance components $X_n$. The allowed $V_{\text{rms}}$ for a capacitor is 110% of the nominal voltage. The actual $V_{\text{rms}}$ is:

$$V_{\text{rms}} = \sqrt{(I_1 X_1)^2 + (I_3 X_3)^2 + \ldots (I_n X_n)^2} \qquad (20.9)$$

The allowable kVAR loading of a capacitor unit is 135% of the nominal rating. The actual loading is given by:

$$\text{kVAR} = V_1^2 \omega C + V_3^2 (3\omega) + \ldots V_n^2 (n\omega) C \qquad (20.10)$$

The allowable peak voltage on a capacitor unit is given by 120% of the nominal voltage. The actual peak voltage is given by:

$$V_{\text{pk}} = \sqrt{2} \, (I_1 X_1 + I_3 X_3 + \ldots I_n X_n) \qquad (20.11)$$

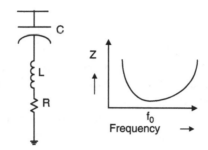

**Figure 20.5** Single-tuned filter with frequency response characteristics.

where the current components $I_n$ and the reactance components $X_n$ are known for a given project.

## 20.5   HARMONIC FILTERS

Filtering the dominant harmonics can reduce the effect of harmonics. There are several filters available to perform this function. The single-tuned notch filter and the high-pass filter are two commonly used devices [1–5].

### 20.5.1   Single-Tuned Filters

A single-tuned or notch filter can be used to filter harmonics at a particular frequency. Figure 20.5 illustrates a common single-tuned notch filter to control a single dominant harmonic component. The impedance characteristics of the filter are also shown in Figure 20.5. The tuned frequency and the operating point may change due to temperature, tolerances, and change in the supply frequency. The single-tuned filter is the simplest device for harmonic control.

### 20.5.2   High-Pass Filter

The frequency of the high-pass filter, the optimal factor $m$, and the MVAR of the capacitor bank are required. A typical high-pass filter and the frequency response are shown in Figure 20.6. As can be seen from the frequency response, the high-pass filter

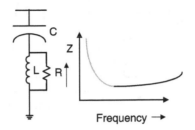

**Figure 20.6** High-pass filter and frequency response characteristics.

reduces the impedance at high harmonic orders to lower values. This filter is more efficient in reducing the harmonics across the entire frequency spectrum. The high-pass filter is not sensitive to the tuned frequency. This type of filter can control harmonic frequencies over a wide range. The resistor produces significant power loss at the fundamental frequency.

### 20.5.3 Multiple Filter Banks

Sometimes there will be several dominant harmonic frequencies in the system. In order to control harmonic frequencies such as the 5th and 7th, single-tuned filters will be used. In order to control the 11th and higher harmonic frequencies, a high-pass filter will be used. Such an arrangement is shown in Figure 20.7. In this example, $L_1$ and $C_1$ are responsible for controlling harmonics at a specific frequency. Similarly $L_2$ and $C_2$ provide harmonic control at another frequency. The $L_3$, $R$, and $C_3$ combination is a high-pass filter that provides harmonic control over a wide range of frequencies.

### 20.6 SYSTEM RESPONSE CONSIDERATIONS

The addition of power factor correction equipment produces parallel resonance between system reactance and the shunt capacitance. In the case of tuned filters, such responses are expected and there is a need to avoid voltage magnifications.

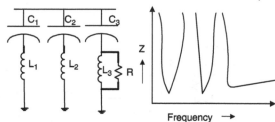

**Figure 20.7** Multiple filters and frequency response characteristics.

## 20.6.1  Resonance Problems

The addition of power factor correction capacitors at low voltage level causes parallel resonance between the source inductance and the capacitance. Harmonic currents close to the parallel resonant frequency are amplified. Higher harmonic currents at the point of common coupling are reduced because the capacitors offer low impedance at high frequencies. At the parallel resonant frequencies, the magnified current can cause excessive voltage distortion and significant harmonic problems within the industrial plant. Consider the simple power system shown in Figure 20.8. The frequency response of the system without shunt capacitors and with 300 kVAR shunt capacitors is shown in Figure 20.9 [6,7]. Without shunt capacitors, the frequency response is a straight line and there is no magnification at any frequency. With shunt capacitors, there is a resonant frequency at 13f that will magnify the harmonics at that

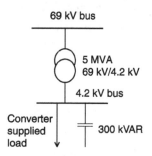

**Figure 20.8** Power system to demonstrate the effect of resonance problem.

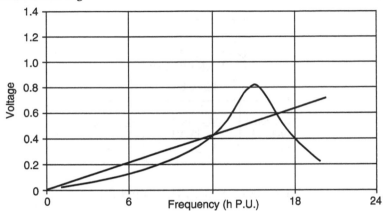

**Figure 20.9** Example power system to show the frequency response without and with 300 kVAR shunt capacitors. (Courtesy of Electrotek Concepts, output from TOP Program.)

frequency. Therefore, with shunt capacitors it is more difficult to meet the harmonic requirements.

### 20.6.2 Power Factor Correction with Tuned Capacitors

Harmonic control and power factor correction are provided simultaneously by installing tuned filters. This prevents voltage magnification at any specific harmonic component from the nonlinear loads. In order to demonstrate this, consider the power system shown in Figure 20.10. The converter load is producing the harmonics. In order to control the harmonics a 5th harmonic filter with 1,200 kVAR and a high-pass filter at the 11th harmonic are used. The corresponding frequency domain analysis is shown in Figure 20.11.

### 20.6.3 Total Harmonic Distortion (THD) of the Voltage

The calculated value of THD of the voltage and the acceptable values are presented in Table 20.8 for all the three cases considered above. The THD of the voltage is acceptable in the base case and with the filters.

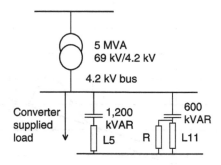

**Figure 20.10** Power system with the tuned filters.

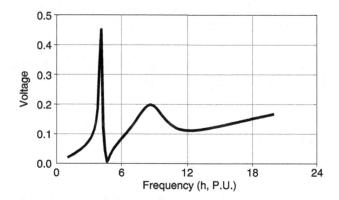

**Figure 20.11** Frequency response characteristics of the power system in Figure 20.10. (Courtesy of Electrotek Concepts, output from TOP Program.)

**TABLE 20.8** Total Harmonic Distortion (THD) at PCC

| Case | Description | THD % | Acceptable THD % |
|------|-------------|-------|------------------|
| 1 | No capacitors present | 2.3% | 5% |
| 2 | With 300 kVAR capacitors | 6.3% | 5% |
| 3 | 5th and 11th harmonic filters | 4.6% | 5% |

**TABLE 20.9** Calculated TDD Components

| Case | $I_5,\%$ | $I_7,\%$ | $I_{11},\%$ | $I_{13},\%$ | TDD % |
|------|------|------|------|------|-------|
| 1 | 1.28 | 0.07 | 22.5 | 14.8 | 6.57 |
| 2 | 1.38 | 2.18 | 42.4 | 42.4 | 11.3 |
| 3 | 1.47 | 0.44 | 2.26 | 1.14 | 5.04 |
| IEEE 519 | 10.00 | 10.00 | 4.50 | 4.50 | 12.00 |

## 20.6.4 Total Demand Distortion (TDD)

From the calculated harmonic currents, the TDD is calculated for all three cases at the point of common coupling. The 5th, 7th, 11th, and 13th harmonic components are dominant and are compared with the allowed values in Table 20.9. The IEEE 519 values are chosen for a ratio of $I_{sc}/I_{Load}$ of 60. From Table 20.9 it can be seen that without shunt capacitors, the individual harmonic components due to dominant 11th and 13th harmonics are above the acceptable limits. With shunt capacitors, the dominant harmonic level increases significantly and is not acceptable. With the 5th harmonic filter and 11th harmonic high-pass filter, all the individual harmonic components and the TDD levels are acceptable.

## 20.7 FILTER DESIGN

The harmonic filter is one of the important components used in the power factor and harmonic control. The main components of a harmonic filter are the capacitors, reactor, and a damping resistance if necessary, as in the case of a high-pass filter. The required kVAR of the capacitor bank can be determined from the given kVA, the existing power factor, and the desired power factor of the load. The design procedure is presented for a simple notch filter and high-pass filter.

Total kVAR of the filter = kVAR
kVAR per phase = kVAR/3
System line-to-line voltage = kV (line)
System voltage per phase = kV (line)/1.732
Select suitable capacitor units from Table 5.2,
    Chapter 5.

Number of cans per phase $=x$
Total number of cans for the bank $=3x$

$$\text{kVAR (delivered)} = \left(\frac{V_{\text{operating}}}{V_{\text{rated}}}\right)^2 \text{kVAR (design)} \qquad (20.12)$$

$$X_c = \frac{kV^2}{MVAR}\,\Omega; \; C = \frac{10^6}{377.7X_c}\,\mu F \qquad (20.13)$$

The resonant frequency $f_0$ of a notch filter is given by:

$$f_0 = \frac{1}{2\pi\sqrt{LC}} \qquad (20.14)$$

where $L$ and $C$ are the inductance and capacitance of the filter.
The inductance of the notch filter is given by:

$$L = \frac{1}{(2\pi)^2(f_0^2)(C)} \qquad (20.15)$$

The fundamental current through the filter $I_1$ is given by:

$$I_1 = \frac{kVAR}{\sqrt{3}\,(kV)} \qquad (20.16)$$

The inductive reactance of the filter reactor is given by:

$$X_L = \omega L \qquad (20.17)$$

The $X_c$ and $X_L$ are related by the harmonic order and are represented by Equation (20.18):

$$X_L = \frac{X_c}{n^2} \qquad (20.18)$$

The use of an inductor in series with a capacitor results in a voltage rise at the capacitor terminals given by:

$$V_c = \left(\frac{n^2}{n^2 - 1}\right)V_{\text{sys}} \qquad (20.19)$$

where $n=$ Tuned impedance harmonic number of the
frequency
$V_{sys}=$ System line-to-line voltage, kV
$V_c=$ Capacitor line-to-line voltage, kV

The fundamental current through a harmonic filter is given by:

$$I_1 = \frac{V(\text{line-line})}{\sqrt{3}\,(X_c - X_L)} \qquad (20.20)$$

The $n$th harmonic current through a filter can be calculated as:

$$If(n) = \frac{I_1}{n} \qquad (20.21)$$

The rms current through the filter is given by:

$$I_{rms} = \sqrt{I_1^2 + \ldots I_n^2} \qquad (20.22)$$

The fundamental voltage component across a capacitor is given by:

$$V_{c1} = I_1 \times X_c \qquad (20.23)$$

The $n$th harmonic voltage across the $n$th harmonic filter is given by:

$$V_{cn} = I_n \frac{X_c}{n} \qquad (20.24)$$

The peak voltage $V_{pk}$ across a capacitor unit can be calculated as:

$$V_{peak} = V_{c1} + \ldots V_{pn} \qquad (20.25)$$

The rms voltage $V_{rms}$ across a capacitor unit can be calculated as:

$$V_{rms} = \sqrt{V_{c1}^2 + \ldots V_{cn}^2} \qquad (20.26)$$

The filter kVAR can be calculated knowing the voltage and the current components:

$$\text{kVAR} = V_{\text{rms}} \times I_{\text{rms}} \qquad (20.27)$$

The presence of the filter reactor changes the effective MVAR delivered. The output MVAR of the filter is:

$$\text{MVAR}_{\text{filter}} = \frac{\text{kV}^2}{(X_{\text{c}} - X_{\text{L}})} \qquad (20.28)$$

where $X_{\text{L}}$ is the inductive reactance of the filter reactor. The kVAR of the inductance is given by:

$$\text{kVAR of the Inductance} = \frac{I^2 \omega L}{1000} \qquad (20.29)$$

where $L$ is in henrys. The connection can be ungrounded wye in order to use the unbalance detection scheme. The reactor rated current is 1.5 times the maximum rms current in order to allow protective coordination with a suitable overcurrent relay.

**Example 20.1**

Design a simple notch filter for a 4.16 kV, three-phase, 60 Hz system where harmonics are produced due to a converter load. The power factor correction approach indicates a need for 460 kVAR shunt capacitors. What is the delivered kVAR? The harmonic is produced due to the presence of a 12-pulse converter supplied load.

Solution

> Total kVAR of the filter (round it) = 600 kVAR
> kVAR per phase (600/3) = 200 kVAR
> System line-to-line voltage = 4.16 kV (line)
> System voltage per phase (4.16 kV/1.732) = 2.4 kV

Select 100 kVAR, 2.77 kV capacitor cans from the capacitor table.
Number of cans per phase $= 2$
Total number of cans for the bank $= 6$

$$\text{kVAR (delivered)} = \left(\frac{2.4\,\text{kV}}{2.77\,\text{kV}}\right)^2 600 = 451\,\text{kVAR}$$

$$X_c = \frac{4.16\,\text{kV}^2}{0.451\,\text{MVAR}} = 38.37\,\Omega; \ C = \frac{10^6}{377.7 \times 38.37} = 69\,\mu\text{F}$$

The dominant harmonic of the 12-pulse converter is the 11th. The desired frequency of the notch filter is 10.7 F (642 Hz). The inductance of the notch filter is given by:

$$L = \frac{10^6}{(2\pi)^2(642)^2(69)} = 0.877\,\text{mH}$$

$I$ (filter), $[451\,\text{kVAR}/(1.732 \times 4.16\,\text{kV})] = 62.5\,\text{A}$
Inductor current $(1.5 \times 62.5\,\text{A}) = 93.9\,\text{A}$
kVAR of the inductor $(93.9^2 \times 377.7 \times 0.877 \times 10^{-3}) = 2.916\,\text{kVAR}$
Inductive reactance $(377.7 \times 0.877 \times 10^{-3}) = 0.33\,\Omega$

$$I_1 = \frac{4{,}160}{\sqrt{3}(38.37 - 0.33)} = 63.1\,\text{A}$$

$I$ $(10.7f) = (63.1/10.7) = 5.9\,\text{A}$

$$I_{\text{rms}} = \sqrt{63.1^2 + 5.9^2} = 63.3\,\text{A}$$

Allowed current through the capacitor $(200\,\text{kVAR}/2.4\,\text{kV})$ $= 72.2\,\text{A}$
$V_{c1}$, $(63.1\,\text{A} \times 38.37\,\Omega) = 2{,}412\,\text{V}$
$V_c(10.7f)$, $(63.1\,\text{A} \times 38.37\,\Omega/10.7) = 226.3\,\text{V}$
Peak voltage $V_{\text{pk}} = 2{,}421 + 226.3 = 2{,}647\,\text{V}$
The allowed peak voltage $(2{,}400 \times 1.414) = 3{,}917\,\text{V}$

$$V_{\text{rms}} = \sqrt{2{,}412^2 + 226.3^2} = 2{,}431\,\text{V}$$

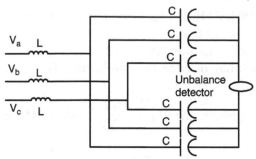

**Figure 20.12**   One-line diagram of the filter circuit.

     Delivered kVAR of the unit $(2{,}412\,\text{V} \times 63.4\,\text{A}/1000) =$
     154.2 kVAR
     Resistance for the high-pass filter $R = 2 \times 10.7 \times 0.33) =$
     7.1 Ω

The one-line diagram of the filter is shown in Figure 20.12.
Use 100 MFD, 2.77 kV capacitors, and a total of six cans.
The inductor is 0.877 mH, 2.916 kVAR, and a total of three
numbers. The BIL of the capacitor bank and the inductor is
75 kV. The filter bank is connected in ungrounded split-wye.
The neutral unbalance current can be sensed in this connection using a current transformer. For the high-pass filter, add a
7.1 Ω resistance across the filter reactance. This is not shown
in the diagram.

### Example 20.2

The measured load current at the 4.16 kV level is 500 A. The
measured 5th and 7th harmonic currents due to a converter
load is 80 A and 60 A, respectively. Calculate the following:
(a) $I_{\text{rms}}$ at the 4.16 kV level; (b) TDD; (c) distortion of 5th harmonic component; (d) distortion of 7th harmonic component.

Solution

    (a)  $I_{\text{rms}}$ at 4.16 kV level

$$I_{\text{rms}} = \sqrt{I_1^2 + I_5^2 + I_7^2} = 509.9 \text{ A}$$

(b)   TDD at 4.16 kV level

$$\text{TDD} = \left(\frac{\sqrt{80^2 + 60^2}}{500}\right)100 = 2.0\%$$

(c)   Distortion of 5th harmonic component

$$\text{Distortion of 5th harmonic} = \left(\frac{I_5}{I_f}\right)100 = \left(\frac{80}{500}\right)100 = 16\%$$

(d)   Distortion of 7th harmonic component

$$\text{Distortion of 7th harmonic} = \left(\frac{I_7}{I_f}\right) \times 100 = \left(\frac{60}{500}\right)100 = 12\%$$

### 20.7.1   Practical Filter Installations

A typical filter installation consists of capacitor banks, fuse cutouts with fuses, a reactor, and surge arresters with proper mounting arrangements. An example filter installation is shown in Figure 20.13 [8]. The capacitor banks and the reactors can

**Figure 20.13**   A typical filter installation at a chemical factory. (Courtesy of IEEE, Reproduced from Reference [8].)

be seen in the rack-mounted installation. The following three photographs shown in Chapter 7 illustrate the pole-mounted, metal-enclosed, and open structure harmonic filters.

> Figure 7.4 A pole-mounted harmonic filter bank
> Figure 7.7 A metal-enclosed harmonic filter bank    ·
> Figure 7.5 An open structure harmonic filter bank

These harmonic filters can be operated using oil switches, vacuum circuit breakers, or vacuum switches.

## 20.8   SUMMARY AND CONCLUSIONS

In this chapter, the harmonic sources and harmonic amplitudes are identified from the system's operational point of view. The acceptable voltage distortion and total demand distortion are presented from IEEE Standard 519. The effect of the shunt capacitor alone on the power system is to introduce new harmonic resonant frequency due to the interaction with source impedance. A tuned harmonic filter is used to control both the power factor and harmonics within acceptable limits. The approach to designing a notch filter is shown.

Filter Capacitor Specifications

| | |
|---|---|
| System line-to-line voltage | kV rms |
| System phase voltage | kV rms |
| Supply frequency | Hz |
| Capacitor nominal voltage | kV rms |
| Capacitor peak voltage | kV peak |
| Filter tuned frequency | Hz |
| Nominal capacitance | μ F |
| Nominal reactive power | MVAR |
| Filter capacitor current | A |
| Harmonic current | A |
| Energization transient | kV peak |

Filter Reactor

| | |
|---|---|
| Rated reactance | mH |
| Tolerance | % |
| Fundamental current | A |

| Harmonic current | A |
|---|---|
| Lightning impulse withstand | kV |
| Energization transient | kV peak |

## PROBLEMS

20.1. Name a few harmonic-producing loads. What makes them nonlinear?

20.2. What is the difference between series resonance and parallel resonance?

20.3. Is there any relation between voltage and current harmonics?

20.4. The industry standard places a limit on individual harmonics and total harmonics. This is applicable both for current and voltage harmonics. Explain why this is necessary.

20.5. The measured fundamental, 5th, and 7th harmonic components of a nonlinear load are 900 A, 100 A, and 40 A, respectively. Calculate the total demand distortion (TDD) and distortion due to 5th and 7th harmonics individually.

20.6. The short-circuit impedance of the source is always used in the harmonic analysis. How will the harmonic change for (a) high source impedance and (b) low source impedance?

20.7. There are 5th and 7th harmonic problems in a power supply connected to a nonlinear load. One of the management decisions based on budget limits is to apply one filter at 5th harmonic level. Will that be sufficient to mitigate the harmonic problem? How can you ensure that the single-tuned filter will solve this harmonic problem satisfactorily?

20.8. Design a notch filter for a 13.8 kV, 60 Hz, three-phase system. The harmonics are produced due to a nonlinear load of 2,000 kVA. The existing facility power factor of 0.72 has to be corrected to 0.95. Select a suitable 5th harmonic notch filter.

# REFERENCES

1. IEEE Standard 519 (1996), Recommended Practices and Requirements for Harmonic Control in Electric Power Systems.

2. Gonzalez, G. A., McCall, J. C. (1987). Design of filters to reduce harmonic distortion in industrial power systems, *IEEE Transactions on Industry Applications*, IA–23(3), 504–511.

3. Hammond, P. W. (1988). A harmonic filter installation to reduce voltage distortion from solid state converters, *IEEE Transactions on Industry Applications*, 24(1), 53–58.

4. Natarajan, R., Nall, A., Ingram, D. (1999). A harmonic filter installation to improve power factor and reduce harmonic distortion from multiple converters, *Proceedings of the 1999 American Power Conference*, Chicago, 6, 680–685.

5. IEEE Standard 18 (1992), IEEE Standard for Shunt Power Capacitors.

6. Electrotek Concepts (1997). SuperHarm — The Harmonic Analysis Program, Electrotek Concepts, Knoxville, TN.

7. Electrotek Concepts (1995). TOP — The Output Processor, Electrotek Concepts, Knoxville, TN.

8. Buddingh, P. C. (2003). Even harmonic resonance — An unusual problem, *IEEE Transactions on Industry Applications*, 39(4), 1181–1186.

# 21

## CAPACITOR TRANSFORMER TRANSIENTS

### 21.1 INTRODUCTION

Use of shunt capacitors in the power system for power factor correction results in significant economic savings. Some of the additional effects of the application of shunt capacitors, such as harmonics, voltage magnification, overvoltage problems, ferroresonance, and related issues, are identified in other chapters. One effect of capacitor switching on a power transformer is to produce high frequency oscillations and in some cases, equipment failure. In order to understand this problem in a system, detailed knowledge of the switching pattern and time/frequency results is needed. In certain applications, the transformer and the shunt capacitors are switched together. The arc furnace is one such load where the furnace transformer and the shunt capacitors are energized or de-energized together. Such a configuration is found to produce harmonic resonance [1]. In EHV and HV systems, the shunt capacitors are installed at the substation and the lines are terminated through stepdown transformers. The switching of these capacitors causes traveling waves and phase-to-phase overvoltages at the transformer termination [2,3]. The transients due to capacitor energizations with a transformer are discussed in this chapter.

## 21.2 TRANSFORMER CAPACITOR ENERGIZATION

Transformers used in industrial facilities such as an arc furnace installation are switched many times during a 24-h period. Under normal conditions, the inrush current does not cause problems. However, overvoltages may occur if the system is sharply tuned at one dominant harmonics produced by the inrush current. This is possible when a large power factor correction capacitor bank is installed at the secondary voltage level of the transformer that is being energized. If the energization occurs 25–50 times a day, the capacitor may be unable to withstand the overvoltages for an extended period of time. A typical arc furnace circuit is shown in Figure 21.1. The power factor correction capacitors interact with the system inductance to produce harmonic resonance at some frequency. That resonant frequency number $h$ can be calculated using the following equation:

$$h = \sqrt{\frac{X_c}{X_L}} \tag{21.1}$$

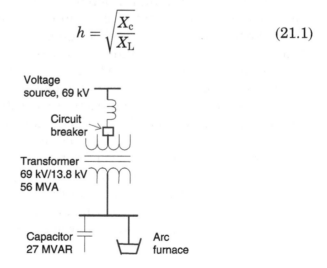

**Figure 21.1**   Arc furnace with capacitor installation.

where $X_c$ and $X_L$ are the reactances of the capacitor bank and the combined reactance of the transformer and the source. The effect of resonance on the energization transient is to amplify the harmonics and produce overvoltages. In order to evaluate the harmonics and overvoltages without and with shunt capacitors, the time domain simulations are performed using the EMTP. The following cases are studied:

1. Time domain plots without shunt capacitors (Figure 21.2).
2. Time domain plots with shunt capacitors (Figure 21.3).
3. Frequency domain plots with shunt capacitors (Figure 21.4).

Figure 21.2 shows the voltage waveform when the transformer alone is energized. The transformer inrush current

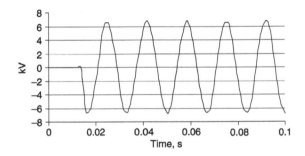

**Figure 21.2** Voltage waveform when energizing the transformer alone.

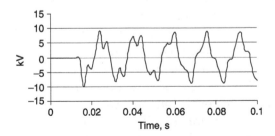

**Figure 21.3** Voltage waveform when energizing the transformer and capacitor bank together.

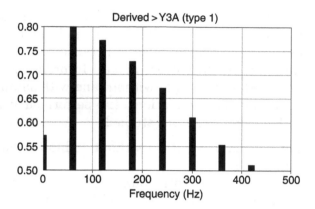

**Figure 21.4**   Harmonic analysis of the voltage waveform presented in Figure 21.3.

**TABLE 21.1**   Dominant Harmonics in the Voltage Waveform in Figure 21.3

| Frequency, Hz | Harmonic Number | Harmonic Magnitude, P.U. |
|---------------|-----------------|--------------------------|
| 60            | 1f              | 0.80                     |
| 120           | 2f              | 0.77                     |
| 180           | 3f              | 0.73                     |
| 240           | 4f              | 0.67                     |
| 300           | 5f              | 0.62                     |

will produce a small amount of voltage distortion. The maximum line-to-ground transient overvoltage is 1.05 P.U. Figure 21.3 shows the voltage waveform during the energization of both the transformer and the capacitance. In this case, the voltage waveform is distorted and the maximum magnitude of the voltage is 1.54 P.U. The distorted voltage is dominant with several harmonics. Figure 21.4 shows the Fast Fourier Transform (FFT) of the voltage waveform shown in Figure 21.3. The dominant harmonics are presented in Table 21.1. As the table shows, the harmonic magnitudes in the voltage waveform are significant. Although 1.54 P.U.

overvoltage is not a concern to the insulation of the electrical system, the duration and frequency of oscillation of the harmonics can cause stress on individual capacitor units and premature failure of the transformer.

## 21.3 CAPACITOR SWITCHING AND TRANSFORMER TRANSIENTS

Energization of the shunt capacitors with long line and transformer termination can produce transients at the remote transformer. Such transients will produce phase-to-phase overvoltages of high frequency. The energization of a capacitor bank at the substation produces traveling waves along the transmission lines toward the transformer. These waves reflect at the transformer with the same sign doubling the incoming waves. It is possible to obtain transients approaching twice the peak system voltage with opposite polarity, resulting in a phase-to-phase transient approaching four times the peak phase-to-ground voltage. These transients can easily exceed the phase-to-phase withstand characteristics of three-phase transformers. This topic has been addressed in many references [2,3]. Some of the results are discussed below.

In Reference [2], a 230 kV phase shifting transformer was installed at the end of a 34.5 mi transmission line. A one-line diagram of the simplified system is shown in Figure 21.5. A 50 MVAR shunt capacitor bank is located at the Klamath

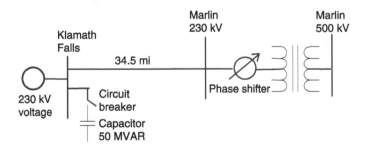

**Figure 21.5** One-line diagram of the system where a phase shifter failed, followed by the energization of the 50 MVAR capacitor bank.

Falls substation. The phase shifter was protected by 180 kV
lightning arresters. The capacitor bank was switched using
the circuit breaker, which was not equipped with closing resistances. The phase shifter failed from what appeared to be
a phase-to-phase insulation failure between B and C phases.
The surge arrester did not conduct during the energization of
the capacitor bank.

In Reference [2], a 115 kV transmission line is shown with
transformer termination at the Riverton 115 kV substation.
A one-line diagram of the simplified system is shown in
Figure 21.6. The 15 MVAR capacitor bank at the Pilot Butte
substation was installed. The circuit breaker 762 was closed
first. Using circuit switcher 764, the capacitor bank was energized. The circuit breaker 762 tripped via the fault bus protection relay. Upon investigation, it was determined that the fault
bus relay was set too sensitively. Coincident with the capacitor
bank switching, the Riverton transformer failed. Inspection of
the transformer indicated a turn-to-turn failure in the B-phase
series-winding group closest to the 115 kV bushing. The transformer manufacturer indicated that the turn-to-turn failure
was due to the overvoltage condition. In addition to the transformer failure, evidence of an overvoltage condition was indicated by a failed internal protective gap on the B-phase of the
CCVT at Riverton.

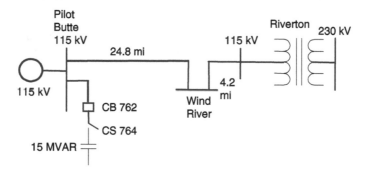

**Figure 21.6** One-line diagram of the system where the Riverton
transformer failed followed by the energization of the 15 MVAR
capacitor bank.

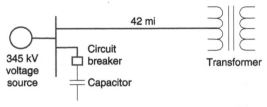

**Figure 21.7** Example circuit diagram used in the study.

An example circuit configuration, which produces such overvoltages, is shown in Figure 21.7. The 345 kV substation is connected to a 42-mi-long transmission line terminating at a three-phase transformer. Switched shunt capacitors are installed at the 345 kV substation. This case is simulated using the EMTP program to examine the voltages at the transformer termination. The energization of the capacitor bank at the substation produces overvoltage and the phase A and C voltages in time domain at the transformer termination are shown in Figures 21.8 and 21.9, respectively. The maximum voltages observed at the transformer termination in phase A and C are 750 kV (2.66 P.U.) and 700 kV (2.48 P.U.) respectively. The phase A to phase C voltage is shown in Figure 21.10. The maximum voltage between phase A and C is 1,250 kV (4.43 P.U.). The allowable phase-to-phase switching surge insulation level for the 345 kV transformer is 1,050 kV or 3.73 P.U. The transient overvoltage between phase A and phase C exceeds the allowable insulation level.

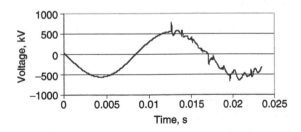

**Figure 21.8** Voltage waveform in phase A at the transformer termination.

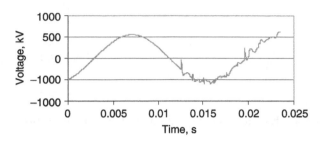

**Figure 21.9** Voltage waveform in phase C at the transformer termination.

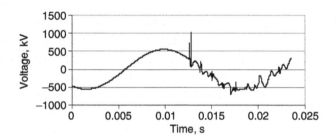

**Figure 21.10** Voltage waveform between phases A and C at the transformer termination.

## 21.4  REDUCING THE VOLTAGE TRANSIENTS

There are several approaches available to reduce the effect of transformer transients. Some of the approaches are discussed below.

### 21.4.1  Closing Resistors

The circuit breakers used to perform capacitor switching can be equipped with closing resistors. An example of a closing resistor used in a circuit breaker in one phase is shown in Figure 21.11. In this arrangement, the auxiliary circuit breaker is connected across the main circuit breaker contacts. A suitable closing resistor is connected in the auxiliary circuit breaker circuit. Figure 21.11 shows two equal resistors. This model is used in the EMTP simulation. The auxiliary contacts are closed first and the main circuit breaker contacts are closed

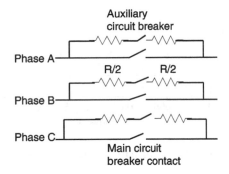

**Figure 21.11**  Illustration of a closing resistance.

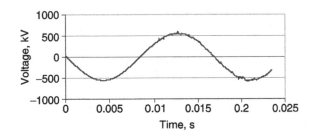

**Figure 21.12**  Voltage waveform at phase A with closing resistor.

after a time delay. The closing resistor in the auxiliary circuit breaker controls the inrush current. The auxiliary contacts are opened after a short duration. The effect of using a closing resistor in the circuit breaker is simulated and the voltage waveforms in phase A, phase C, and the voltage between phases A and C are shown in Figures 21.12, 21.13, and 21.14, respectively. The voltages in phases A, C, and between phases A and C are shown in Table 21.2. These transient overvoltages at the transformer location are acceptable.

### 21.4.2  Controlled Closing

With certain circuit breakers, it is possible to control the point of closing of the poles in order to minimize the transient overvoltages. If the contacts close when the voltage across them

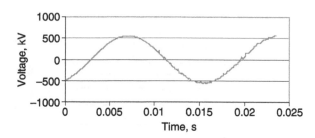

**Figure 21.13**  Voltage waveform at phase C with closing resistor.

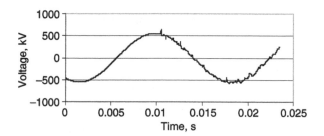

**Figure 21.14**  Voltage waveform between phase A and phase C with closing resistor in the circuit breaker.

**TABLE 21.2**  Transient Overvoltages with Closing Resistor

| Location | Voltage, kV | Voltage, P.U. with Closing R | Voltage, P.U. without Closing R |
|---|---|---|---|
| Phase A | 600 | 2.12 | 2.66 |
| Phase C | 580 | 2.06 | 2.48 |
| Phase-to-phase (A – C) | 620 | 2.19 | 4.43 |

is nearly zero, the transients will be reduced. An example of a synchronized control switching circuit breaker suitable for a 230 kV, 40 kA, three-phase application is shown in Figure 21.15.

### 21.4.3  Staggered Closing of the Circuit Breaker Contacts

In grounded wye capacitor banks, it is possible to reduce the phase-to-phase transients by staggering the closing times to

**Figure 21.15** Synchronized switching SF$_6$ circuit breaker 120 kV, three-phase, 40 kA. (Courtesy of Mitsubishi Electric Products Inc., Warrendale, PA.)

ensure that no two phases close together. This does not help in the case of ungrounded wye banks and also does not reduce the internal resonance condition.

### 21.4.4 Capacitor Bank Reactors

A reactor in each phase of the capacitor bank can also act to reduce the transient voltage. Such series reactors are commonly used in shunt capacitor banks to reduce the inrush currents and transient overvoltages.

### 21.4.5 Surge Arresters

Surge arresters at the transformer terminal will also act to reduce the transient. Using the required MOV arresters, the line-to-ground voltage on each phase can be limited to nearly 2.0 P.U. For the worst case, the phase-to-phase condition of +2.0 per unit on one phase and −2.0 per unit on another phase, the phase-to-phase voltage can reach 4.0 per unit. If surge arresters of lowest rating were applied phase to phase, protection can be provided to 87% of that for phase-to-ground arresters. This would reduce the maximum transient from 4.0 per unit to 3.5 per unit.

### 21.4.6    Capacitor Bank Connection

For switching devices that are random in closing times, ungrounded wye capacitor banks will tend to have a higher percentage of phase-to-phase voltages while grounded wye banks will have a significantly lower percentage. For a grounded wye bank, the transients on each phase are nearly independent of each other. A high phase-to-phase voltage will occur on a grounded wye bank.

On an ungrounded wye bank, the first phase to close does not generate a transient. The second phase will then generate a significant phase-to-phase voltage depending on the exact closing time. This is the main reason why ungrounded wye banks will tend to result in high phase-to-phase voltages as compared to grounded banks.

### 21.5    CONCLUSIONS

The transients due to the energization of a transformer and shunt capacitor bank are identified. The energization voltages contain significant harmonics and the voltage waveform is distorted. Therefore, there is need to control such energizations using a closing resistor in the circuit breaker or similar approaches.

The energization of a capacitor bank at a substation can produce significant overvoltages at the transformer terminated through a transmission line.

The transients produced at the terminated transformer can be reduced by using a closing resistor in the circuit breaker, controlled closing, staggered closing, capacitor bank reactors, and surge arresters at the transformer terminals. Some capacitor bank connections, such as wye grounded, produce limited transients.

Other conditions not discussed in this chapter are the transformer terminated through a transmission line with temporary faults. Single or double line-to-ground faults tend to increase the voltage on the unfaulted phase of the transformer. If the line is compensated using shunt capacitor banks, then the voltage rise will be higher and ferroresonance may be likely to occur. In such faulted cases, the damage may occur to both the transformer and capacitor equipment. In the case of one open conductor in the transmission line, the voltage on the open conductor can rise to unacceptable values. There may be interaction between the shunt capacitors and the magnetizing reactance of the transformer causing ferroresonance. The transformer connection on both the ends of the transmission line plays an important role in such cases. The grounding of the transformer and the capacitor bank is another important factor.

## PROBLEMS

21.1. Energization of a transformer and capacitor bank together produces damaging transients in installations such as an arc furnace. Is it possible to energize the transformer first and then energize the capacitor bank?

21.2. How can you reduce the transients due to the energization of a transformer and capacitor bank together?

21.3. It is seen that the energization of a capacitor bank produces transients at the transformer terminated at the end of the transmission line. Is there any other circuit configuration to avoid this condition?

21.4. What are the approaches available to reduce the transients at the transformer termination due to the energization of the capacitor banks?

21.5. Are there any similar line/transformer/capacitor configurations that can produce damaging overvoltages?

## REFERENCES

1. Witte, J. F., DeCesaro, F. P., Mendis, S. R. (1994). Damaging long-term overvoltages on industrial capacitor banks due to transformer energization inrush currents, *IEEE Transactions on Industry Applications*, 30(4), 1107–1115.

2. Bayless, R. S., Selman, J. D., Truax, D. E., Reid, N. E. (1988). Capacitor switching and transformer transients, *IEEE Transactions on Power Delivery*, 3(3), 349–357.

3. Mikhail, S. S., McGranaghan, M. F. (1986). Evaluation of switching concerns associated with 345 kV shunt capacitor applications, *IEEE Transactions on Power Delivery*, 1(2), 231–240.

# 22

## CAPACITOR SWITCHING

### 22.1 INTRODUCTION

It is common practice to install shunt capacitors to improve the power factor and voltage profile at all voltage levels in the power system. It is also a well-accepted practice to improve the power factor of the industrial systems using local capacitor banks. If the industrial load is fed from converter equipment, then notch filters are used to control the power factor and harmonics simultaneously. These shunt capacitor banks are switched in and out as needed. The switching operations include energizing, de-energizing, fault clearing, backup fault clearing, and reclosing. Sometimes a restrike occurs due to excessive voltage across the circuit breaker blades. Further, in the capacitor banks there may be bus fault conditions responsible for significant outrush current. There is also energization of a capacitor bank when another capacitor bank is charged, i.e., the back-to-back switching. Sometimes the EHV or HV systems are connected to the low voltage system using stepdown transformers. It is known that the voltage magnification will occur in the low voltage capacitor banks at certain conditions when the switched HV capacitors are significantly larger in size. It can be seen that there are several technical issues involved in switching of the capacitor banks.

Such capacitor switching transients are analyzed in References [1–3]. The switching studies are performed using the Electro Magnetic Transients Program [4].

## 22.2 SWITCHING OPERATIONS

Cases of energization, de-energization, fault clearing, backup fault clearing, reclosing, and restriking are examined for a radial power system. In this study, a 230 kV, three-phase, 60 Hz source is used with a peak voltage of 188.8 kV. The impedance of the radial line is $(0.10 + 1.02)$ $\Omega$/phase. The shunt capacitance is $3\,\mu F$/phase. The size of the shunt capacitor is 60 MVAR. The circuit breaker is represented as a time-dependent switch that is opened or closed as needed.

### 22.2.1 Energization

The circuit diagram used in this study is shown in Figure 22.1. In order to show the overvoltage transients due to the energization of a capacitor bank, the circuit breaker is closed at a predetermined time. The time domain waveform at the capacitor location is shown in Figure 22.2 for phase A. The maximum overvoltage during the energization is 2.1 P.U. and the frequency of oscillation is 588 Hz. The inrush current magnitude and frequency of the inrush current during energization of the capacitor bank are given by:

$$I_{pk} = \frac{V}{Z_s} \tag{22.1}$$

where

$$Z_s = \sqrt{\frac{L}{C}} \tag{22.2}$$

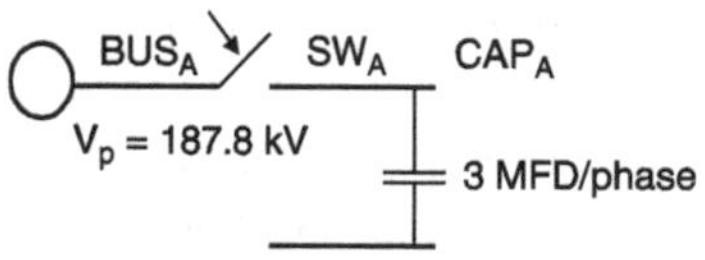

**Figure 22.1**   One-line diagram of the system for energization study for phase A.

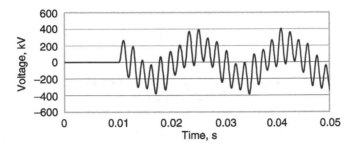

**Figure 22.2** Time domain waveform of the voltage at the capacitor during energization.

> $V =$ Peak system voltage (line-to-ground), kV/phase
> $L =$ Circuit inductance, H/phase
> $C =$ Circuit capacitance, F/phase
> $Z_s =$ Circuit surge impedance, Ω/phase

The frequency of oscillation ($f_0$) is given by:

$$f_0 = \frac{1}{2\pi\sqrt{LC}} \tag{22.3}$$

**Example 22.1**

For the 230 kV system the source impedance is 9.5 Ω/phase. The shunt capacitance is 3 MFD/phase. Calculate the magnitude and frequency of oscillation of the inrush current if the capacitor banks are energized.

Solution

> $V$ (line-line) $= 230\,$kV
> $X = 9.5\,$Ω/phase
> $C = 3\,$MFD/phase
> $L = 9.5/377.7 = 0.025\,$H/phase

$$Z_s = \sqrt{\frac{L}{C}} = \sqrt{\frac{0.025}{3 \times 10^{-6}}} = 91.3\,\Omega/\text{phase}$$

$$I_{\text{pk}} = \frac{V_{\text{pk}}}{Z_s} = \frac{230\,\text{kV} \times \left(\sqrt{2}/\sqrt{3}\right)}{91.3\,\Omega} = 2.07\,\text{kA}$$

$$f = \frac{1}{2\pi\sqrt{0.025 \times 3\,\mu\text{F}}} = 581\,\text{Hz}$$

The calculated frequency using the EMTP program is 588 Hz as per Figure 22.2.

### 22.2.2 De-Energization

The circuit diagram used in the study is shown in Figure 22.3. The circuit breaker is opened at a specific time and the electrical circuit opens at respective current zeros in various phases. The voltage wave at the open end of the line is presented in Figure 22.4. It can be seen that the voltages go to the DC mode and the charges stay on the lines. This is due to the presence of the shunt capacitance at the end of the open

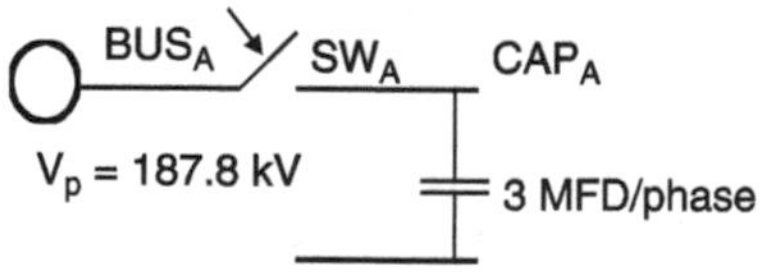

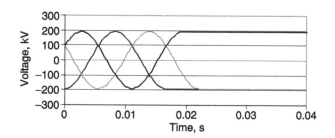

**Figure 22.3** One-line diagram of the system for de-energization study for phase A.

**Figure 22.4** Voltage waveform at the capacitance during de-energization.

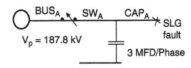

**Figure 22.5** TRV waveforms at the switch during de-energization.

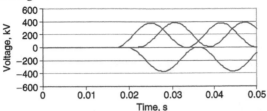

**Figure 22.6** One-line diagram of the system for fault clearing study for phase A.

line. Another waveform of importance is the transient recovery voltage (TRV) across the circuit breaker. The TRV waveforms are shown in Figure 22.5. The maximum overvoltage and the TRV during de-energization are 1 and 2 P.U., respectively. If the TRV magnitudes exceed the allowed ratings of the circuit breaker, then there may be a restrike at the circuit breaker blades.

### 22.2.3 Fault Clearing

The circuit diagram used in the study is shown in Figure 22.6. It is assumed that a single line-to-ground fault occurs in phase A. The circuit breaker is opened and the voltage waveforms are studied. The phase voltages at the end of the line are illustrated in Figure 22.7. It can be seen that the voltage of the faulted line is zero and the unfaulted lines have a maximum of 1.0 P.U. The circuit breaker TRV voltage waveforms are shown in Figure 22.8. The maximum TRV is 2.3 P.U. The frequency of oscillation of the TRV is 5,000 Hz.

### 22.2.4 Reclosing

This is the process of reconnecting a three-phase source to the transmission line with trapped charges in the line by closing

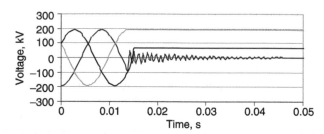

**Figure 22.7**  Time domain waveforms at the end of the line during fault clearing.

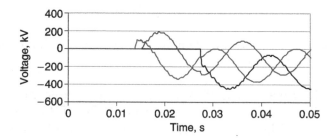

**Figure 22.8**  TRV waveforms at the switch during fault clearing.

the circuit breaker. In this case, there is a shunt capacitor bank connected at the end of the open line. The one-line diagram used in the study is shown in Figure 22.1. Such an operation is performed in order to minimize the time of discontinuity in the service. In this case, the circuit breaker is assumed to be closed and is opened at 10 ms. This leaves the residual charges in the line. The circuit beaker is reclosed at 20 ms. The voltage at the source is presented in Figure 22.9. The maximum overvoltage magnitude due to the presence of trapped charges is 2.5 P.U. and the frequency of oscillation is approximately 400 Hz.

## 22.2.5  Backup Fault Clearing

Consider two transmission lines connected to the same source. If there is a fault at the open end of the line and if the primary circuit breaker fails to clear, then the backup circuit breaker is operated to clear the fault. Such an operation is called backup

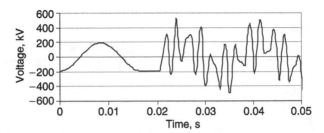

**Figure 22.9** Voltage waveforms at the source during reclosing for phase A.

fault clearing and takes place at a delayed clearing time. The switching surge results are similar to the fault clearing cases and are not shown.

### 22.2.6 Restriking

Sometimes during the de-energization of the capacitive circuits an arc is re-established if the voltage between the circuit breaker blades exceeds the breakdown voltage. The worst case restriking occurs one half cycle after initial clearing, when the system voltage is at a peak, with polarity opposite to that of the trapped charge on the capacitor. The simulated waveforms are presented for the circuit shown in Figure 22.3. The restrike is simulated using a voltage-dependent switch, which will close at a specified overvoltage. The voltage at the source during restrike is shown in Figure 22.10. The circuit breaker

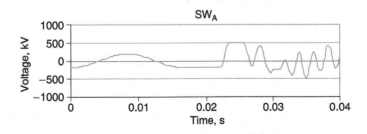

**Figure 22.10** Voltage waveforms at the source during restriking for phase A.

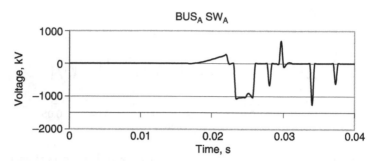

**Figure 22.11** TRV across the circuit breaker during restriking for phase A.

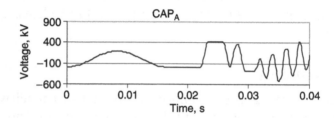

**Figure 22.12** Voltage waveforms at the capacitor during restriking for phase A.

TRV is shown in Figure 22.11. The capacitor voltage is shown in Figure 22.12. The restriking transient produces an overvoltage of 3.2 P.U. in the phase, 2.2 P.U. in the capacitor circuit, and a TRV of 6.4 P.U. in the circuit breaker. These are significantly higher overvoltages that will produce severe stress on various equipment, including the surge arrester.

### 22.2.7  Prestrike

During energization of the capacitor banks, an arc can establish even before the physical circuit breaker blades are closed. This phenomenon is called prestrike. During the prestrike, normal high frequency current flow occurs. When the circuit breaker blades actually establish physical contact, the transients are much higher than the ordinary energization transients. Such prestrikes are well known in vacuum circuit

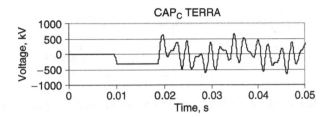

**Figure 22.13** Voltage at the capacitor bank during prestrike for phase C.

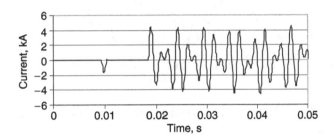

**Figure 22.14** Current through the capacitor during prestrike for phase C.

breakers. An example is presented based on the circuit shown in Figure 22.1. The capacitor voltage and the capacitor current are shown in Figures 22.13 and 22.14, respectively. The peak transient voltage in the capacitor circuit has increased from 2.0 P.U. to 3.28 P.U., which will result in a surge arrester operation. The capacitor inrush current also is significantly higher, which may damage the capacitor bank.

## 22.2.8 Back-to-Back Switching

Energizing a capacitor bank with an adjacent capacitor bank already in service is known as back-to-back switching. High magnitude and frequency currents can be associated with back-to-back switching. The current must be limited to acceptable levels for switching devices and current transformer burdens. Usually, series reactors are used with individual capacitor banks to limit the current magnitude and frequency.

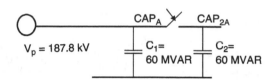

**Figure 22.15**  Back-to-back capacitor switching for phase A.

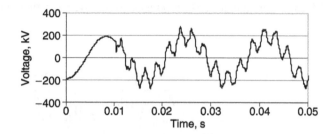

**Figure 22.16**  Voltage waveform at the source during back-to-back switching.

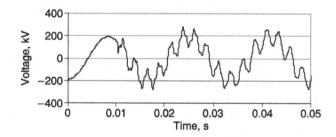

**Figure 22.17**  Voltage waveform at the 230 kV capacitor in service.

Figure 22.15 illustrates the simplified circuit that can be used to analyze the inrush currents during back-to-back switching without current limiting series reactors. The voltage waveforms at the source, at the 230 kV capacitor in service, and at the capacitor bank to be connected are shown in Figures 22.16, 22.17, and 22.18, respectively. The peak voltages at the source, existing capacitor in service, and the switched capacitor are 1.48 P.U. Note that capacitors $C_1$ and $C_2$ are equal in size.

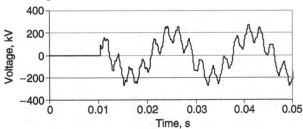

**Figure 22.18** Voltage waveform at the 230 kV capacitor to be connected.

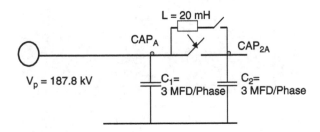

**Figure 22.19** Back-to-back switching with series reactor for phase A.

### 22.2.8.1 With Series Reactor in the Capacitor Bank Circuit

In order to control the inrush current during the energization of the second bank, when the first bank is already in service, a current limiting reactor is required. A circuit switcher can be used for this application with 20 mH in the series circuit during the energization. The inductor is short circuited immediately after the energization. A typical circuit diagram for the back-to-back switching using a series reactor is shown in Figure 22.19. A typical three-phase circuit switcher suitable for capacitor switching is shown in Figure 22.20. The three closing reactors are clearly visible in this photograph. Such circuit switchers are less expensive than circuit breakers and very effective in controlling switching surges [6]. The voltage waveform at the source with series reactor switching is shown in Figure 22.21. The peak transient voltage is 1.32 P.U.

**Figure 22.20**   Mark V three-phase circuit switcher for capacitor applications. (Courtesy of S&C Electric Company, Chicago.)

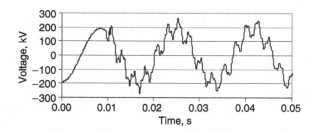

**Figure 22.21**   Source voltage during back-to-back switching with series reactor.

### 22.2.9   Voltage Magnification

Consider the presence of capacitor banks at the high and low voltage power systems as shown in Figure 22.22. Severe

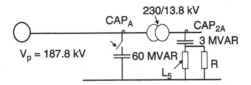

**Figure 22.22** Circuit for the analysis of voltage magnification for phase A.

overvoltages can result in the switching of high voltage capacitors, due to voltage magnification at the low voltage capacitors at a remote location. The magnification will be highest when:

- The natural frequencies of the coupled inductive-capacitive circuits are equal; $L_1C_1 = L_2C_2$.
- The capacitive MVAR of the high voltage capacitor is significantly higher than the low voltage capacitor bank; MVAR of $C_1 > 25$ times of $C_2$.
- There is not enough damping at the low voltage capacitor bank.

In order to show the voltage magnification effect, the low voltage capacitors in Figure 22.22 are already in service and the high voltage capacitors are switched in. In a normal electrical circuit, the switching transients are expected to be present only at the high voltage capacitors. In this case, the inductive-capacitive coupling produces oscillations in the low voltage capacitor circuits as well. Voltage magnification will occur at the low voltage capacitor banks when the natural frequencies $f_1$ and $f_2$ are equal, the switched capacitive MVAR is significantly higher than the MVAR of the remote capacitor and the equivalent source of the remote source is weak. The voltage plot at the switched capacitor is shown in Figure 22.23. The corresponding voltage at the 13.8 kV filter capacitor is shown in Figure 22.24. It can be seen in both plots that the oscillations are sustained and are not decaying. The overvoltage magnitudes at the 230 kV capacitor bank and the 13.8 kV capacitor are 2.1 P.U. and 1.6 P.U., respectively. This type of oscillation at the remote low voltage bus is not acceptable.

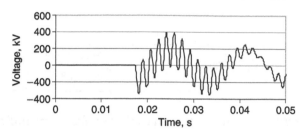

**Figure 22.23**   Voltage waveform at the 230 kV bus; $R$ is absent.

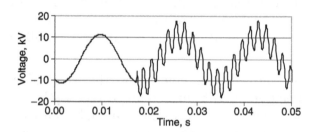

**Figure 22.24**   Voltage waveform at the 13.8 kV bus; $R$ is absent.

### 22.2.9.1   Corrective Action to Control the Voltage Amplification

One way to control the voltage magnification at the low voltage bus is to provide adequate damping. Converting the fifth harmonic notch filter to a high-pass filter can achieve this. The high-pass filter contains a damping resistance across the reactor. In the example case it is 9 $\Omega$/phase. The voltage waveforms at the 230 kV capacitor bank and the 13.8 kV capacitor during the energization of the high voltage capacitor bank are shown in Figures 22.25 and 22.26, respectively. The overvoltage magnitudes at the 230 kV capacitor bank and the 13.8 kV capacitor are 1.5 P.U. each. There is a significant reduction in the magnitudes and the oscillations are decaying rapidly.

### 22.2.10   Outrush Currents due to Close-In Faults

Consider a circuit breaker closing on a nearby bus fault as shown in Figure 22.27a. This produces a severe duty on the

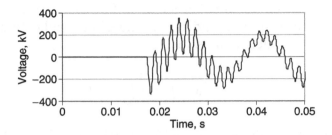

**Figure 22.25** Voltage waveform at the 230 kV capacitor; $R = 9\ \Omega$/phase.

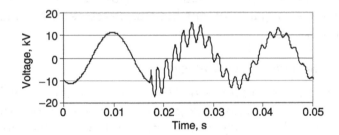

**Figure 22.26** Voltage waveform at the 13.8 kV capacitor; $R = 9\ \Omega$/phase.

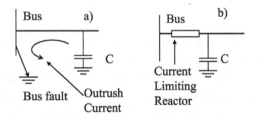

**Figure 22.27** (a) Outrush current from a capacitor bank to a bus fault. (b) Current limiting reactor at the bus location to limit the outrush current.

circuit breaker. If there is a capacitor involved in such a fault, then there will be significant outrush current from the capacitor bank. This fault current has to be limited such that the product of the peak current and the frequency is less than

$2 \times 10^7$ in the circuit-breaker location. For the specific purpose of circuit breakers intended for capacitor applications, the $I_{pk} \times f$ can be calculated as [3]:

$$I_{pk} \times f = \frac{V_p}{2\pi L} \tag{22.4}$$

where $V_{pk}$ is the peak voltage of the system and $L$ is the inductance between the capacitor bank and the fault location. The outrush currents can be limited by adding current limiting reactors. An example current limiting reactor is shown at a bus and shunt capacitor bank in Figure 22.27b. There are several ways to place current limiting reactors in the multiple capacitor banks.

## 22.2.11 Sustained Overvoltages

The possibility of high steady-state and resonant overvoltages caused by the capacitor energization is a matter of practical importance. The steady-state voltage following the capacitor energization can be calculated using the following equation [3]:

$$V_{bus} = V_p\left(1 + \frac{X_s}{X_c - X_s}\right) \tag{22.5}$$

where $X_s$ is the reactance of the source, $X_c$ is the reactance of the capacitor bank, and $V_p$ is the bus voltage before energization of the capacitors. Consider a three-phase, short-circuit rating at the 230 kV substation without shunt capacitors as 4,092 MVA. The corresponding source reactance is 12.92 Ω. For a shunt capacitor rating of 105 MVAR at the 230 kV substation, the steady-state voltage is 1.0407 P.U. At times, increased voltages may exceed the rating of the capacitors and fuses in the filter bank. The surge arrester rating at 230 kV is 180 kV. The MCOV rating of the arrester is 144 kV. This rating is very close to the steady-state voltage of the system with switched capacitors at no-load. Suitable overvoltage relaying is recommended to avoid this operating condition.

## 22.2.12 Resonance Effects

The addition of a shunt capacitor to a bus produces a resonant frequency due to the interaction with the source impedance. If the resonant frequency coincides with a harmonic frequency of the power supply, there will be oscillations in the voltage and current waveforms. The resonant frequency number $h_1$ due to the capacitor banks can be calculated using the following equation:

$$h_1 = \sqrt{\frac{MVA_{sc}}{MVAR_c}} \qquad (22.6)$$

where $MVA_{sc}$ is the short-circuit rating of the source and $MVAR_c$ is the rating of the capacitor bank.

### Example 22.2

The short-circuit current at a 230 kV bus is 16,900 MVA. The power factor at this location has to be improved by installing 100 MVAR shunt capacitors. What is the new resonant frequency at this bus location?

Solution

$$MVA_{sc} = 16,900\,MVA, \qquad MVAR_c = 100$$

Using Equation (22.6), the resonant frequency number is given by:

$$h_1 = \sqrt{\frac{16,900}{100}} = 13$$

## 22.3  VOLTAGE ACCEPTANCE CRITERIA

Equipment overvoltage withstand capability is related to the magnitude and duration of the overvoltages. The following types of overvoltages are considered.

### 22.3.1  Transient Overvoltage

These are the overvoltages caused by switching operations. They last for a few milliseconds up to a few cycles. The switching surge voltage withstand capability of the equipment is determined based on the switching surge withstand voltages specified in the applicable industry standards.

| | | |
|---|---|---|
| Transformer | C57.12.00 | BIL and switching surge |
| Shunt reactor | C57.21.00 | BIL and switching surge |
| Shunt capacitors | IEEE Std 18 | BIL and switching surge |
| GIS | C37.122 | BIL and switching surge |
| Circuit breaker | C37.06 | BIL and switching surge |
| Circuit breaker TRV | C37.09 | Circuit breaker TRV |
| Insulator [1] | | BIL and switching surge |

The allowed peak switching surge voltages are compared with the calculated switching surge voltages. In all cases, the peak switching surge voltage has to be less than the allowed switching surge voltage.

### 22.3.2  Transient Recovery Voltage (TRV)

The TRVs are the voltages measured across the circuit breaker poles during opening. The severity of a TRV depends on both the magnitude and the rate of rise of voltage across the opening circuit breaker poles. Based on ANSI Standard C37.09, the allowable TRV values for various circuits are given in Table 22.1.

### 22.3.3  Surge Arrester Transient Overvoltage Capability and Protective Levels

The surge arrester should have adequate protective margins. The maximum switching surge spark overvoltage has to be higher than the calculated switching surge voltage in all cases. The calculated switching surge voltages for all three phases in all critical locations and the TRV are compared with equipment capabilities. The TRV of circuit breakers, switching surge spark overvoltage of surge arresters, insulation, and switching surge withstand capabilities of critical equipment are presented in Appendix C.

### 22.4  INSULATION COORDINATION

Based on the switching surge study, an insulation coordination study can be conducted for the 230 kV example system. The

TABLE **22.1**   Allowable TRV Values

| Type of Capacitor Circuit | Allowed Maximum TRV |
|---|---|
| Grounded shunt capacitor | 2.0 P.U. |
| Unloaded cable | 2.0 P.U. |
| Unloaded transmission line | 2.4 P.U. |

TABLE **22.2**   Switching Surge Voltage Limitations for Selected Cases

| Equipment | BIL kV | Impulse Withstand Level kV | Switching Surge Limit P.U. | Related IEEE or ANSI Standard |
|---|---|---|---|---|
| Transformer | 900 | 745 | 3.9 | C57.12 |
| Capacitor | 1050 | 750 | 3.9 | IEEE 824 |
| Circuit breaker | 900 | 675 | 3.6 | C37.06 |
| CB TRV | | | 2.4 | C37.09 |

equipment overvoltage withstand capability is related to the magnitude and duration of the overvoltages. These overvoltages are caused by switching operations and last for a few milliseconds up to a few cycles. The switching surge withstand capability can be assessed using the switching surge withstand voltages specified by the applicable industry standards. The resulting surge voltage limitations are listed in Table 22.2.

### 22.4.1   The Surge Arrester Transient Overvoltage Capability and Protective Levels

A 180 kV surge arrester is marginal with regard to its maximum continuous operating voltage (MCOV) capability. The arrester protective margins are well within those recommended in ANSI Standard C62.2. The maximum switching surge protective level of the 180 kV surge arrester is 351 kV (1.87 P.U. on 230 kV base).

### 22.4.2    Transient Recovery Voltage (TRV)

The TRVs are the voltages measured across the circuit breaker poles after opening. The severity of the TRV depends on both the magnitude and the rate of rise of the voltage across the opening circuit breaker poles. Based on ANSI Standard C37.09, the circuit breaker switching capability is tested at the maximum TRV of 2.4 P.U. The TRV voltages experienced in the fault clearing are less than 2.4 P.U.

### 22.5    CONCLUSIONS

In this chapter, the effects of switching the power factor correction capacitors are studied. The following aspects of capacitor installation and switching are examined:

| | |
|---|---|
| Energization | De-energization |
| Fault clearing | Reclosing |
| Backup fault clearing | Restriking |
| Prestrike | Back-to-back capacitor switching |
| Voltage magnification | High frequency overvoltages |
| Outrush currents | Sustained overvoltages |

The nature of the transient overvoltages and the impact on the insulation coordination are identified.

### PROBLEMS

22.1.  What is the difference between energization and de-energization?

22.2.  What are the various capacitor switching problems?

22.3.  In a 230 kV, three-phase power system, the short-circuit rating is 18,000 MVA. 110 MVAR shunt capacitors are added to improve the power factor. In the distribution system close to this high voltage system, there is a filter bank at 13.8 kV tuned to the 6.7th harmonic. Are there any technical problems due to the installation of the high voltage capacitor bank? Explain.

22.4. What is TRV? Is there any relation between over-voltages in the phase circuit and the TRV of the circuit breaker?

22.5. Why are the reclosing transients higher than the energization transient?

22.6. Define restrike. How can restrike be avoided?

22.7. Define prestrike. Is there any difference between restrike and prestrike?

22.8. Is there a need to perform back-to-back capacitor switching? Is it possible to avoid such an operation in a power system?

22.9. Define voltage magnification. The HV capacitor system is large, and it is a high-pass filter. The low voltage system contains a capacitor bank. Is it still possible to see oscillations in the low voltage capacitor bank? Explain.

22.10. Define outrush current. What is the difference between inrush and outrush currents?

## REFERENCES

1. Natarajan, R., Hale, E., Ashmore, S., Larsson, K. (1999). A 230 kV power factor correction installation taking into account the low voltage filters, *Proceedings of the 1999 American Power Conference*, 61, 686–691.

2. Hammond, P. W. (1988). A harmonic filter installation to reduce voltage distortion from solid state converters, *IEEE Transactions on Industry Applications*, 24(1), 53–58.

3. Mikhail, S. S., McGranaghan, M. F. (1986). Evaluation of switching concerns associated with 345 kV shunt capacitor applications, *IEEE Transactions on Power Delivery*, 1(2), 231–240.

4. Dommel, H. W. (1976). *EMTP User's Manual*, University of British Columbia, Greater Vancouver.

5. ANSI Standard C37.09 (1998), IEEE Standard Test Procedures for AC High Voltage Circuit Breakers Rated on a Symmetrical Current Basis.

6. S&C Electric Company (1995). S&C Circuit Switches — Mark V. Data Bulletin 711–95, S&C Electric Company, Chicago.

7. Eletric Power Research Institute (1975). Transmission Line Reference Book 345 kV and Above, Electric Power Research Institute, Palo Alto, CA.

# 23

## INDUCED VOLTAGES IN CONTROL CABLES

### 23.1 INTRODUCTION

In high voltage substations, there are different kinds of conductors close to one another, such as high voltage buses, current transformers, potential transformers, carrier couplers, bushing, control cables, substation ground conductors, and equipment ground connections. The control cables are used to carry potential transformer outputs, current transformer outputs, circuit breaker control signals, relaying, and other communication signals. Increasingly, electronic equipment is used in switchyards and control houses. The induced voltage produced inside a substation can couple into low voltage control cables and electronic equipment unless it is suitably protected. Parallel conductors exhibit both mutual inductance and capacitance. Since the power conductors carry relatively large currents and operate at higher voltages as compared to control cables, power frequency voltages may appear on the control cables through this coupling and cause considerable noise problems. In addition, if care is not taken to ground the system properly, ground currents at these frequencies may be coupled with the instrumentation and control system resistively, capacitively, or inductively, producing nuisance trips [1–3]. This chapter deals with the calculation of the induced voltages in the control cables due to switching and lightning surges. In the presence of shunt capacitor banks in the substation, the magnitude and frequency of the switching surges increase.

## 23.2   SOURCES OF INDUCED VOLTAGES

### 23.2.1   Capacitive Coupling

The electrostatic coupling between the power system conductor and the control cable can produce voltage induction at the supply frequency. An example capacitance coupling between a power conductor and control cable is shown in Figure 23.1. The capacitance acts as a voltage divider. During transient switching operations, there will be induced currents in the control cable, given by Equation (23.1):

$$i = C\frac{dV}{dt} \tag{23.1}$$

Increased distance between the power conductor and the control cable can reduce the induced voltage in the control cable.

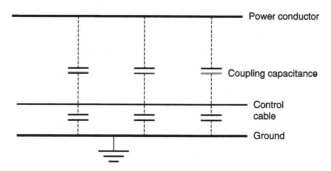

**Figure 23.1**   Capacitive coupling between a power conductor and control cable.

## 23.2.2 Inductive Coupling

The presence of a power conductor close to a control cable can produce inductive coupling between the two. The current through the power conductor produces magnetic flux as shown in Figure 23.2. If a control cable is present in the magnetic field, then there will be induced voltage at the power frequency. The magnitude of the induced voltage depends on the mutual coupling between the conductors and the current through the conductor. The induced voltage in the control cable is given by:

$$e \text{ (control cable)} = M \frac{di}{dt} \quad (23.2)$$

where $M$ is the mutual inductance between the power conductor and the control cable and $i$ is the current through the conductor.

The electric field is proportional to the charge per unit length $\rho$ on the bus and is inversely proportional to the shortest distance $r$ between the field point on the bus given by:

$$E = \frac{\rho}{2\pi\varepsilon_0 r} \quad (23.3)$$

$$\rho = CV_{\text{ph}} \quad (23.4)$$

$$C = \frac{1}{Z_{\text{s}} \times c} \quad (23.5)$$

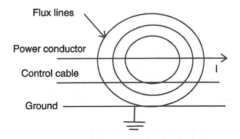

**Figure 23.2** Inductive coupling between the power conductor and the control cable.

where  $c =$ Velocity of light $= 3 \times 10^8$ m/s
$\quad\quad C =$ Capacitance of the bus
$\quad\quad V_{ph} =$ Voltage per phase
$\quad\quad Z_s =$ Surge impedance, $\Omega$/phase

Re-arranging the above equations:

$$E = \frac{377 \times V_{ph}}{2\pi Z_s\, h} \tag{23.6}$$

where

$$\eta = \sqrt{\frac{\mu_0}{\varepsilon_0}} = \frac{1}{\varepsilon_0 c} = 377\,\Omega \tag{23.6}$$

The radius $r$ is equal to the bus height $h$. The vertical electric field is doubled upon reflection from the ground. For 1.0 per unit switching transients, the electric field is given by:

$$E = \frac{377 \times V_{ph}}{\pi Z_s\, h} \tag{23.7}$$

## Example 23.1

Calculate the electric field at a distance of 8 m from the phase conductor of a 230 kV system. Assume a surge impedance of 350 $\Omega$/phase.

Solution

$\quad\quad h = 8$ m,$\quad\quad Z_s = 350\,\Omega$/phase

$\quad\quad$System line-to-line voltage $= 230$ kV

$\quad\quad$Phase voltage $(230\,\mathrm{kV}/1.732) = 132.8$ kV

$$E = \frac{(377\,\Omega)(132.8\,\mathrm{kV})}{\pi(350\,\Omega)(8\,\mathrm{m})} = 5.7\,\mathrm{kV/m}$$

The transient induced voltages in the control cables are due to circuit breaker switching operations and to the traveling waves produced by the lightning strike. The transient current amplitudes depend on the surge impedance of the conductor and the peak instantaneous phase-to-ground system voltage.

### 23.2.3 Switching Transients due to Circuit Breaker Operations

The moving contacts of the circuit breakers not only allow multiple breakdowns of the insulating medium between the components of the high voltage system, but also allow the breakdown potentials to exceed the system operating voltage due to trapped charges. The oscillation frequencies may vary from the nominal supply frequency to several kHz. When control cables are present, there will be induced voltages due to mutual coupling.

### 23.2.4 Lightning Transients

Lightning strikes can also cause arcing in substation equipment and produce transients. When control cables are laid in parallel to the power line conductors transmitting such transients there will be induced voltages.

In a substation, the induced voltages in the control cables may be due to conducted coupling, radiated coupling such as electrostatic coupling, or inductive coupling. The induced voltage through the control cables can cause damage to electronic equipment.

### 23.3 ACCEPTABLE INDUCED VOLTAGES

The acceptable induced voltages in control cables are adopted from IEC Standard 801-4 due to fast electrical transients. The following four levels of environmental conditions are identified:

Level 1   Well protected
Level 2   Protected
Level 3   Typical industrial
Level 4   Severe industrial

**TABLE 23.1**  Severity and Acceptable Induced Voltage Levels

| Level | On Power Supply | | On Signal and Control Cable | |
|---|---|---|---|---|
| | $V_{oc}$ kV | $I_{sc}$ A | $V_{oc}$ kV | $I_{sc}$ A |
| 1 | 0.5 | 10 | 0.25 | 5 |
| 2 | 1.0 | 20 | 0.50 | 10 |
| 3 | 2.0 | 40 | 1.00 | 20 |
| 4 | 4.0 | 80 | 2.00 | 40 |

The acceptable peak amplitudes for various severity levels are presented in Table 23.1. The open circuit voltage for each severity level for both the power supplies and data lines are given in Table 23.1. The short-circuit values are estimated by dividing the open circuit voltage by 50 Ω source impedance. This value represents the worst-case voltage seen by the surge suppression element. A typical 230 kV substation is situated with open bus bars and other equipment, which can be identified as a level 4 severity. The corresponding acceptable peak-to-peak open circuit voltage during switching operations is 4 kV. The acceptable induced open circuit voltage in data lines is 2 kV.

## 23.4  EXAMPLE SYSTEM

The 230 kV and the 115 kV substation used in this study is operated with the entire cross-bays closed. The substation uses a one and a half circuit breaker scheme with double bus system. The 60 MVAR capacitor bank is connected to the 230 kV system. The specifications of the capacitor bank are:

Nominal system voltage = 230 kV
Maximum system voltage = 253 kV (+10%)
Rating of capacitor bank at 230 kV = 60 MVAR
Frequency = 60 Hz
Connection = Grounded wye

### 23.4.1 Circuit Breaker for Capacitance Switching

Maximum voltage = 242 kV
Interrupting current = 40 kA
Continuous current = 3,000 A
BIL level = 1,050 kV
Bus diameter = 12.5 cm
Diameter of the control cable = 0.8 cm
Diameter of the shield = 0.3 cm

The 230 kV and 115 kV circuit breakers along with the 60 MVAR capacitor banks are shown in Figure 23.3. The electromagnetic interference was suspected to be responsible for equipment failure or nuisance tripping in the other 115 kV power factor capacitor installation project within the utility. A study was performed to identify the related issues and apply suitable mitigation measures. The induced voltages in the control cables can emanate in a substation due to the

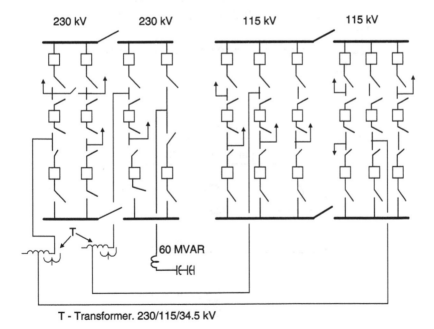

T - Transformer. 230/115/34.5 kV

**Figure 23.3** 230 kV and 115 kV substations and capacitor location.

switching operations (energizing, de-energizing, reclosing, fault clearing, backup fault clearing) and lightning strikes.

### 23.5  CALCULATION OF INDUCED VOLTAGES

The Electro Magnetic Transients Program (EMTP) is the commonly used software for studying the steady state and transients in time domain. This program is used to simulate the induced voltages in the control cables due to the switching of 230 kV lines in the substation. The buses in the substation and the representative control cable are shown in Figure 23.4. The equivalent circuit of the system used in the transient analysis is shown in Figure 23.5. The following assumptions are made in this analysis [4]:

1. The load resistance in the 230 kV system is taken as 351 Ω. This produces the required current flow through the 230 kV bus.
2. Only one control cable is used in this study.
3. The induced voltages due to energization are dominant.
4. The shield of the control cable is represented as a conductor.

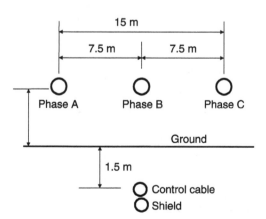

**Figure 23.4**   The 230 kV bus, control cable, and the shield.

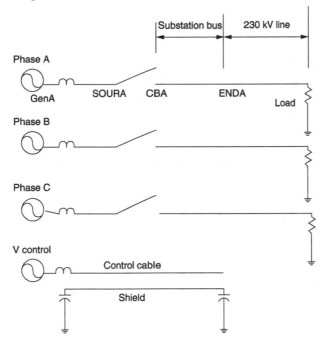

**Figure 23.5**  Circuit diagram used in the EMTP simulation.

5. The voltage source of the control cable is taken as 24 V.
6. A load of 150 Ω is assumed in the control cable. This is a typical load impedance in a current transformer circuit.

The following five cases are simulated in order to evaluate the induced voltages in the cable by studying the voltage at the open end or load of the control cable, shield voltage at the left end, and shield voltage at the right end. The time domain plots of the induced voltages in the control cables are shown in Figures 23.6 and 23.7 for case 1 and in Figures 23.8 and 23.9 for case 4. The summary of the peak voltages is presented in Table 23.2. The descriptions of the cases and the observations are listed below.

1. *Energize the 230 kV line with the control cable load end open.* The shield end is not grounded. The induced

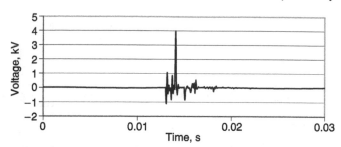

**Figure 23.6**  Voltage at the load end of the control cable, case 1.

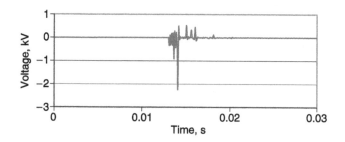

**Figure 23.7**  Voltage at the left end of the shield, case 1.

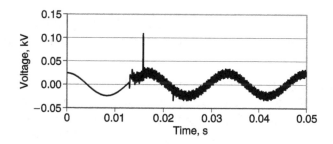

**Figure 23.8**  Voltage at the 150 Ω load of the cable, case 4.

voltage in the control cable (6 kV) is higher than the
allowed open circuit voltage per IEC 801-4 [3], which
is 4 kV. The induced voltages at the ungrounded shield
at both ends are 4.0 and 3.2 kV, respectively.

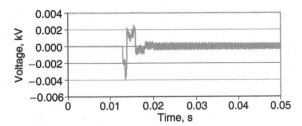

**Figure 23.9**  Voltage at the left end of the shield, case 4.

**Table 23.2**  Induced Voltage in the Control Cable

| | V at Control Cable Load/Open End | | Shield Voltage Left End kV | Shield Voltage Right End kV |
|------|------|------|------|------|
| Case | EMTP kV | IEC kV | | |
| 1 | 6.00 | 4.0 | 3.20 | 3.200 |
| 2 | 0.60 | – | 1.20 | 1.200 |
| 3 | 1.00 | – | 0.03 | 2.500 |
| 4 | 0.07 | – | 5.00 | 0.007 |
| 5 | 3.50 | 4.0 | 6.00 | 0.007 |

2. *Energize the 230 kV line with a 150 Ω load on the control cable.* The shield ends are not grounded. With a 150 Ω load resistance and ungrounded shields, the induced voltage in the control cable circuit is 600 V.
3. *Energize the 230 kV line with 150 Ω load on the control cable.* The left end of the shield is grounded and the right end is not grounded. The induced voltage in the control cable is 1.3 kV.
4. *Energize the 230 kV line with 150 Ω load on the control cable.* Both ends of the shield are grounded. The induced voltage in the control cable is 70 V. This is acceptable.
5. *Energize the 230 kV line with no load on the control cable circuit.* Both ends of the shield are grounded. The voltage in the control cable is 3.5 kV. This is acceptable as per IEC 801-4, but this induced voltage magnitude is not acceptable in data cables.

When both ends of the shield are not grounded, the induced voltages on the shield are high. If the shield is grounded at only one end, the induced voltage on the grounded end is small, but the induced voltage on the ungrounded end is still very high. When the shield is grounded at both ends, the induced voltages on the shields and control cable are very small.

When the control cable shields are grounded on both ends, an alternative ground path should be provided for power fault currents. During a power system fault involving the substation ground, voltage differences exist between points on the ground. The shield of the control cable will be subject to large currents and will be fused. A separate ground cable of at least #2/0 (approximately $68\,mm^2$) copper cable should be included to provide an alternate ground fault current path. Due to the complex nature of the switching arc during the transients, the calculated induced voltages do not reflect the lower or upper boundary values. These values are in line with the other published simulation results and measurements.

## 23.6 EFFECT OF LIGHTNING ON CONTROL CABLES

Lightning is one of the major causes of equipment failure in the power system. During a lightning strike, there are induced voltages in the control cables due to the coupling between the phase conductors and control cables. The lightning surge can be modeled as a current source or as a voltage source. In this study, the lightning surge is modeled as a voltage source with 2,000 kV surge voltage as shown in Figure 23.10. It is assumed that the lightning strikes at the end of the line. From a practical point of view, lightning can strike at any location in the transmission line or substation. The equivalent circuit of the system used in the transient analysis is shown in Figure 23.11. The effect of a lightning arrester is not included in the study. With the lightning arrester, the magnitude of the traveling waves will be small and hence the induced voltages in the control cables may be less [4].

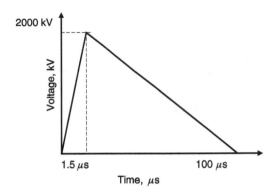

**Figure 23.10**   Waveform of the lightning surge voltage.

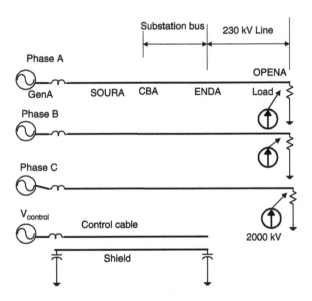

**Figure 23.11**   Circuit diagram used in the lightning surge study.

The following cases are simulated in order to evaluate the induced voltages in the cable. The voltage at the open end or load of the control cable, the shield voltage at the left end, and the shield voltage at the right end are studied. The time domain plots of the induced voltages in the control cables are shown in Figures 23.12 and 23.13 for case $L_1$ and in

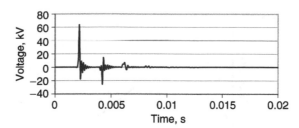

**Figure 23.12**  Voltage at the cable end, case $L_1$.

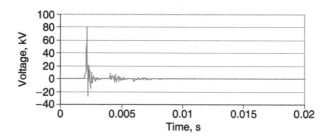

**Figure 23.13**  Voltage at the shield end, case $L_1$.

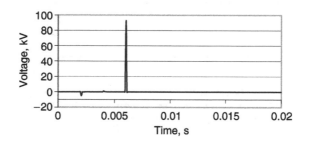

**Figure 23.14**  Voltage at the open end of the cable, case $L_4$.

Figures 23.14 and 23.15 for case $L_4$. The summary of the peak voltages is presented in Table 23.3. A description of the cases and observations are given below.

L1. *Lightning strike to the 230 kV line when the control cable load end is open.* The shield ends are not grounded. The induced voltage in the control cable (50 kV) is higher than the allowed open circuit

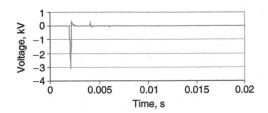

**Figure 23.15** Voltage at the left end of the shield, case L₄.

TABLE 23.3 Induced Voltages in the Control Cables due to Lightning

| | V at Control Cable Load/Open End | | Shield Voltage | Shield Voltage |
|---|---|---|---|---|
| Case | EMTP kV | IEC kV | Left End kV | Right End kV |
| 1 | 50.0 | 4.0 | 120.0 | 120.0 |
| 2 | 65.0 | – | 40.0 | 40.0 |
| 3 | 18.0 | – | 1.5 | 110.0 |
| 4 | 3.5 | – | 2.5 | 2.5 |
| 5 | 135.0 | 4.0 | 2.5 | 2.5 |

voltage as per IEC 801-4 [3], which is 4 kV. The induced voltage at the ungrounded shield is 120 kV.

L2. *Lightning strike to the 230 kV line when the load on the control cable is 150 Ω.* The shield ends are not grounded. With 150 Ω load resistance and with ungrounded shields, the induced voltage in the control cable circuit is 65 kV. This voltage is very high for a control cable.

L3. *Lightning strike to the 230 kV line when the load on the control cable circuit is 150 Ω.* The left end of the shield is grounded and the right end is not grounded. With one end of the shield grounded, the induced voltage in the control cable is 18 kV. Again, this voltage is very high for a control cable.

L4. *Lightning strike to the 230 kV line when the load on the control cable circuit is 150 Ω with both ends of the shield grounded.* With both ends of the shield grounded, the induced voltage in the control cable is 3.5 kV. This is acceptable.

L5. *Lightning strike to the 230 kV line.* There is no load on the control cable circuit. Both ends of the shield are grounded. The induced voltage in the control cable is 135 kV. This voltage is very high as per IEC 801-4.

In any substation location, there are control cables between the control room and the substation equipment. A control wire conduit in a substation location is shown in Figure 23.16 [5]. Due to the nature of the lightning strike, lightning arresters are required at the substation terminals. A measured waveform of an induced voltage in a control cable is shown in Figure 23.17 [1]. The maximum measured voltage in a current transformer control cable is 3 kV in a 500 kV system.

## 23.7 CONCLUSIONS

The allowable induced voltage magnitude for control cables in a 230 kV substation is 4 kV. The corresponding allowable induced voltage in the data cables is 2 kV as per IEC Standard 801-4. The induced voltage in the control cables is studied during switching and lightning using the transients program.

**Figure 23.16** Control wire conduit in a substation location. (Courtesy of U.S. Department of Labor, OSHA website, Reference [5].)

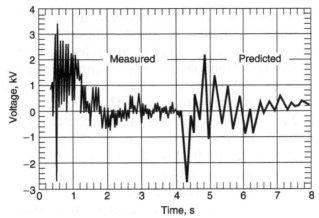

**Figure 23.17** Measured induced voltage in a control cable in a 500 kV system.

With one end of the shield grounded, the induced voltage in the control cable is 1.3 kV. With both ends of the shield grounded, the voltage in the control cable is 70 V. This is acceptable as per IEC 801-4.

A similar study was performed to evaluate the induced voltages due to a lightning strike and it was found that the induced voltages are acceptable with both ends of the shield grounded.

The control cables to be used in the 230 kV substation are equipped with shields. When the cable shields are grounded on both ends, an alternative ground path should be provided for power fault currents.

## PROBLEMS

23.1. Why are voltages induced in the control cables in the substation environment? What are the sources of these induced voltages?

23.2. Explain the basic reason for identifying the induced voltages based on the severity of the environment. Can you give an example of a well-protected environment?

23.3. Is there a difference between an ordinary substation and one with capacitor banks?

23.4. Name some of the control cables used in the substation. What is the approximate length of these cables? Where do these cables originate and where are they terminated?

23.5. Most of the power system conductors are on overhead towers and the control cables are in underground ducts. Is it possible to get voltage induction in the control cables? Draw the flux lines showing the induction mechanism.

23.6. Consider a control cable in a substation where a lightning strike occurs. Give the relative magnitude of induced voltages under the following conditions: (a) with no lightning arrester present in the substation; (b) with a lightning arrester present in the substation.

23.7. Calculate the electric field at a distance of 10 m from the phase conductor of a 500 kV system. Assume a surge impedance of 375 Ω.

# REFERENCES

1. Thomas, D. E., Wiggins, C. M., Salas, T. M., Nickel, F. S., Wright, S. E. (1994). Induced transients in substation cables: Measurements and models, *IEEE Transactions on Power Delivery*, 9(4), 1861–1868.

2. IEEE Standard 1050 (1989), Guide for Instrumentation and Control Equipment Grounding in Generating Stations.

3. Rashid, M., Russell, W. (1995). IEC 801: The transient immunity standard defined, *International Journal of EMC–Item*, 18–28.

4. Natarajan, R., Martin, D., Sanjayan, R. (2000). Analysis of induced voltages in 230 kV substation control cables, *Proceedings of the American Power Conference*, 62, 122–126.

5. www.osha.gov/sltc/etools/electric_power/glossary.html, website of the Occupational and Health Administration, U.S. Department of Labor.

# 24

## ECONOMIC ANALYSIS

### 24.1 INTRODUCTION

In all engineering applications, it is necessary to justify the installation of new equipment. In weighing the various alternatives, the economic analysis problems can be classified as fixed input type, fixed output type, or situations where neither input nor output is fixed. Whatever the nature of the problem, the proper economic criteria are to optimize the benefit/cost ratio. Such an approach may not be applicable to government projects, where the funds often are allotted based on political reasoning. The present worth approach is useful for evaluating the investment, cost of losses, operating cost, and savings [1,2].

### 24.2 BASIS OF ENGINEERING ECONOMICS

#### 24.2.1 Time Value of Money

When cash flow-related consequences occur in a very short duration, the income and expenditures can be summed and the net cash balance can be calculated. When the time span is longer, the effect of interest on the investment needs to be calculated. For example, an investment of $100 at 6% interest for 1 year will result in $6 interest and the total will be $106

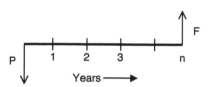

**Figure 24.1**   Present worth value $P$ of future money $F$.

after a year. This example illustrates the time value of money. When considering projects with a useful life of several years, the time value of money has to be taken into account.

## 24.2.2   Present Worth Value of Money

The process of calculating the future value of money in today's value is called present worth. The present worth $P$ is given by:

$$P = F\left[\frac{1}{(1+i)^n}\right] \tag{24.1}$$

where $F$ is the value of money after $n$ years and $i$ is the interest rate. This concept is illustrated in Figure 24.1. $F$, $n$, and $i$ are known in this problem and $P$ is to be calculated.

## Example 24.1

What is the present value of $10,000 that will be received after 3 years from today? Assume an interest rate of 8%.

Solution

$$n = 3 \text{ years}, \qquad i = 8\%$$
$$P = \$10,000, \quad (1+0.08)^{-3} = \$7,938 \text{ (using Equation (24.1))}$$

## 24.2.3   Annual Cash Flow Analysis

Suppose an amount $A$ is deposited at the end of every year for $n$ years. If $i$ is the interest rate compounded annually,

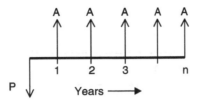

**Figure 24.2** Present worth value of money, for a uniform annual amount $A$.

then the present worth amount $P$ can be represented as shown in Figure 24.2. The mathematical representation is:

$$P = A\left[\frac{(1+i)^n - 1}{i(i+1)^n}\right] \qquad (24.2)$$

**Example 24.2**

If a $500 annual deposit is made for 5 years, what is the present worth value of those deposits? The interest rate is 6%.

Solution

$$n = 5 \text{ years}, \qquad i = 6\%$$

$$P = 500\left[\frac{(1+0.06)^5 - 1}{0.06(0.06+1)^5}\right] = \$2,106 \text{ (using Equation (24.2))}$$

### 24.2.4 Future Worth Value of Money

The process of calculating the future distribution of money is called future worth. Consider a one-time investment of $P$ at today's value and find the future worth $F$ of the money after $n$ years. The concept is displayed in Figure 24.1, where $P$ is given and $F$ is to be evaluated. For a given payment of $P$ at today's value, the future worth $F$ is given by:

$$F = P(1+i)^n \qquad (24.3)$$

This is the inverse of the present worth approach given by Equation (24.1).

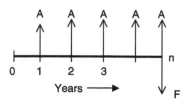

**Figure 24.3** Evaluation of the future worth $F$, for a uniform investment of $A$/year.

## Example 24.3

Calculate the future worth, $F$, for a deposit of $6,000. The number of investment years is 3, and the interest rate is 8%.

Solution

$n = 3$ years,     $i = 8\%$
$F = 6,000,$   $(1 + 0.08)^3 = \$7,558$ (using Equation (24.3))

### 24.2.5  Annual Cash Flow Analysis Using Future Worth

Suppose an amount $A$ is deposited at the end of every year for $n$ years. If $i$ is the interest rate compounded annually, then the future worth amount $F$ can be represented as shown in Figure 24.3. The mathematical representation is:

$$F = A\left[\frac{(1+i)^n - 1}{i}\right] \qquad (24.4)$$

## Example 24.4

If a $6,000 annual deposit is made for 5 years, what is the future worth value of those deposits? The interest rate is 8%.

Solution

$n = 5$ years,     $i = 8\%$

$$F = 6,000\left[\frac{(1+0.08)^5 - 1}{0.08}\right] = \$35,200 \text{ (using Equation (24.4))}$$

## 24.3 COST COMPONENTS

Installation of shunt capacitors for power factor correction and harmonic control is an efficient way to improve power system performance. When electric equipment such as a capacitor bank, a circuit breaker, or a transformer is purchased, there are several associated cost components. These include the capital cost of the equipment, installation cost, maintenance cost, salvage value, and the annual returns. There are other factors related to income tax, depreciation, tax credit, property tax, and insurance. All these factors are taken into account in order to evaluate the economic efficiency of the investment. The following description is for power factor correction capacitors, but it is applicable to similar equipment such as transformers or circuit breakers.

1. *Capital cost of the capacitor banks, C(cap).* The capacitor units are available in 50–400 kVAR ratings and can be purchased from the manufacturer. These units can be connected in series parallel combination to achieve the desired MVAR rating. In addition to the capacitor units, reactors, circuit breakers, protective relaying, fuses, mounting racks, surge arresters, and cables are required for the assembly.

2. *Cost of installation, C(inst).* The capacitor units are installed on pole-mounted structures, in substations, or in metal-enclosed cabinets. In substations, suitable foundations have to be erected. The capacitor units are installed in the racks and the necessary series–parallel connections are performed in the field. The installation cost is application dependent and varies from project to project.

3. *Maintenance cost, C(maint).* The cost of maintenance depends on the type of design and the location of installation.

4. *Salvage value, C(sal).* Usually any electrical equipment is designed for certain useful life. This depends on the type of equipment, design life, maintenance of the equipment and operating conditions. Sometimes the equipment may have zero value at the end of useful life.

## 24.3.1  Benefits Resulting from the Installation of Capacitor Banks

As identified in Chapter 9, the benefits resulting from the installation of capacitor banks may include one or more of the following:

1. Savings due to release of transformer capacity.
2. Savings due to improved power factor.
3. Savings due to reduced losses in the transformer and feeder.
4. Cost involved with reactive energy.
5. Savings due to avoiding power factor penalty.
6. Savings due to decrease in the kVA demand.
7. Savings due to the reduction in the kVAR demand.

It should be noted that the specific savings at a location depend on the tariff structure. All the above savings will not be available in any specific application.

1. *Savings due to release of transformer capacity.* If the power factor of the load is corrected by shunt capacitors, then there is less kVA demand from the transformer. The release in the transformer kVA demand can be calculated as:

$$\text{kVA}_R = (P/\text{PF}_1 - P/\text{PF}_2) \tag{24.5}$$

where $P$ is the real power demand and $\text{PF}_1$ and $\text{PF}_2$ are the power factors before and after the power factor correction. If $C_{TR}$ is the charge on the released transformer capacity per kVA per year, then the total savings due to the released transformer capacity $C_{TRS}$ is given by:

$$C_{TRS} = \text{kVA}_R \times C_{TR} \tag{24.6}$$

2. *Savings due to improved power factor.* If the power factor of the load is improved from $\text{PF}_1$ to $\text{PF}_2$, then the savings in the kVAR demand is given by:

$$\text{kVAR}_S = P(\tan\theta_1 - \tan\theta_2) \tag{24.7}$$

If $C_{QD}$ is the cost of the reactive power per unit, then the cost savings $C_{kVAR}$ due to reduction in the kVAR demand is given by:

$$C_{kVAR} = C_{QD} \times kVAR_S \qquad (24.8)$$

3. *Savings due to reduced losses in the transformer and feeder.* This is an indirect system cost reflected on the total tariff. If the reactive compensation is available at the terminals, then the current in the feeder and transformer circuit is reduced, and that leads to reduced $I^2R$ losses in both of them. If $I_1$ and $I_2$ are the respective rms current magnitudes before and after the compensation and if $R$ is the combined transformer and feeder resistance per phase, then the total average cost savings $C_L$ due to reduced ohmic losses per year can be expressed as:

$$C_L = 3(C_W)\,(I_1^2 - I_2^2)\,(R \times 8{,}760 \times LF \times 10^{-3}) \qquad (24.9)$$

where LF is the loss factor, $C_W$ is the cost of energy per kWh, and 8760 is the number of hours in a year. The load factor is defined as the ratio of the average real power loss to the maximum real power loss over a specified period of time.

4. *Cost involved with reactive energy.* In the absence of suitable power factor correction, the utility has to supply the necessary reactive power to the low power factor load. If $C_Q$ is the cost per unit of reactive energy and if the $kVAR_h$ is known, then the total cost of reactive energy is:

$$C_{QT} = (C_Q)\,(kVAR_h) \qquad (24.10)$$

5. *Savings due to avoiding power factor penalty.* If the power factor of the load is improved to acceptable levels of the utility such as 0.95 or above, then

the power factor penalty can be avoided. If $C_{kVAR}$ is the cost of the reactive power demand per unit and if the kVAR saved is known, then cost savings due to power factor improvement $C_{PF}$ can be calculated:

$$C_{PF} = (C_{kVAR})\ (kVAR) \qquad (24.11)$$

6. *Savings due to decrease in the kVA demand.* If the decrease in kVA demand is known due to the power factor improvement from Equation (24.5), then the cost savings due to the decrease in the kVA demand $C_S$ can be calculated:

$$C_S = C_{SD} \times kVA_R \qquad (24.12)$$

where $C_{SD}$ is the maximum demand cost per kVA.

7. *Savings due to the reduction in the kVAR demand.* If the kVARs saved due to power correction are known from Equation (24.7), and if $C_{QD}$ is the cost of reactive power demand, then the cost savings $C_{QDT}$ is given by:

$$C_{QDT} = C_{QD} \times kVAR_s \qquad (24.13)$$

Each utility has its own tariff structure and billing approaches. The cost savings due to improved power factor has to be calculated taking into account all the related tariffs appropriately. The cost functions can be identified as follows:

$$C(INV) = C(cap) + C(inst) + C(maint) - C(sal) \qquad (24.14)$$

$$C(BEN) = C_{TRS} + C_{kVAR} + C_{QT} + C_L + C_{PF} + C_S + C_{QDT}$$
$$(24.15)$$

Depending on the type of tariff, the components in the benefits will be selected.

## 24.4 EFFECT OF INCOME TAX AND RELATED FACTORS

In order to perform a realistic economic analysis, the effects of federal income tax, depreciation, tax credits, and the inflation have to be included in the calculations.

### 24.4.1 Taxes

Federal, state, and city taxes are part of the business expenditure. Income tax forms a major portion of the taxes.

### 24.4.2 Tax credits

Federal and state governments provide tax credits in order to encourage investments in energy-related projects. For example, in 1978, the U.S. Congress enacted the Energy Tax Act to provide a 10% energy tax credit in addition to the 10% regular investment tax credit on alternate energy sources. In passing the Crude Oil Windfall Tax Act of 1980, Congress increased the energy tax credit to 15% for renewable energy sources. These tax credits will come and go and the prevailing rates are to be used for the academic (= current) year. Some of the specific projects for tax credits are:

| | |
|---|---|
| Alternative energy sources | Wind or solar energy |
| Recycling equipment | Shale oil equipment |
| Cogeneration projects | Hydroelectric generation equipment |

### 24.4.3 Depreciation

The depreciation approach is a useful method for the recovery of an investment through income tax over a period of time. Usually, the depreciation is accounted on capital assets such as a house, car, and industrial equipment. Depreciation is not applicable for items such as land, salvage value, and interest amounts. There are two types of depreciation, book and tax depreciation. With book depreciation, a certain amount is written off each year and credited to a depreciation reserve for the purpose of reinvesting in a new plant. Tax depreciation is based on the existing tax laws and is used by the government as a tool to encourage investment. The rate is basically decided by the government and can be changed at any time. The following depreciation methods are used in engineering economic analysis:

> Straight-line method
> Sum of year of digits method
> Declining balance method
> The accelerated cost recovery system

Sometimes the recovery of depreciation on a specific asset may be related to the use rather than time. In such situations, the depreciation will be calculated by unit of production depreciation method. This is not applicable to electrical equipment.

### 24.4.3.1   The Straight-Line Method

If the investment amount and the number of years of depreciation are known, then:

$$\text{Straight-line depreciation per year} = \frac{\text{Investment}}{\text{Life of the equipment, years}} \quad (24.16)$$

Typically, this approach is used for book purposes and is not used for tax applications. As an example, if the investment is $50,000 and if the useful economic life is 5 years, then the annual depreciation by the straight-line method is ($50,000/5) or $10,000.

### 24.4.3.2   Sum of Years of Digit Method

With this method, a greater depreciation amount is used in earlier years and there is lesser depreciation in the later years compared to the straight-line method. The sum of years of digits (SYD) is found by using the formula:

$$\text{SYD} = \frac{n(n+1)}{2} \quad (24.17)$$

For a 5-year economic life, the SYD is $[5 \times (5+1)/2]$, which equals 15 years. The depreciation for the example case year by year is given in Table 24.1.

TABLE 24.1  SYD of Depreciation for the Example Case

| Year | Fraction | Depreciation, $ |
|---|---|---|
| 1 | 5/15 | 16,667 |
| 2 | 4/15 | 13,333 |
| 3 | 3/15 | 10,000 |
| 4 | 2/15 | 6,667 |
| 5 | 1/15 | 3,333 |
| Total | | 50,000 |

### 24.4.3.3 Declining Balance Depreciation

Using this method, a constant depreciation rate is applied to the book value of the property. This depreciation rate is based on the type of property and when it was acquired. The three rates described in the Economic Recovery Tax Act of 1981 are 200%, 175%, and 150% of the straight-line rate method. Since 200% is twice the straight line rate, it is called the double declining balance method and is given by the formula:

Double declining balance depreciation in any year
$$= 2 \times \text{Book Value}/n \tag{24.18}$$

The book value equals the cost depreciation to date.

Double declining depreciation in any year
$$= 2\,(\text{Cost} - \text{depreciation to date})/n \tag{24.19}$$

### Example 24.5

The investment in equipment is $50,000. Calculate the double declining depreciation schedule. Assume $n = 5$ years.

Solution

Using the formula given by Equation (24.19).

First year depreciation $= 2(\$50,000 - \$0)/5 = \$20,000$
Second year depreciation $= 2(\$50,000 - \$20,000)/5 = \$12,000$
Third year depreciation $= 2(\$50,000 - \$32,000)/5 = \$7,200$

TABLE 24.2   ACRS Classes of Depreciable Property

| Personal Property | Property Class |
|---|---|
| Auto and light duty trucks | 3 year |
| Machinery and equipment used for research | |
| Special tools and other personal property with a life of 4 years or less | |
| Most machinery and equipment | 5 year |
| Office furniture and equipment | |
| Heavy duty trucks, ships, aircraft | |
| Public utility property with a life of 25 years or less | 10 year |
| Railroad tank class, manufactured homes | |
| Public utility property with a life of more than 30 years | 15 year |
| Real property (real estate), all buildings | |

Fourth year depreciation = 2($50,000 − $39,200)/5 = $4,320

Fifth year depreciation = 2($50,000 − $43,520)/5 = $2,592

Salvage value = $3,888

Total = $50,000

### 24.4.3.4   Accelerated Cost Recovery Depreciation System (ACRS)

The Economic Recovery Act of 1981 established the ACRS depreciation. With this method, the depreciation life is far less than the actual equipment life. The salvage value is assumed to be zero. The first step in calculating the depreciation is to identify the property class from Table 24.2. The next step is to read the depreciation schedule from Table 24.3.

### Example 24.6

A desktop computer is personal property and costs $3,000. Compute the ACSR for tax purposes.

Solution

A desktop computer is personal property. As Table 24.2 shows, this equipment is in a 5-year ACRS depreciation class. The depreciation schedule is calculated and is shown in Table 24.4.

TABLE 24.3  ACRS Depreciation for Real Property

| Year | 3 Year (%) | 5 Year (%) | 10 Year (%) | 15 Year (%) |
|---|---|---|---|---|
| 1 | 33 | 20 | 10 | 7 |
| 2 | 45 | 32 | 18 | 12 |
| 3 | 22 | 24 | 16 | 12 |
| 4 | | 16 | 14 | 11 |
| 5 | | 8 | 12 | 10 |
| 6 | | | 10 | 9 |
| 7 | | | 8 | 8 |
| 8 | | | 6 | 7 |
| 9 | | | 4 | 6 |
| 10 | | | 2 | 5 |
| 11 | | | | 4 |
| 12 | | | | 3 |
| 13 | | | | 3 |
| 14 | | | | 2 |
| 15 | | | | 1 |

TABLE 24.4  Depreciation Using 5-Year ACRS Method

| Year | ACRS% | Cost ($) | Depreciation ($) |
|---|---|---|---|
| 1 | 20 | 3,000 | 600 |
| 2 | 32 | 3,000 | 960 |
| 3 | 24 | 3,000 | 720 |
| 4 | 16 | 3,000 | 480 |
| 5 | 8 | 3,000 | 240 |

## 24.5  ECONOMIC EVALUATIONS

There are many approaches available for the evaluation of the economic efficiency of one or more project options, including the following:

Payback method
Rate of return method
Benefit–cost ratio analysis
Breakeven analysis

Generally, the following assumptions are made in this type of analysis:

- One of the parameters in the present worth analysis is assumed. It may be either the interest rate or the useful life.
- The salvage value is usually taken as zero.
- Even if the interest rate is provided, it is likely to change in the future.
- When comparing products from a superior technology versus the older technology, the latter will be the attractive choice. In such cases, the economic advantages and the technical merits are to be weighted together.

### 24.5.1  Payback Period

The payback period is the time required for the benefits of an investment to equal the cost of the investment.

#### 24.5.1.1  Simple Payback Period

The simple payback period can be calculated as:

$$\text{Simple payback period} = \frac{\text{Investment}}{\text{Benefits per year}} \qquad (24.20)$$

The simple payback period calculation ignores the time value of money. One of the ways of comparing mutually exclusive economic aspects is to resolve the consequences to the present time using the present worth method. The present worth of the benefits per year is obtained by multiplying the benefits with the present worth factor given by:

$$\text{Present worth factor (PWF)} = \frac{1}{(1+i)^n} \qquad (24.21)$$

where $n$ is the number of years. The cumulative value of benefits is given by:

$$P = A \sum_{n=1}^{\infty} \left[ \frac{1}{(1+i)^n} \right] = A \left[ \frac{(1+i)^n - 1}{i(i+1)^n} \right] \qquad (24.22)$$

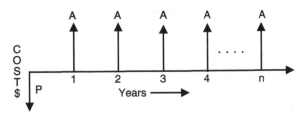

**Figure 24.4** Application of present worth analysis for payback period.

where $A$ is the benefits per year. The investment and the economic returns are expressed on a time-varying basis as shown in Figure 24.4. For a given project, the calculation of the payback period using the present worth analysis involves the evaluation of the present worth factor, benefits per year, and the present worth of the cumulative benefits.

In order to perform the economic analysis, the effect of income tax must be included. Sometimes certain tax credits may be available for energy-related projects and they need careful consideration. Using this provision, the business can deduct a percentage of its new equipment purchases as a tax credit, usually after computing the tax return. At the same time the basis for the depreciation remains the full cost of the equipment. The main components taking into account the tax effects are:

- Cash flow before taxes
- Depreciation = Investment × depreciation rate
- Change in taxable income = Cash flow before taxes − depreciation
- Income taxes = Taxable income × incremental income tax
- After tax cash flow = Before tax cash flow − income tax

### 24.5.1.1 Effect of Inflation on the Payback Period

In general, the prices and cost of services change with time and hence the inflationary trend should be included in the payback period calculations. The following steps are included in the analysis.

- Cash flow before taxes
- Actual cash = Cash flow before taxes $\times (1+f)^n$
- Depreciation = Investment $\times$ depreciation rate
- Change in taxable income = Actual cash flow − depreciation
- Income taxes = Change in taxable income $\times$ incremental income tax
- After tax cash flow = After tax cash flow $\times (1+f)^n$

The calculation approach is demonstrated using an example.

## Example 24.7

A power factor controller was installed in order to correct the power factor of an induction generator [1]. A one-line diagram of the system is shown in Figure 24.5. The cost of the device was $10,000. The Federal tax credit during 1984 on wind electric generation was 25%. The California State tax credit was 25%. The incremental tax rate was 46%. The system-related and cost data are:

| | |
|---|---|
| Transformer impedance | $= (0.0046 + j\ 0.0036)$ $\Omega$/ phase |
| Feeder impedance | $= (0.0359 + j\ 0.0652)$ $\Omega$/ phase |
| Total circuit | |
|   resistance $(R)$ | $= 0.0405\ \Omega$ |

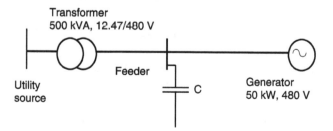

**Figure 24.5**   One-line diagram of the system for Example 24.7.

Generator output $= 50\,\text{kW}$

Power factor before
compensation $(PF_1)$ $= 0.60$

Power factor after
compensation $(PF_2)$ $= 0.95$

Current before
compensation $(I_1)$ $= 80\,\text{A}$

Current after
compensation $(I_2)$ $= 51\,\text{A}$

Loss factor (LF) $= 0.4$

Maximum kVAR $= 40.3\,\text{kVAR}$

Cost of the power factor
correction device $= \$10,000$

Cost of released
transformer kVA $(C_{TR})$ $= \$12$ per kVA/year

Cost of real energy $(C_W)$ $= \$0.06/\text{kWh}$

Cost of reactive
energy $(C_Q)$ $= \$0.0025/\text{kVARh}$

Maximum demand cost
of kW, kVAR, kVA $= \$3.80/\text{kW/month}$

Interest rate, $(i)$ $= 10\%$

Calculate (a) the simple payback period, (b) the payback period taking into account the present worth value of money along with the tax credits, and (c) the payback period taking into account the inflation. Consider the accelerated depreciation scheme of 5 years with rates at 20%, 32%, 24%, 16%, and 8%.

Solution

$P = \$10,000$

$i = 10\%$

Total tax credits $= 50\%$

Savings due to power factor penalty per year $= \$1,064$

$$(C_{PFP} = 23.3\ \text{kVAR} \times \$3.80/\text{kVAR} \times 12)$$

Savings due to reactive power factor reactive power
cost $= \$23$

$$(C_{QT} = 93,500\ \text{kVARh} \times 0.4 \times \$0.0025/\text{kVARh})$$

**TABLE 24.5** Benefits and Payback Period Calculations Using Present Worth Analysis

| Year | Benefits $ | PWF | Present Worth of Benefits $ | Cumulative PWV $ |
|------|-----------|--------|------------------|------------------|
| 1 | 1,479 | 0.9091 | 1,345 | 1,345 |
| 2 | 1,479 | 0.8264 | 1,222 | 2,567 |
| 3 | 1,479 | 0.7513 | 1,111 | 3,678 |
| 4 | 1,479 | 0.6830 | 1,010 | 4,688 |
| 5 | 1,479 | 0.6209 | 918 | 5,607 |
| 6 | 1,479 | 0.5645 | 835 | 6,441 |
| 7 | 1,479 | 0.5132 | 759 | 7,200 |
| 8 | 1,479 | 0.4665 | 690 | 7,890 |
| 9 | 1,479 | 0.4241 | 627 | 8,518 |
| 10 | 1,479 | 0.3855 | 570 | 9,088 |
| 11 | 1,479 | 0.3505 | 518 | 9,606 |
| 12 | 1,479 | 0.3186 | 471 | 10,077 |

Savings due to transformer capacity ($C_{TRS} = 23.6 \times$ $12/kVA) = $295

Savings due to loss reduction ($C_L = 3 \times \$0.06(80^2 - 51^2) \times 0.0405 \times 8,760 \times 0.4 \times 10^{-3} = \$97$

$$A = \$1,064 + \$23 + \$295 + \$97 = \$1,479$$

(a) Simple payback period $= \dfrac{10,000}{1,479} = 6.8$ years.

(b) The present worth value calculations are shown for 12 years using Equations (24.21) and (24.22) in Table 24.5. The investment of $10,000 will be equal to the PWV of benefits at 11.2 years. The calculation of the cash flow, taking into account depreciation and income tax, is shown in Table 24.6. The effective investment taking the tax credit into account is $5,000. The payback period is 3.1 years.

(c) The calculation of the cash flow taking the inflation into account is shown in Table 24.7. The effective investment after tax credit is $5,000. The payback

**TABLE 24.6** Cash Flow Analysis Taking into Account Tax, Tax Credits, and Depreciation

| Year | Benefits $ | Depreciation $ | Taxable Income $ | Income Tax $ | After Tax $ | PWF | Present Worth of Benefits $ | Cumulative PWV $ |
|---|---|---|---|---|---|---|---|---|
| 1 | 1,479 | 2,000 | −521 | −239.66 | 1,718.66 | 0.9091 | 1,562 | 1,562 |
| 2 | 1,479 | 3,200 | −1,721 | −791.66 | 2,270.66 | 0.8264 | 1,877 | 3,439 |
| 3 | 1,479 | 2,400 | −921 | −423.66 | 1,902.66 | 0.7513 | 1,429 | 4,868 |
| 4 | 1,479 | 1,600 | −121 | −55.66 | 1,534.66 | 0.6830 | 1,048 | 5,917 |
| 5 | 1,479 | 800 | 679 | −312.34 | 1,166.66 | 0.6209 | 724 | 6,641 |

**TABLE 24.7** Cash Flow Calculations Taking into Account Inflation

| Year | Benefits $ | Depreciation $ | Taxable Income $ | Income Tax $ | After Tax $ | PWF | Present Worth of Benefits $ | Cumulative PWV $ |
|------|-----------|----------------|------------------|--------------|-------------|------|------------------------------|-------------------|
| 1 | 1,422.1154 | 2,000 | −577.8846 | −265.8269 | 1,687.942 | 0.9091 | 1,475 | 1,475 |
| 2 | 1,367.4186 | 3,200 | −1,832.581 | −842.9874 | 2,210.406 | 0.8264 | 1,689 | 3,164 |
| 3 | 1,314.8256 | 2,400 | −1,085.174 | −499.1802 | 1,814.006 | 0.7513 | 1,212 | 4,376 |
| 4 | 1,264.2554 | 1,600 | −335.7446 | −154.4425 | 1,418.698 | 0.6830 | 828 | 5,204 |
| 5 | 1,215.6302 | 800 | 415.6302 | 191.1899 | 1,024.440 | 0.6209 | 523 | 5,727 |

period taking depreciation, tax, and inflation into account is 3.8 years.

## 24.5.2  Rate of Return Analysis

The rate of return may be defined as the interest rate paid on the unpaid balance such that the loan is exactly repaid within the schedule of payments. This type of analysis is useful when the economic activity is very slow and the interest rate is too low (e.g., years 2002 and 2003). When the interest rate is around 2%, the effect of compounding is insignificant. To calculate the rate of return on an investment, use:

$$\frac{\text{Present Worth of Benefits}}{\text{Present Worth of Costs}} = 1 \qquad (24.23)$$

In Equation (24.23) the only unknown is the interest rate, $i$. An example is shown below.

## Example 24.8

An investment of $57,400 was made to install a capacitor bank. The expected project lifetime is 10 years. It was estimated that the cost savings due to kVA reduction and loss reduction would be $8,173 per year. Calculate the rate of return on this investment.

Solution

$n = 10$ years,     $A = \$8,173,$     $P = \$57,400$
Using Equation (24.22) and solving for $i$:

$$57,400 = 8,173\left[\frac{(1+i)^{10} - 1}{i(i+1)^{10}}\right]$$

Solving this equation by trial and error, $i = 7\%$.

The rate of return analysis is probably the most frequently used analysis in engineering projects. The rate of return gives a measure of the desirability of the project in terms that are easily understood. Sometimes the rate of return problem cannot be solved with a positive interest rate due to the nature of the cash flow. Therefore, there is a need to be careful when choosing this technique.

### 24.5.3 Benefit–Cost Ratio Analysis

At a given minimum attractive rate of return, an alternative can be acceptable if:

$$\text{Benefit–Cost ratio} = \frac{\text{Present Worth of Benefits}}{\text{Present Worth of Costs}} \qquad (24.24)$$

This analysis can be used for all the three types of economic problems namely, fixed input, fixed output, and where neither input nor output is fixed. In this case, obtain the benefit–cost ratio. If it is greater than 1.0, then the project is viable.

### Example 24.9

An investment of $57,400 was made to install a capacitor bank. The prevailing interest rate is 6%. The projected life duration is 20 years according to the manufacturer. The life duration is to be de-rated due to the environmental operating conditions. Life duration of 10 or 15 years is under consideration. Calculate the benefit–cost ratio for each of the life durations. The estimated annual benefits due to this project are $8,173.

Solution

$$P = \$57,400, \qquad A = \$8,173, \qquad i = 6\%$$

$$\text{For } n = 10 \text{ years:} \quad \frac{8{,}173\left[\dfrac{(1+0.06)^{10} - 1}{0.06(1+0.06)^{10}}\right]}{57{,}400} = 1.048$$

$$\text{For } n = 15 \text{ years:} \quad \frac{8{,}173\left[\dfrac{(1+0.06)^{15} - 1}{0.06(1+0.06)^{15}}\right]}{57{,}400} = 1.383$$

$$\text{For } n = 20 \text{ years}: \quad \frac{8{,}173\left[\dfrac{(1+0.06)^{20}-1}{0.06(1+0.06)^{20}}\right]}{57{,}400} = 1.633$$

All the project life duration (10, 15, and 20) years provide benefit–cost ratios greater than 1.

### 24.5.4  Breakeven Analysis

Breakeven analysis is a form of sensitivity approach to the economic analysis problem. This type of analysis is useful in engineering problems involving stage construction. Should a project be constructed now to meet the present and part of the future requirement, or should it be constructed in stages as the need for increased capacity arises? A power factor correction project falls into such a category in an expanding load growth situation. Consider a situation with an existing power factor of 0.80. The desired power factor of the utility to meet the standard requirements is 0.95. The required power factor correction equipment is 100 MVAR. Then there is a new transmission project to be completed in 3 years at the same location and the expected power factor correction equipment is another 100 MVAR. The question is whether to install all the 200 MVAR equipment in one or two stages. Such a problem can be analyzed using the breakeven analysis.

### Example 24.10

Consider a typical power factor correction project that can be installed in one or two stages. The expected installation cost is as follows:

> Full project install in one stage, 200 MVAR $= \$350{,}000$
> Install 100 MVAR in stage 1 now $= \$190{,}000$
> Install another 100 MVAR after $n$ years $= \$200{,}000$

The expected life duration of the project is 20 years, irrespective of whether it is a one- or two-stage construction. The interest rate is 6%. Perform a breakeven analysis.

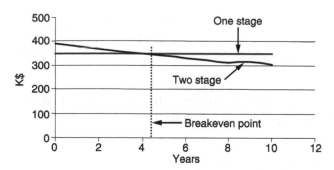

**Figure 24.6**   Breakeven chart for Example 24.10.

Solution

One-stage construction cost $= \$350{,}000$.

*Two-stage construction.* The first stage construction to be done now and the second stage after $n$ years. Calculate the present worth value for several values of $n$.

PW of cost $= \$19{,}000 + \$200{,}000(1 + 0.06)^{-n}$

$n = 2$ years;  PW of cost $= 19{,}000 + 177{,}999 = \$367{,}999$

$n = 3$ years;  PW of cost $= 19{,}000 + 167{,}923 = \$357{,}923$

$n = 4$ years;  PW of cost $= 19{,}000 + 158{,}418 = \$348{,}418$

$n = 5$ years;  PW of cost $= 19{,}000 + 149{,}451 = \$339{,}451$

$n = 10$ years; PW of cost $= 19{,}000 + 111{,}678 = \$301{,}678$

The data is plotted in the form of a graph to find the breakeven point and is shown in Figure 24.6. The breakeven point is the time at which both the project alternatives have equal costs. This is around 4 years in this example. However, the project decision has to consider the capacity requirement as well.

## 24.6   CONCLUSIONS

Economic analysis is an important aspect of the decision making process. The fundamentals of present worth and future worth analyses are discussed. The cost components involved in a power factor correction project are identified. The effects of taxes, depreciation, tax credits, and inflation on economic analysis are discussed. The payback period approach is analyzed for the simple case, taking into account the present worth value of money, taxes, depreciation, tax credits, and inflation. The other important approaches suitable for the analysis of the power factor correction project are benefit–cost analysis and rate of return analysis. Finally, the breakeven analysis is presented for construction of a project in stages. It should be noted that there are other economic analysis techniques available for the power factor correction problem. These include incremental rate of return analysis and gradient method, where the benefits are nonuniform every year.

## PROBLEMS

24.1. What are the different approaches available for the economic analysis of engineering projects? What is the basis for the selection of a suitable approach for a specific project?

24.2. A man bought $4,000 worth of office furniture. He expects it to last for 12 years. What will be the equivalent uniform annual cost if the interest rate is 6%?

24.3. An investor purchased a lot for $3,500. Each year he paid $80 for the property taxes. At the end of 4 years he sold the lot and received $4,400. What is the rate of return on his investment?

24.4. Describe the rate of return method.

24.5. Which of the four methods (payback period, rate of return, benefit–cost, and breakeven) of analyses is suitable for power factor correction projects? Why?

24.6. Consider the installation of a solar electric panel during the tax year 2003. The cost of the solar panels and installation is 1,900 per kW. The rating of the panel is 30 kW. Assume a tax credit of 30%. The expected benefits due to the sale of electricity per year are \$4,900. Calculate (a) the simple payback period; (b) the payback period taking into account the present worth value of money; (c) the payback period using the PWV method along with the tax credits; and (d) the payback period for (c) taking into account the inflation. Consider the accelerated depreciation scheme of 5 years with rates at 20%, 32%, 24%, 16%, and 8%.

24.7. There are two alternatives available for a power system project. The first alternative, to install the entire equipment in one stage, would cost \$220,000. The maintenance cost would be \$100,000 per year. The equipment life would be 25 years and the interest rate is 4%. The second alternative is to install stage one now at a cost of \$142,000. The maintenance cost would be \$75,000 per year. The cost of the second stage of construction would be \$126,000. Calculate the present worth of the cost and select a suitable alternative.

24.8. Compare the straight-line, sum of digits, 200% double declining balance, and ACRS methods of depreciation for a \$10,000 investment and a 5-year useful life.

# REFERENCES

1. Natarajan, R., Venkata, S. S., El-Sharkawi, M. A., Butler, N. G. (1987). Economic feasibility analysis of intermediate size wind electric energy conversion systems employing induction generators, *International Journal of Energy Systems*, 7(2), 90–94.

2. Newnan, D. G. (1983). *Engineering Economic Analysis*, Second Edition, Engineering Press, San Jose.

# Appendix A

## CAPACITOR CIRCUIT FUNDAMENTALS

There are several equations commonly used for power system calculations when dealing with capacitor circuits. Some of the important relationships are listed below.

Equivalent capacitance $C$ when several capacitors are connected in parallel is given by:

$$C = C_1 + C_2 + C_3 + \ldots \tag{A.1}$$

Equivalent kVAR when several capacitor banks are connected in parallel is given by:

$$kVAR = kVAR_1 + kVAR_2 + kVAR_3 + \ldots \tag{A.2}$$

Equivalent capacitance $C$ when several capacitors are connected in series:

$$C = \frac{1}{(1/C_1) + (1/C_2) + (1/C_3)} \ldots \tag{A.3}$$

Equivalent kVAR when several capacitor banks are connected in series:

$$\text{kVAR} = \frac{1}{(1/\text{kVAR}_1) + (1/\text{kVAR}_2) + (1/\text{kVAR}_3)} \cdots \quad \text{(A.4)}$$

The capacitive reactance $X_C$ for a capacitance $C$ is in microfarads:

$$X_C = \frac{10^6}{(2\pi f)C} = \frac{2,653}{C} \quad \text{at } 60\,\text{Hz} \quad \text{(A.5)}$$

In terms of kVAR, the capacitance is given by:

$$C = \frac{10^6}{(2\pi f)X_C} = \frac{(1,000)\,\text{kVAR}}{(2\pi f)(\text{kV})^2} \quad \text{(A.6)}$$

$$\text{kVAR} = \frac{(2\pi f c)(\text{kV})^2}{1,000} = \frac{(1,000)(\text{kV})^2}{X_C} \quad \text{(A.7)}$$

For a three-phase system with voltage in kV and current in A the kVA is:

$$\text{kVA} = \sqrt{3}\,(\text{kV})(I) \quad \text{(A.8)}$$

The following relations can be used for motor-related calculations:

Induction motor kVA = Motor hp rating (approximately)
Synchronous motor at 0.8 power factor, kVA = Motor hp rating (approximately)
Synchronous motor at 1.0 power factor, kVA = (0.8) Motor hp rating (approximately)

The short circuit current $I_{\text{sc}}$ at the terminal of a transformer is given by:

$$I_{\text{sc}} = \frac{\text{Full load current}}{\text{Transformer impedance}} \quad \text{(A.9)}$$

Calculate the $I$(RMS) in the harmonic environment:

$$I(\text{RMS}) = \sqrt{\sum_{n=1}^{\infty} I_n^2} = \sqrt{I_1^2 + I_3^2 + I_5^2} \qquad (A.10)$$

The total demand distortion (TDD) is calculated using Equation (A.11):

$$\text{TDD} = \frac{\sqrt{\sum_{n=2}^{\infty} I_n^2}}{I_L} \qquad (A.11)$$

where $I_L$ is the load current and $I_n$ is the harmonic current component. The TDD due to a specific harmonic is calculated using Equation (A.12):

$$\text{TDD} = \frac{I_n}{I_L} \qquad (A.12)$$

**Example A.1**

Three capacitors have capacitances 5, 10, and 15 MFD, respectively. The three elements are connected in parallel across a 100 V DC supply. (a) Calculate the charge on each capacitor. (b) Calculate the total capacitance. (c) Connect the three capacitances in series and calculate the voltage across each unit.

Solution

(a) The three capacitor units are connected in parallel as shown in Figure A.1:

$Q_1 = C_1 \quad V = (5 \times 10^{-6})\,(100\,\text{V}) = 500\,\mu\text{C}$
$Q_2 = C_2 \quad V = (10 \times 10^{-6})\,(100\,\text{V}) = 1,000\,\mu\text{C}$
$Q_3 = C_3 \quad V = (15 \times 10^{-6})\,(100\,\text{V}) = 1,500\,\mu\text{C}$

(b) Total capacitance:

$C = C_1 + C_2 + C_3 = (5 + 10 + 15) \times 10^{-6} = 30\,\mu\text{F}$

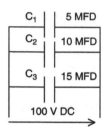

**Figure A.1**  Three capacitors in parallel.

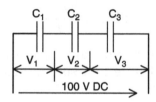

**Figure A.2**  Three capacitors in series.

(c) The series connection is shown in Figure A.2:

$$\frac{1}{C} = \frac{1}{C_1} + \frac{1}{C_2} + \frac{1}{C_3}, \qquad \frac{1}{C} = \frac{1}{5 \times 10^{-6}} + \frac{1}{10 \times 10^{-6}} + \frac{1}{15 \times 10^{-6}}$$

Solving $C = 2.727 \, \mu\text{F}$:

$$Q = CV = (2.727 \times 10^{-6} \, \text{F})\,(100 \text{ V}) = 272.7 \, \mu\text{C}$$

The charge is the same in all three capacitor units:

$$V_1 = \frac{Q}{C_1} = \frac{272.7 \, \mu\text{C}}{5 \, \mu\text{F}} = 54.5 \text{ V}$$

$$V_2 = \frac{Q}{C_2} = \frac{272.7 \, \mu\text{C}}{10 \, \mu\text{F}} = 27.3 \text{ V}$$

$$V_3 = \frac{Q}{C_3} = \frac{272.7 \, \mu\text{C}}{15 \, \mu\text{F}} = 18.2 \text{ V}$$

$$V = V_1 + V_2 + V_3 = 54.5 \text{ V} + 27.3 \text{ V} + 18.2 \text{ V} = 100 \text{ V}$$

## Example A.2

The value of shunt capacitance used in a 230 kV, 60 Hz, three-phase system is 3 µF per phase. Calculate the reactance per phase. What is the total kVAR delivered by the shunt capacitor bank?

Solution

$$C = 3\,\text{MFD/phase}$$

$$X_C = \frac{2{,}653}{3} = 884.3\ \Omega/\text{phase, using Equation (A.5)}$$

$$\text{kVAR (total)} = \frac{(1{,}000)(230\ \text{kV})^2}{884.3}$$
$$= 59{,}821\ \text{kVAR, using Equation (A.7)}$$

## Example A.3

Consider a 1,000 kVA, three-phase transformer, 6% impedance, 33 kV/4.16 kV. Find the short-circuit current at the terminal of the transformer.

Solution

$$I(\text{full load current}) = \frac{1{,}000\ \text{kVA}}{\sqrt{3}\ (4.16\ \text{kV})} = 138.8\,\text{A}$$

$$I_{sc} = \frac{138.8}{0.06} = 2{,}312\,\text{A, using Equation (A.9)}$$

## Example A.4

The fundamental, 5th, and 7th harmonic current magnitudes are 100 A, 11 A, and 5 A, respectively. Calculate the total RMS current, TDD, and TDD of the specific harmonics.

Solution

$$I(\text{RMS}) = \sqrt{100^2 + 11^2 + 5^2} = 100.72\,\text{A, using Equation (A.10)}$$

$$\text{TDD} = \frac{\sqrt{11^2 + 5^2}}{100} \times 100 = 12\%, \text{ using Equation (A.11)}$$

$$\text{TDD (at 5th harmonic)} = \frac{11}{100} \times 100 = 11\%$$

$$\text{TDD (at 9th harmonic)} = \frac{5}{100} \times 100 = 5\%$$

## PROBLEM

A.1. Two capacitors, $C_1$ and $C_2$, are connected in series. The capacitors are connected to a 100 V DC voltage. The measured voltage across $C_1$ is 70 V and the voltage across $C_2$ is 30 V. A 1.0 μF is connected across $C_1$. Then the measured voltages across ($C_1$ and 1.0 MFD) and $C_2$ are 20 V and 80 V, respectively. Calculate the values of $C_1$ and $C_2$. Draw the necessary circuit diagram.

# Appendix B

## DEFINITIONS

**Active power**  The in-phase component of volt-amperes in an electric circuit. Also called real power.

**Adjustable speed drive (ASD)**  An electric drive designed to provide easily operable means for speed adjustment of the motor.

**Alternating current (AC)**  An electric current or voltage that reverses direction of flow periodically, as opposed to direct current, and that has alternately positive and negative values.

**Ampere (A)**  A unit of measurement of electric current, which is the rate at which electrons flow in a wire; one ampere is $6.023 \times 10^{23}$ electrons per second.

**Apparent power**  The mathematical product of volts and amperes.

**Back-to-back switching**  The switching of a capacitor bank that is connected in parallel with one or more other capacitor banks.

**Background harmonic**  The harmonic voltage distortion at a location, which is not caused by the voltage distortion levels.

**Bus**  A conductor or group of conductors that serves as a common connection for two or more circuits and is used to interconnect equipment of the same voltage.

**Capacitance**  The property of an arrangement of conductors and dielectrics that stores energy in the form of an electrical charge when potential differences exist between the conductors.

**Capacitor**  In a power system, the capacitor is installed to supply reactive power. The power system capacitor consists of metal foil plates separated by paper or plastic insulation with suitable insulating fluid and sealed in a metal tank.

**Capacitor control**  The device required for automatically switching shunt power capacitor banks.

**Capacitor inrush current**  The transient charging current that flows in a capacitor when a capacitor bank is initially connected to a voltage source.

**Capacitor fuse**  A fuse applied to disconnect a faulted phase in its capacitor bank from a (capacitor group fuse) power system.

**Capacitor outrush current**  The high frequency, high magnitude current discharge of one or more capacitors into a short circuit—such as into a failed capacitor unit connected in parallel with the discharging units, or into a breaker closing into a fault.

**Characteristic harmonic**  One of the predominant frequencies present in a harmonic source. For example, the characteristic harmonics of a six-pulse converter are 5, 7, 11, 13, 17, 19, . . . .

**Circuit breaker**  A switching device for closing, carrying, and interrupting currents under normal circuit conditions, and closing, carrying for a specified time, and interrupting currents under abnormal circuit conditions such as those under faults or short circuits. The medium in which circuit interruption is performed may be designated as oil circuit breakers, air-blast circuit breakers, gas or sulfur hexafluoride circuit breakers, or vacuum circuit breakers.

**Converter** A device that changes alternating current to direct current.

**Crest factor ($C_f$)** The ratio of the peak value of a periodic waveform to the rms value. The crest factor of a sine wave is 1.414. For distorted waveforms, the crest factor can vary.

**Current transformer** A device used to sense currents in high voltage circuits for instrumentation or control purposes.

**Damping** A power system characteristic that tends to retard the effects of transient conditions or harmonics.

**Delta connected** A connection of the windings of a three-phase transformer or three single-phase transformers making up a three-phase bank that is in series, to form a closed path. Delta connections are also used for three-phase shunt reactor banks, shunt capacitor banks, or generator windings.

**Displacement power factor (DPF)** The ratio of the active power to the apparent power between the fundamental (60 Hz) components of the voltage and current, neglecting the influence of harmonics.

**Distortion power** The harmonic content of apparent power not contributing toward real power.

**Filter** In an electric system, a device that blocks certain frequencies while allowing other frequencies to pass.

**Filter capacitors** Utilized with inductors and/or resistors for controlling harmonic problems in the power system, such as reducing voltage distortion due to large rectifier loads or arc furnaces.

**Fixed capacitor bank** A capacitor bank that is left in circuit, without any automatic controls.

**Frequency** The repetition rate of a periodically recurring quantity, commonly stated in hertz (Hz). The standard frequency of alternating current in the U.S. is 60 Hz.

**Frequency response** A formula or a graph that depicts the impedance of an electrical network or devices versus driving frequency.

**Fundamental frequency**   The lowest frequency component of a periodically recurring complex wave.

**Fuse**   An overcurrent protective device with a circuit-opening fusible part that is heated and melted by the passage of excessive current through it.

**Harmonic**   A sinusoidal wave having a frequency that is an integral multiple of a fundamental frequency.

**Harmonic distortion**   Distortion in the wave shape of voltage or current caused by the presence of harmonics.

**Harmonic resonance**   Conditions where circuit inductive and capacitive reactance negate each other at a particular harmonic frequency.

**Hertz (Hz)**   The unit of frequency in cycles per second.

*I*   Symbol for current

$I^2R$ **losses**   Heating losses due to the flow of current through a resistance in a transformer, motor, or conductor.

**Isolated capacitor bank**   A capacitor bank that is not in parallel with other capacitor banks.

**Individual capacitor fuse**   A fuse applied to disconnect an individual faulted capacitor from its bank.

**Impedance**   A characteristic of an electric circuit that determines its hindrance to the flow of electricity. The higher the impedance, the lower the current. The unit of measure is $\Omega$.

**Inductance**   The property of an electric circuit that causes it to store energy in the form of a magnetic field, because of which a varying current in a circuit induces an electromotive force (voltage) in that circuit.

**kilo (k)**   A prefix indicating 1,000.

**kVA**   Kilo-Volt-Amperes.

**kVAR**   Kilo-Volt-Amperes Reactive.

**kW**   Kilowatts.

**Lagging**   An electrical current whose phasing is behind the voltage phasing is said to lag.

**Leading** An electrical current whose phasing is ahead of the voltage phasing is said to lead.

**Line-to-line voltage** The voltage reference for three-phase power systems.

**Line-to-neutral voltage** The voltage reference for single-phase power systems.

**Linear load** An electrical load that draws a sinusoidal current from a sinusoidal voltage source.

**Load** A device that receives electrical power.

**Losses** The term applied to energy (kilowatt-hours) and power (kilowatts) lost when operating an electric system, occurring mainly as energy turning to waste heat in electrical conductors and apparatus.

**Maximum demand** The greatest demand of a load occurring during a specified time period.

**Mega (M)** Prefix meaning 1,000,000.

**Micro (μ)** Prefix meaning 1/1,000,000th.

**Milli (m)** Prefix meaning 1/1,000th.

**Modeling** Technique of system analysis and design using mathematical or physical idealizations of all or a portion of the system.

**MVA** Megavolt amperes.

**Neutral** The common connection between three phases of a power system.

**Nominal system voltage** The voltage by which the system is designated and to which certain operating characteristics are related, near the voltage level at which the system operates.

**Nonlinear load** An electrical device drawing power that does not draw a sinusoidal current.

**One-line diagram** A diagram that, by means of a single-phase representation, shows the structure of electric circuits and components.

**Overcurrent** Current in excess of the rated capacity of a circuit.

**Overvoltage**  A voltage above the normal rated voltage or above the maximum statutory limits.

**Parallel**  Two circuit elements connected across the same two points.

**Parallel resonance**  A resonance occurs when the capacitive reactance and inductive reactance of a device are adjusted so that the device either maximizes or minimizes current flow at a specific frequency.

**Phase angle**  In a power system, the displacement in time of the phase of one quantity (voltage or current) from the phase of the same quantity at a different or reference location.

**Positive sequence**  The component of three-phase voltages or currents that are balanced.

**Power**  Power is expressed in watts, the product of applied voltage and resulting in-phase current.

**Power factor**  The ratio of power in watts to the apparent power (product of volts times amperes) in volt-amperes.

**Power factor correction**  A method to improve the power factor toward unity.

**Pulse width modulated**  A power electronic switching technique where voltage (or current) switches on and off many times during the fundamental period.

**Quality factor ($Q$)**  The electrical quality of a coil, capacitor, or circuit. Mathematically, $Q$ is the ratio of reactance to resistance. The higher the $Q$, the greater the selectivity of the circuit.

**Reactive power (VAR)**  The out-of-phase component of the total volt-amperes in an electric circuit, usually expressed in VAR (volt-amperes reactive).

**Reactor**  In an electrical system, a device used to introduce inductive reactance into a circuit.

**Rectifier**  A device that converts alternating current into direct current.

**rms**  The value of an alternating current or voltage that produces the same amount of heat at a certain resistance as an equal direct current or voltage is called the effective, or the rms value.

**Series connected**   Two electrical circuit elements are series connected when the same identical current must flow through both.

**Series resonance**   If the circuit capacitance and inductance are in series, the device exhibits low impedance and maximized current flow at resonance.

**Short circuit**   An unintentional connection between a point in a circuit and ground, or between two points in a circuit.

**Short circuit capacity**   The kVA capability of a circuit during short circuit.

**Sinusoidal**   The shape of a wave form such as an alternating current or voltage that has alternating positive and negative cycles.

**Six-pulse converter**   A three-phase rectifier that utilizes six diodes or other power electronic switches to convert AC voltage to DC voltage.

**Source impedance**   The strength of electrical supply, as characterized by the equivalent impedance, seen by a short circuit to ground.

**Steady state**   When the operating condition of a system is consistent.

**Stiff system**   A system with low source impedance.

**Surge arrester**   A device that is used to protect other electrical equipment from overvoltage by discharging or bypassing surge current.

**Switched capacitor bank**   A power factor correction bank that has automatic controls to vary the amount of power factor correction, according to the site power factor or voltage.

**Synchronous machine**   An electric motor with an excitation system that allows it to rotate at synchronous frequency.

**Thyristor**   A semiconductor switch with on and off states. May be unidirectional or bidirectional, and may be a triggered three-terminal device (a controlled rectifier).

**Total harmonic distortion (THD)** The ratio of the rms value of the harmonic content to the rms value of the fundamental. THD is usually expressed as a percent of the fundamental.

**Transformer** A device for transferring electrical energy from one circuit to another by magnetic induction, usually between circuits of different voltages. Consists of a magnetic core on which there are two or more windings.

**Twelve-pulse converter** A three-phase rectifier that utilizes two six-pulse converters connected to cancel the 5th and 7th harmonic components.

**Volt (V)** The unit of electromotive force, or voltage, that if steadily applied to a circuit with a resistance of one ohm will produce a current of one ampere.

**Voltage (*V*)** The driving force that causes a current to flow in an electric circuit. Voltage and volt are often used interchangeably.

**Volt-Ampere (VA)** The mathematical product of volt and ampere.

**Watt (W)** The electrical unit of power.

**Wye connected** A connection of the windings of a three-phase transformer or three single-phase transformers making up a three-phase bank such that one end of each winding is connected to a common point.

**Z** Symbol for impedance

**Zero sequence** The component of unbalance in a three-phase system that is in phase for all phases.

# Appendix C

## CIRCUIT BREAKER DATA

The circuit breakers used in the power system applications are classified into two categories in the ANSI Standard C37.06, 1979 as general purpose and definite purpose.

### General Purpose Circuit Breakers

These are used for the switching of lines, transformers, reactors, and buses. The preferred ratings of such circuit breakers are from ANSI Standard C37.06, Tables C-1, C-2, and C-3 representing the indoor, outdoor, and gas insulated switchgear as follows.

Table C-1 Preferred Ratings for Oilless Indoor Circuit Breakers (4.76 kV–38 kV).

Table C-2 Preferred Ratings for Outdoor Circuit Breakers 72.5 kV and Below, Including Circuit Breakers in Gas Insulated Substations (15.5 kV–72.5 kV).

Table C-3 Preferred Ratings for Outdoor Circuit Breakers 121 kV and Above, Including Circuit Breakers Applied in Gas Insulated Substations (121 kV–800 kV).

**TABLE C-1**  Preferred Ratings for Indoor Oilless Circuit Breakers

| Rated Maximum Voltage kV, rms | Rated Voltage Range Factor K | Rated Continuous Current at 60 Hz A, rms | Rated Short Circuit Current (at Rated Maximum kV) kA, rms | Rate Interrupting Time Cycles | Rated Maximum Voltage Divided by K kV, rms | Maximum Symmetrical Interrupting Capability and Rated Short Time Current kA, rms | Closing and Latching Capability 2.7K Times Rated Short Circuit Current kA, Crest |
|---|---|---|---|---|---|---|---|
| 4.76 | 1.36 | 1,200 | 8.8 | 5 | 3.50 | 12 | 32 |
| 4.76 | 1.24 | 1,200, 2,000 | 29.0 | 5 | 3.85 | 36 | 97 |
| 4.76 | 1.19 | 1,200, 2,000, 3,000 | 41.0 | 5 | 4.00 | 49 | 132 |
| 8.25 | 1.25 | 1,200, 2,000 | 33.0 | 5 | 6.60 | 41 | 111 |
| 15.00 | 1.30 | 1,200, 2,000 | 18.0 | 5 | 11.50 | 23 | 62 |
| 15.00 | 1.30 | 1,200, 2,000 | 28.0 | 5 | 11.50 | 36 | 97 |
| 15.00 | 1.30 | 1,200, 2,000, 3,000 | 37.0 | 5 | 11.50 | 48 | 130 |
| 38.00 | 1.65 | 1,200, 2,000, 3,000 | 21.0 | 5 | 23.00 | 35 | 95 |
| 38.00 | 1.00 | 1,200, 3,000 | 40.0 | 5 | 38.00 | 40 | 108 |

**TABLE C-2** Preferred Ratings for Outdoor Circuit Breakers 72.5 kV and Below, Including Circuit Breakers Applied in Gas Insulated Substations

| Rated Maximum Voltage kV, rms | Rated Voltage Range Factor K | Rated Continuous Current at 60 Hz A, rms | Rated Short Circuit Current (at Rated Maximum kV) kA, rms | Rated Maximum Voltage Divided by K kV, rms | Maximum Symmetrical Interrupting Capability and Rated Short Time Current kA, rms | Closing and Latching Capability 2.7 K Times Rated Short Circuit Current kA, Crest |
|---|---|---|---|---|---|---|
| 15.5 | 1.0 | 600, 1,200 | 12.5 | 15.5 | 12.5 | 34 |
| 15.5 | 1.0 | 1,200, 2,000 | 20.2 | 15.5 | 20.2 | 54 |
| 15.5 | 1.0 | 1,200, 2,000 | 25.0 | 15.5 | 25.0 | 68 |
| 15.5 | 1.0 | 1,200, 2,000, 3,000 | 40.0 | 15.5 | 40.0 | 108 |
| 25.8 | 1.0 | 1,200, 2,000 | 12.5 | 25.8 | 12.5 | 34 |
| 25.8 | 1.0 | 1,200, 2,000 | 25.0 | 25.8 | 25.0 | 68 |
| 38.0 | 1.0 | 1,200, 2,000 | 16.0 | 38.0 | 16.0 | 43 |
| 38.0 | 1.0 | 1,200, 2,000 | 20.0 | 38.0 | 20.0 | 54 |
| 38.0 | 1.0 | 1,200, 2,000 | 25.0 | 38.0 | 25.0 | 68 |
| 38.0 | 1.0 | 1,200, 2,000 | 31.5 | 38.0 | 31.5 | 85 |
| 38.0 | 1.0 | 1,200, 2,000, 3,000 | 40.0 | 38.0 | 40.0 | 108 |
| 48.3 | 1.0 | 1,200, 2,000 | 20.0 | 48.3 | 20.0 | 54 |
| 48.3 | 1.0 | 1,200, 2,000 | 31.5 | 48.3 | 31.5 | 85 |
| 48.3 | 1.0 | 1,200, 2,000, 3,000 | 40.0 | 48.3 | 40.0 | 108 |
| 72.5 | 1.0 | 1,200, 2,000 | 20.0 | 72.5 | 20.0 | 54 |
| 72.5 | 1.0 | 1,200, 2,000, | 31.5 | 72.5 | 31.5 | 85 |
| 72.5 | 1.0 | 1,200, 2,000, 3,000 | 40.0 | 72.5 | 40.0 | 108 |

**TABLE C-3** Preferred Ratings for Outdoor Circuit Breakers 121 kV and Above, Including Circuit Breakers Applied in Gas Insulated Substations

| Rated Maximum Voltage kV, rms | Rated Voltage Range Factor K | Rated Continuous Current at 60 Hz A, rms | Rated Short Circuit Current (at Rated Maximum kV) kA, rms | Rate Interrupting Time Cycles | Rated Maximum Voltage Divided by K kV, rms | Maximum Symmetrical Interrupting Capability and Rated Short Time Current kA, rms | Closing and Latching Capability 2.7 K Times Rated Short Circuit Current kA, Crest |
|---|---|---|---|---|---|---|---|
| 121 | 1.0 | 1,200 | 20 | 3 | 121 | 20 | 54 |
| 121 | 1.0 | 1,600, 2,000, 3,000 | 40 | 3 | 121 | 40 | 1,808 |
| 121 | 1.0 | 2,000, 3,000 | 63 | 3 | 121 | 63 | 170 |
| 145 | 1.0 | 1,200 | 20 | 3 | 145 | 20 | 54 |
| 145 | 1.0 | 1,600, 2,000, 3,000 | 40 | 3 | 145 | 40 | 108 |

| | | | | | | | |
|---|---|---|---|---|---|---|---|
| 145 | 1.0 | 2,000, 3,000 | 63 | 3 | 145 | 63 | 170 |
| 145 | 1.0 | 2,000, 3,000 | 80 | 3 | 145 | 80 | 216 |
| 169 | 1.0 | 1,200 | 16 | 3 | 169 | 16 | 43 |
| 169 | 1.0 | 1,600 | 31.5 | 3 | 169 | 31.5 | 85 |
| 169 | 1.0 | 2,000 | 40 | 3 | 169 | 40 | 108 |
| 169 | 1.0 | 2,000 | 50 | 3 | 169 | 50 | 135 |
| 169 | 1.0 | 2,000 | 63 | 3 | 169 | 63 | 170 |
| 242 | 1.0 | 1,600, 2,000, 3,000 | 31.5 | 3 | 242 | 31.5 | 85 |
| 242 | 1.0 | 2,000, 3,000 | 40 | 3 | 242 | 40 | 108 |
| 242 | 1.0 | 2,000 | 50 | 3 | 242 | 50 | 135 |
| 242 | 1.0 | 2,000, 3,000 | 63 | 3 | 242 | 63 | 170 |
| 362 | 1.0 | 2,000, 3,000 | 40 | 2 | 362 | 40 | 108 |
| 362 | 1.0 | 2,000 | 63 | 2 | 362 | 63 | 170 |
| 550 | 1.0 | 2,000, 3,000 | 40 | 2 | 550 | 40 | 108 |
| 550 | 1.0 | 3,000 | 63 | 2 | 550 | 63 | 170 |
| 800 | 1.0 | 2,000, 3,000 | 40 | 2 | 800 | 40 | 108 |
| 800 | 1.0 | 3,000 | 63 | 2 | 800 | 63 | 170 |

## Definite Purpose Circuit Breakers

These are used for the switching of shunt capacitors. Preferred ratings of such circuit breakers from ANSI Standard C37.06, Tables C-1A, C-2A, and C-3A, for indoor, outdoor, and gas insulated switchgear are:

Table C-1A Preferred Capacitance Current Switching Ratings for Indoor Circuit Breakers (4.76 kV–38 kV).

Table C-2A Preferred Capacitance Current Switching Ratings for Outdoor Circuit Breakers 72.5 kV and below Including Circuit Breakers in Gas Insulated Substations (15.5 kV–72.5 kV).

Table C-3A Preferred Capacitance Current Switching Ratings for Outdoor Circuit Breakers 121 kV and Above Including Circuit Breakers Applied in Gas Insulated Substations (121 kV–800 kV).

## REFERENCES

1. ANSI Standard C37.06 (2000), Preferred Ratings and Related Required Capabilities for AC High Voltage Breakers.

2. IEEE Standard 141 (1993), Recommended Practice for Electric Power Distribution for Industrial Plants.

TABLE C-1A  Preferred Capacitance Current Switching Ratings for Indoor Oilless Circuit Breakers

| Rated Maximum Voltage kV, rms | Rated Short Circuit Current kA, rms | Rated Continuous Current A, rms | General Purpose Circuit Breakers Rated Capacitance Switching Current Shunt Capacitor Bank or Cable Isolated Current A, rms | Definite Purpose Circuit Breakers Rated Capacitance Switching Current | | | |
|---|---|---|---|---|---|---|---|
| | | | | Overhead Line Current A, rms | Shunt Capacitor Bank or Cable Isolated Current A, rms | Back-to-Back Inrush Current Peak Current kA | Frequency Hz |
| 4.76 | 8.80 | 1,200 | 400 | 630 | 630 | 15 | 2,000 |
| 4.76 | 29.00 | 1,200 | 400 | 630 | 630 | 15 | 2,000 |
| 4.76 | 29.00 | 2,000 | 400 | 1,000 | 1,000 | 15 | 1,270 |
| 4.76 | 41.00 | 1,200, 2,000 | 400 | 630 | 630 | 15 | 2,000 |
| 4.76 | 41.00 | 3,000 | 400 | 1,000 | 1,000 | 15 | 1,270 |
| 8.25 | 33.00 | 1,200 | 250 | 630 | 630 | 15 | 2,000 |
| 8.25 | 33.00 | 2,000 | 250 | 1,000 | 1,000 | 15 | 1,270 |
| 15.00 | 18.00 | 1,200 | 250 | 630 | 630 | 15 | 2,000 |
| 15.00 | 18.00 | 2,000 | 250 | 1,000 | 1,000 | 15 | 1,270 |
| 15.00 | 28.00 | 1,200 | 250 | 630 | 630 | 15 | 2,000 |
| 15.00 | 28.00 | 2,000 | 250 | 1,000 | 1,000 | 15 | 1,270 |
| 15.00 | 37.00 | 1,200 | 250 | 630 | 630 | 15 | 2,000 |
| 15.00 | 37.00 | 2,000 | 250 | 1,000 | 1,000 | 18 | 2,400 |
| 15.00 | 37.00 | 3,000 | 250 | 1,600 | 1,600 | 25 | 1,330 |
| 38.00 | 21.00 | 1,200, 2,000, 3,000 | 50 | 250 | 250 | 18 | 6,000 |
| 38.00 | 40.00 | 1,200, 3,000 | 50 | 250 | 250 | 25 | 8,480 |

**TABLE C-2A** Preferred Capacitance Current Switching Ratings for Outdoor Circuit Breakers 72.5 kV and Below, Including Circuit Breakers Applied in Gas Insulated Substations

| Rated Maximum Voltage kV, rms | Rated Short Circuit Current kA, rms | Rated Continuous Current A, rms | General Purpose Circuit Breakers Rated Capacitance Switching Current — Shunt Capacitor Bank or Cable — Isolated Current A, rms | Definite Purpose Circuit Breakers Rated Capacitance Switching Current | | | | |
|---|---|---|---|---|---|---|---|---|
| | | | | Shunt Capacitor Bank or Cable | | Back-to-Back | | |
| | | | | Overhead Line Current A, rms | Isolated Current A, rms | Inrush Current | | |
| | | | | | | Peak Current kA | Frequency Hz | |
| 15.5 | 12.5 | 600, 1,200 | 250 | 100 | 400 | 20 | 4,240 | |
| 15.5 | 20.2 | 1,200, 2,000 | 250 | 100 | 400 | 20 | 4,240 | |
| 15.5 | 25.0 | 1,200, 2,000 | 250 | 100 | 400 | 20 | 4,240 | |

| 15.5 | 40.0 | 1,200, 2,000, 3,000 | 250 | 100 | 400 | 20 | 4,240 |
|---|---|---|---|---|---|---|---|
| 25.8 | 12.5 | 1,200, 2,000 | 160 | 100 | 400 | 20 | 4,240 |
| 25.8 | 25.0 | 1,200, 2,000 | 160 | 100 | 400 | 20 | 4,240 |
| 38.0 | 16.0 | 1,200, 2,000 | 100 | 100 | 250 | 20 | 4,240 |
| 38.0 | 20.0 | 1,200, 2,000 | 100 | 100 | 250 | 20 | 4,240 |
| 38.0 | 25.0 | 1,200, 2,000 | 100 | 100 | 250 | 20 | 4,240 |
| 38.0 | 31.5 | 1,200, 2,000 | 100 | 100 | 250 | 20 | 4,240 |
| 38.0 | 40.0 | 1,200, 2,000, 3,000 | 100 | 100 | 250 | 20 | 4,240 |
| 48.3 | 20.0 | 1,200, 2,000 | 10 | 100 | 250 | 20 | 6,800 |
| 48.3 | 31.5 | 1,200, 2,000 | 10 | 100 | 250 | 20 | 6,800 |
| 48.3 | 40.0 | 1,200, 2,000, 3,000 | 10 | 100 | 250 | 20 | 6,800 |
| 72.5 | 20.0 | 1,200, 2,000 | 20 | 100 | 630 | 25 | 3,360 |
| 72.5 | 31.5 | 1,200, 2,000 | 20 | 100 | 630 | 25 | 3,360 |
| 72.5 | 40.0 | 1,200, 2,000, 3,000 | 20 | 100 | 630 | 25 | 3,360 |

**TABLE C-3A** Preferred Capacitance Current Switching Ratings for Outdoor Circuit Breakers 121 kV and Above, Including Circuit Breakers Applied in Gas Insulated Substations

| Rated Maximum Voltage kV, rms | Rated Short Circuit Current kA, rms | Rated Continuous Current A, rms | General Purpose Circuit Breakers Rated Capacitance Switching Current — Shunt Capacitor Bank or Cable | | Definite Purpose Circuit Breakers Rated Capacitance Switching Current | | | | |
|---|---|---|---|---|---|---|---|---|---|
| | | | | | Shunt Capacitor Bank or Cable | | Back-to-Back | Inrush Current | |
| | | | Overhead Line Current A, rms | Isolated Current A, rms | Overhead Line Current A, rms | Isolated Current A, rms | Isolated Current A, rms | Peak kA | Frequency Hz |
| 121 | 20 | 1,200 | 50 | 50 | 160 | 315 | 315 | 16 | 4,250 |
| 121 | 40 | 1,600, 2,000, 3,000 | 50 | 50 | 160 | 315 | 315 | 16 | 4,250 |
| 121 | 63 | 2,000, 3,000 | 50 | 50 | 160 | 315 | 315 | 16 | 4,250 |

| | | | | | | | | | |
|---|---|---|---|---|---|---|---|---|---|
| 145 | 20 | 1,200, 2,000 | 63 | 63 | 160 | 315 | 315 | 16 | 4,250 |
| 145 | 40 | 1,600, 2,000, 3,000 | 80 | 80 | 160 | 315 | 315 | 16 | 4,250 |
| 145 | 63 | 2,000, 3,000 | 80 | 80 | 160 | 315 | 315 | 16 | 4,250 |
| 145 | 80 | 2,000, 3,000 | 80 | 80 | 160 | 315 | 315 | 16 | 4,250 |
| 169 | 16 | 1,200 | 100 | 100 | 160 | 400 | 400 | 20 | 4,250 |
| 169 | 31.5 | 1,600 | 100 | 100 | 160 | 400 | 400 | 20 | 4,250 |
| 169 | 40 | 2,000 | 100 | 100 | 160 | 400 | 400 | 20 | 4,250 |
| 169 | 50 | 2,000 | 100 | 100 | 160 | 400 | 400 | 20 | 4,250 |
| 169 | 63 | 2,000 | 100 | 100 | 160 | 400 | 400 | 20 | 4,250 |
| 242 | 31.5 | 1,600, 2,000, 3,000 | 160 | 160 | 200 | 400 | 400 | 20 | 4,250 |
| 242 | 40 | 2,000, 3,000 | 160 | 160 | 200 | 400 | 400 | 20 | 4,250 |
| 242 | 50 | 2,000 | 160 | 160 | 200 | 400 | 400 | 20 | 4,250 |
| 242 | 63 | 2,000, 3,000 | 160 | 160 | 200 | 400 | 400 | 20 | 4,250 |
| 362 | 40 | 2,000, 3,000 | 250 | 250 | 315 | 500 | 500 | 25 | 4,250 |
| 362 | 63 | 2,000 | 250 | 250 | 315 | 500 | 500 | 25 | 4,250 |
| 550 | 40 | 2,000, 3,000 | 400 | 400 | 500 | 500 | 500 | | |
| 550 | 63 | 3,000 | 400 | 400 | 500 | 500 | 500 | | |
| 800 | 40 | 2,000, 3,000 | 500 | 500 | 500 | 500 | 500 | | |
| 800 | 63 | 3,000 | 500 | 500 | 500 | 500 | 500 | | |

# Appendix D

## SURGE ARRESTER DATA

Surge arresters are used to protect the power system equipment from overvoltages produced due to switching and lightning. Before the 1980s, gaped silicon carbide arresters were used. Metal Oxide Varistors (MOV) were introduced in the 1980s for the same applications. Presently, MOV arresters are used in the protection of overhead lines, underground cables, transformers, circuit breakers, shunt capacitors, and other power system equipment. The following MOV surge arrester ratings are reproduced from IEEE Standard 141.

Table D-1 Station Class MOV Arrester Characteristics
Table D-2 Intermediate Class MOV Arrester
  Characteristics
Table D-3 Distribution Class MOV Arrester
  Characteristics, Normal Duty
Table D-4 Distribution Class MOV Arrester
  Characteristics, Heavy Duty
Table D-5 Distribution Class MOV Arrester
  Characteristics, Riser Pole

**TABLE D-1**   Station Class MOV Arrester Characteristics

| kV, rms | MCOV kV, rms | FOW kV, Peak | Discharge Peak kV at Indicated Impulse Current for an 8/20 Wave | | | | | | SSP kV, Peak |
|---|---|---|---|---|---|---|---|---|---|
| | | | 1.5 kA | 3 kA | 5 kA | 10 kA | 20 kA | 40 kA | |
| 3 | 2.55 | 9.1 | 6.9 | 7.2 | 7.5 | 8.0 | 9.0 | 10.3 | 6.3 |
| 6 | 5.10 | 17.9 | 13.6 | 14.2 | 14.8 | 15.8 | 17.7 | 20.3 | 12.4 |
| 9 | 7.65 | 26.6 | 20.2 | 21.2 | 22.0 | 23.5 | 26.4 | 30.2 | 18.4 |
| 10 | 8.40 | 29.3 | 22.2 | 23.3 | 24.2 | 25.9 | 29.1 | 33.3 | 20.3 |
| 12 | 10.20 | 35.5 | 26.9 | 28.2 | 29.4 | 31.4 | 35.2 | 40.4 | 24.6 |
| 15 | 12.70 | 44.2 | 33.5 | 35.1 | 36.6 | 39.1 | 43.9 | 50.3 | 30.6 |
| 18 | 15.30 | 53.3 | 40.4 | 42.3 | 44.1 | 47.1 | 52.8 | 60.6 | 36.8 |
| 21 | 17.00 | 59.1 | 44.8 | 46.9 | 48.9 | 52.3 | 58.7 | 67.2 | 40.9 |
| 24 | 19.50 | 67.8 | 51.4 | 53.8 | 56.1 | 60.0 | 67.3 | 77.1 | 46.9 |
| 27 | 22.00 | 76.5 | 58.0 | 60.8 | 63.3 | 67.7 | 75.9 | 87.0 | 52.9 |
| 30 | 24.40 | 84.9 | 64.3 | 67.4 | 70.3 | 75.1 | 84.2 | 96.5 | 58.7 |
| 36 | 29.00 | 101.0 | 76.4 | 80.0 | 83.4 | 8.2 | 100.0 | 115.0 | 69.7 |
| 39 | 31.50 | 110.0 | 83.0 | 86.9 | 90.6 | 96.9 | 109.0 | 125.0 | 75.8 |
| 45 | 36.50 | 128.0 | 96.8 | 102.0 | 106.0 | 113.0 | 127.0 | 146.0 | 88.3 |
| 48 | 39.00 | 136.0 | 103.0 | 108.0 | 113.0 | 120.0 | 135.0 | 155.0 | 93.8 |
| 54 | 42.00 | 135.0 | 105.0 | 112.0 | 115.0 | 122.0 | 136.0 | 151.0 | 98.0 |
| 60 | 48.00 | 154.0 | 120.0 | 127.0 | 131.0 | 139.0 | 155.0 | 173.0 | 110.0 |
| 72 | 57.00 | 183.0 | 142.0 | 151.0 | 156.0 | 165.0 | 184.0 | 205.0 | 131.0 |
| 90 | 70.00 | 223.0 | 174.0 | 184.0 | 190.0 | 202.0 | 226.0 | 251.0 | 161.0 |
| 90 | 70.00 | 236.0 | 185.0 | 195.0 | 202.0 | 214.0 | 237.0 | 266.0 | 169.0 |
| 96 | 74.00 | 242.0 | 190.0 | 201.0 | 208.0 | 220.0 | 245.0 | 274.0 | 175.0 |
| 108 | 76.00 | 267.0 | 209.0 | 221.0 | 229.0 | 243.0 | 271.0 | 301.0 | 193.0 |
| 108 | 84.00 | 279.0 | 219.0 | 232.0 | 239.0 | 254.0 | 284.0 | 316.0 | 202.0 |
| 120 | 98.00 | 311.0 | 244.0 | 257.0 | 266.0 | 283.0 | 315.0 | 351.0 | 231.0 |
| 132 | 106.00 | 340.0 | 264.0 | 280.0 | 289.0 | 306.0 | 342.0 | 381.0 | 249.0 |
| 144 | 115.00 | 368.0 | 287.0 | 303.0 | 314.0 | 332.0 | 369.0 | 413.0 | 271.0 |
| 168 | 131.00 | 418.0 | 326.0 | 345.0 | 357.0 | 379.4 | 421.0 | 470.0 | 308.0 |
| 172 | 140.00 | 446.0 | 348.0 | 368.0 | 381.0 | 404.0 | 448.0 | 502.0 | 330.0 |
| 180 | 144.00 | 458.0 | 359.0 | 380.0 | 392.0 | 417.0 | 463.0 | 517.0 | 339.0 |
| 192 | 152.00 | 483.0 | 379.0 | 401.0 | 414.0 | 440.0 | 488.0 | 546.0 | 360.0 |
| 228 | 182.00 | 571.0 | 447.0 | 474.0 | 489.0 | 520.0 | 578.0 | 645.0 | 424.0 |

Notes:

MCOV = Maximum continuous overvoltage.

FOW = Front of wave protective level.

SSP = Maximum switching surge protective level.

TABLE **D-2** Intermediate Class MOV Arrester Characteristics

| kV, rms | MCOV kV, rms | FOW kV, Peak | Discharge Peak kV at Indicated Impulse Current for an 8/20 Wave | | | | | | SSP kV, Peak |
|---|---|---|---|---|---|---|---|---|---|
| | | | 1.5 kA | 3 kA | 5 kA | 10 kA | 20 kA | 40 kA | |
| 3 | 2.55 | 10.4 | 6.6 | 7.2 | 7.5 | 8.0 | 9.3 | 10.8 | 5.9 |
| 6 | 5.10 | 18.9 | 13.1 | 14.2 | 14.8 | 16.2 | 18.2 | 21.2 | 11.7 |
| 9 | 7.65 | 30.5 | 22.0 | 23.5 | 23.5 | 26.0 | 31.5 | 38.0 | 20.0 |
| 10 | 8.40 | 33.5 | 24.5 | 28.0 | 28.0 | 29.0 | 35.0 | 42.0 | 22.5 |
| 12 | 10.20 | 41.0 | 30.0 | 31.5 | 31.5 | 35.5 | 42.5 | 51.0 | 27.5 |
| 15 | 12.70 | 61.0 | 37.0 | 39.5 | 39.5 | 44.0 | 52.5 | 61.5 | 34.0 |
| 18 | 15.30 | 61.0 | 44.5 | 48.0 | 48.0 | 52.0 | 63.0 | 77.0 | 40.5 |
| 21 | 17.00 | 68.5 | 49.5 | 53.5 | 53.5 | 59.0 | 70.5 | 95.5 | 45.5 |
| 24 | 19.50 | 78.0 | 57.0 | 60.0 | 60.0 | 67.0 | 81.0 | 98.0 | 52.0 |
| 27 | 22.00 | 88.0 | 64.0 | 68.5 | 68.5 | 76.0 | 91.0 | 110.0 | 58.5 |
| 30 | 24.40 | 97.5 | 71.0 | 76.0 | 76.0 | 84.5 | 101.0 | 122.0 | 66.0 |
| 36 | 29.00 | 116.0 | 84.0 | 91.0 | 91.0 | 101.0 | 121.0 | 145.0 | 78.0 |
| 39 | 31.50 | 126.0 | 91.5 | 98.0 | 98.0 | 109.0 | 131.0 | 158.0 | 84.0 |
| 45 | 36.50 | 146.0 | 106.0 | 114.0 | 114.0 | 126.0 | 152.0 | 183.0 | 97.0 |
| 48 | 39.00 | 156.0 | 113.0 | 122.0 | 122.0 | 135.0 | 163.0 | 195.0 | 104.0 |
| 54 | 42.00 | 168.0 | 122.0 | 130.0 | 130.0 | 145.0 | 174.0 | 210.0 | 112.5 |
| 60 | 48.00 | 191.0 | 13.0 | 149.0 | 149.0 | 165.0 | 198.0 | 239.0 | 127.0 |
| 72 | 57.00 | 227.0 | 165.0 | 177.0 | 177.0 | 196.0 | 236.0 | 284.0 | 151.0 |
| 90 | 70.00 | 280.0 | 203.0 | 218.0 | 218.0 | 242.0 | 290.0 | 351.0 | 186.0 |
| 90 | 70.00 | 294.0 | 214.0 | 230.0 | 230.0 | 255.0 | 306.0 | 370.0 | 196.0 |
| 96 | 74.00 | 303.0 | 220.0 | 236.0 | 236.0 | 262.0 | 314.0 | 379.0 | 201.0 |
| 108 | 76.00 | 335.0 | 244.0 | 261.0 | 261.0 | 290.0 | 348.0 | 420.0 | 223.0 |
| 108 | 84.00 | 350.0 | 254.0 | 273.0 | 273.0 | 303.0 | 364.0 | 439.0 | 233.0 |
| 120 | 98.00 | 390.0 | 284.0 | 304.0 | 321.0 | 336.0 | 406.0 | 490.0 | 260.0 |

Notes:

MCOV = Maximum continuous overvoltage.

FOW = Front of wave protective level.

SSP = Maximum switching surge protective level.

**TABLE D-3**  Distribution  Class  MOV  Arrester  Characteristics, Normal Duty

| kV, rms | MCOV kV, rms | FOW kV, Peak | Discharge Peak kV at Indicated Impulse Current for an 8/20 Wave | | | | | | SSP kV, Peak |
|---|---|---|---|---|---|---|---|---|---|
| | | | 1.5 kA | 3 kA | 5 kA | 10 kA | 20 kA | 40 kA | |
| 3 | 2.55 | 12.5 | 9.8 | 10.3 | 11.0 | 12.3 | 14.3 | 18.5 | 8.5 |
| 6 | 5.10 | 25.0 | 19.5 | 20.5 | 22.0 | 24.5 | 28.5 | 37.0 | 17.0 |
| 9 | 7.65 | 33.5 | 26.0 | 28.0 | 30.0 | 33.0 | 39.0 | 50.5 | 23.0 |
| 10 | 8.40 | 36.0 | 27.0 | 29.5 | 31.5 | 36.0 | 41.5 | 53.0 | 24.0 |
| 12 | 10.20 | 50.0 | 39.0 | 41.0 | 44.0 | 49.0 | 57.0 | 74.0 | 34.0 |
| 15 | 12.70 | 58.5 | 45.5 | 48.5 | 52.0 | 57.5 | 67.5 | 87.5 | 40.0 |
| 18 | 15.30 | 67.0 | 52.0 | 56.0 | 60.0 | 66.0 | 76.0 | 101.0 | 46.0 |
| 21 | 17.00 | 73.0 | 55.0 | 60.0 | 64.0 | 73.0 | 84.0 | 107.0 | 49.0 |
| 24 | 19.50 | 92.0 | 71.5 | 76.5 | 82.0 | 90.5 | 106.5 | 138.0 | 63.0 |
| 27 | 22.00 | 100.5 | 78.0 | 84.0 | 90.0 | 99.0 | 117.0 | 151.5 | 69.0 |
| 30 | 24.40 | 108.0 | 81.0 | 88.5 | 94.5 | 108.0 | 124.5 | 159.0 | 72.0 |

Notes:

MCOV = Maximum continuous overvoltage.

FOW = Front of wave protective level.

SSP = Maximum switching surge protective level.

**TABLE D-4**  Distribution  Class  MOV  Arrester  Characteristics, Heavy Duty

| kV, rms | MCOV kV, rms | FOW kV, Peak | Discharge Peak kV at Indicated Impulse Current for an 8/20 Wave | | | | | | SSP kV, Peak |
|---|---|---|---|---|---|---|---|---|---|
| | | | 1.5 kA | 3 kA | 5 kA | 10 kA | 20 kA | 40 kA | |
| 3 | 2.55 | 12.5 | 9.5 | 10.0 | 10.5 | 11.0 | 13.0 | 15.3 | 8.0 |
| 6 | 5.10 | 25.0 | 19.0 | 20.0 | 21.0 | 22.0 | 26.0 | 30.5 | 16.0 |
| 9 | 7.65 | 34.0 | 24.5 | 26.0 | 27.5 | 30.0 | 35.0 | 41.0 | 22.5 |
| 10 | 8.40 | 36.5 | 26.0 | 28.0 | 29.5 | 32.0 | 37.5 | 43.5 | 23.5 |
| 12 | 10.20 | 50.0 | 38.0 | 40.0 | 42.0 | 44.0 | 52.0 | 61.0 | 32.0 |
| 15 | 12.70 | 59.0 | 43.5 | 46.0 | 48.5 | 52.0 | 61.0 | 71.5 | 38.5 |
| 18 | 15.30 | 68.0 | 49.0 | 52.0 | 55.0 | 60.0 | 70.0 | 82.0 | 45.0 |
| 21 | 17.00 | 75.0 | 53.0 | 57.0 | 60.0 | 65.0 | 76.0 | 88.5 | 48.0 |
| 24 | 19.50 | 93.0 | 68.0 | 72.0 | 76.0 | 82.0 | 96.0 | 112.5 | 61.0 |
| 27 | 22.00 | 102.0 | 73.5 | 78.0 | 82.5 | 90.0 | 105.0 | 123.0 | 67.5 |
| 30 | 24.40 | 109.5 | 78.0 | 84.0 | 88.5 | 96.0 | 112.5 | 130.5 | 70.5 |
| 36 | 29.00 | 136.0 | 98.0 | 104.0 | 110.0 | 120.0 | 140.0 | 164 | 90.0 |

Notes:

MCOV = Maximum continuous overvoltage.

FOW = Front of wave protective level.

SSP = Maximum switching surge protective level.

TABLE **D-5** Distribution Class MOV Arrester Characteristics, Riser Pole

| kV, rms | MCOV kV, rms | FOW kV, Peak | Discharge Peak kV at Indicated Impulse Current for an 8/20 Wave | | | | | | SSP kV, Peak |
|---|---|---|---|---|---|---|---|---|---|
| | | | 1.5 kA | 3 kA | 5 kA | 10 kA | 20 kA | 40 kA | |
| 3 | 2.55 | – | – | – | – | – | – | – | – |
| 6 | 5.10 | 17.4 | 13.0 | 14.0 | 14.7 | 16.2 | 18.1 | 21.1 | 11.7 |
| 9 | 7.65 | 25.7 | 19.3 | 21.0 | 21.9 | 24.0 | 27.0 | 31.6 | 17.5 |
| 10 | 8.40 | 28.5 | 21.2 | 23.0 | 24.0 | 26.5 | 29.8 | 34.8 | 19.2 |
| 12 | 10.20 | 34.8 | 25.9 | 28.0 | 29.4 | 32.3 | 36.2 | 42.2 | 23.3 |
| 15 | 12.70 | 43.1 | 32.3 | 36.0 | 36.6 | 40.2 | 46.1 | 52.7 | 29.1 |
| 18 | 15.30 | 51.4 | 38.6 | 41.9 | 43.8 | 48.0 | 54.0 | 63.2 | 34.9 |
| 21 | 17.00 | 57.6 | 42.8 | 46.4 | 48.6 | 53.6 | 60.2 | 70.5 | 38.7 |
| 24 | 19.50 | 68.6 | 51.6 | 55.9 | 58.5 | 64.2 | 72.1 | 84.3 | 46.6 |
| 27 | 22.00 | 77.1 | 57.9 | 62.9 | 65.7 | 72.0 | 81.0 | 94.8 | 52.4 |
| 30 | 24.40 | 88.5 | 63.5 | 69.0 | 72.0 | 79.5 | 89.4 | 104.4 | 57.6 |
| 36 | 29.00 | 102.8 | 77.2 | 83.8 | 87.6 | 96.0 | 108.8 | 126.4 | 69.8 |

Notes:
MCOV = Maximum continuous overvoltage.
FOW = Front of wave protective level.
SSP = Maximum switching surge protective level.

## REFERENCES

1. IEEE Standard 141 (1993), Recommended Practice for Electric Power Distribution for Industrial Plants.

2. IEEE Standard 1036 (1992), IEEE Guide for Application of Shunt Capacitors.

3. IEEE Standard C62.22 (1991), IEEE Guide for the Application of Metal Oxide Surge Arresters for Alternating-Current Systems.

# Appendix E

## FUSE CHARACTERISTICS

### E-1  EXPULSION FUSES

Two types of expulsion fuses are available for capacitor bank protection:

- K-link fuses for faster speed
- T-link fuses for slower speed

The needed information for the expulsion type of fuses such as cut outs, voltage, current, and testing methods are presented in ANSI Standard C37.42. The minimum melting time-current characteristics of the expulsion fuses are presented in this appendix in two figures and two tables:

Figure E-1  Minimum melting time-current characteristics of the K-link fuses

Figure E-2  Minimum melting time-current characteristics of the T-link fuses

Table E-1  Melting Currents for Type K (Fast) Fuse Links

Table E-2  Melting Currents for Type T (Slow) Fuse Links

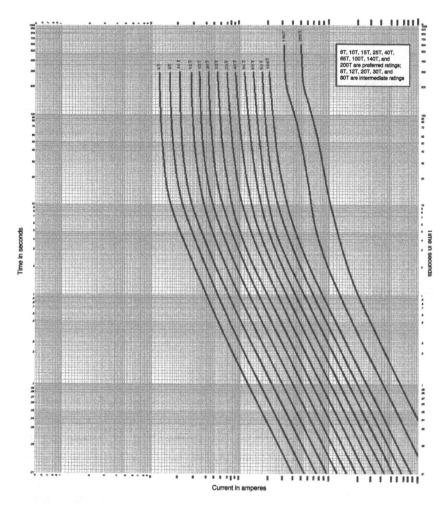

6T, 10T, 15T, 25T, 40T, 65T, 100T, 140T, and 200T are preferred ratings; 8T, 12T, 20T, 30T, and 80T are intermediate ratings

Time in seconds

Time in seconds

Current in amperes

**Figure E-1** Typical minimum melting time-current characteristic curves of the K-link fuses. (Courtesy of S&C Electric Company, Chicago.)

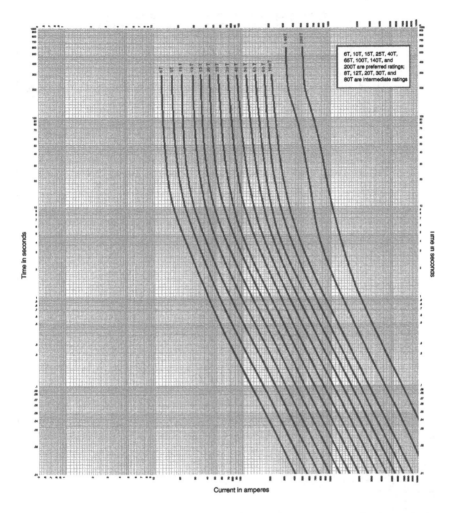

**Figure E-2** Typical minimum melting time-current characteristic curves of the T-link fuses. (Courtesy of S&C Electric Company, Chicago.)

**TABLE E-1** Melting Currents for Type K (Fast) Fuse Links

| Rated Continuous Current A | 300 or 600 s Melting Current A | | 10 s Melting Current A | | 0.1 s Melting Current A | | Speed Ratio |
|---|---|---|---|---|---|---|---|
| | Minimum | Maximum | Minimum | Maximum | Minimum | Maximum | |
| **Preferred Ratings** | | | | | | | |
| 6 | 12 | 14.4 | 13.5 | 20.5 | 72 | 86 | 6.0 |
| 10 | 19.5 | 23.4 | 22.5 | 34 | 128 | 154 | 6.6 |
| 15 | 31 | 37.2 | 37 | 55 | 215 | 258 | 6.9 |
| 25 | 50 | 60 | 60 | 90 | 360 | 420 | 7.0 |
| 40 | 80 | 96 | 98 | 148 | 565 | 680 | 7.1 |
| 65 | 128 | 153 | 159 | 237 | 918 | 1100 | 7.2 |
| 100 | 200 | 240 | 258 | 388 | 1520 | 1820 | 7.6 |
| 140 | 310 | 372 | 430 | 650 | 2470 | 2970 | 8.0 |
| 200 | 480 | 576 | 760 | 1150 | 3880 | 4650 | 8.1 |
| **Intermediate Ratings** | | | | | | | |
| 8 | 15 | 18 | 18 | 27 | 97 | 116 | 6.5 |
| 12 | 25 | 30 | 29.5 | 44 | 166 | 199 | 6.6 |
| 20 | 39 | 47 | 48 | 71 | 273 | 328 | 7.0 |
| 30 | 63 | 76 | 77.5 | 115 | 447 | 546 | 7.1 |
| 50 | 101 | 121 | 126 | 188 | 719 | 862 | 7.1 |
| 80 | 160 | 192 | 205 | 307 | 1180 | 1420 | 7.4 |
| **Ratings below 6 A** | | | | | | | |
| 1 | 2 | 2.4 | 1 | 10 | 1 | 58 | – |
| 2 | 4 | 4.8 | 1 | 10 | 1 | 58 | – |
| 3 | 6 | 7.2 | 1 | 10 | 1 | 58 | – |

TABLE E-2 Melting Currents for Type T (Slow) Fuse Links

| Rated Continuous Current A | 300 or 600 s Melting Current A | | 10 s Melting Current A | | 0.1 s Melting Current A | | Speed Ratio |
|---|---|---|---|---|---|---|---|
| | Minimum | Maximum | Minimum | Maximum | Minimum | Maximum | |
| Preferred Ratings | | | | | | | |
| 6 | 12 | 14.4 | 15.3 | 23 | 120 | 144 | 10.0 |
| 10 | 19.5 | 23.4 | 23.4 | 40 | 224 | 269 | 11.5 |
| 15 | 31 | 37.2 | 37.2 | 67 | 388 | 466 | 12.5 |
| 25 | 50 | 60 | 60 | 109 | 635 | 762 | 12.7 |
| 40 | 80 | 96 | 96 | 178 | 1040 | 1240 | 13.0 |
| 65 | 128 | 153 | 153 | 291 | 1650 | 1975 | 12.9 |
| 100 | 200 | 240 | 240 | 475 | 2620 | 3150 | 12.1 |
| 140 | 310 | 372 | 372 | 775 | 4000 | 4800 | 12.9 |
| 200 | 480 | 576 | 576 | 1275 | 6250 | 7470 | 13.0 |
| Intermediate Ratings | | | | | | | |
| 8 | 15 | 18 | 20.5 | 31 | 166 | 199 | 11.1 |
| 12 | 25 | 30 | 34.5 | 52 | 296 | 355 | 11.8 |
| 20 | 39 | 47 | 57 | 85 | 496 | 595 | 12.7 |
| 30 | 63 | 76 | 93 | 138 | 812 | 975 | 12.9 |
| 50 | 101 | 121 | 152 | 226 | 1310 | 1570 | 13.0 |
| 80 | 160 | 192 | 248 | 370 | 2080 | 2500 | 13.0 |
| Ratings below 6 A | | | | | | | |
| 1 | 2 | 2.4 | 1 | 11 | 1 | 100 | – |
| 2 | 4 | 4.8 | 1 | 11 | 1 | 100 | – |
| 3 | 6 | 7.2 | 1 | 11 | 1 | 100 | – |

# REFERENCE

1. ANSI Standard 37.42 (1981), Specifications for Distribution Cutouts and Fuse Links.

# BIBLIOGRAPHY

## General

Alexander, R., Synchronous closing control for shunt capacitor banks, *IEEE Transactions on Power Apparatus and Systems*, Vol. 104, No. 8, September 1985.

Andrei, R., Ahmed, M., Tumageanian, H.K., and Smith, J.C., World's first commercial bridge capacitor bank installation on the American electric power system, *IEEE Transactions on Power Delivery*, Vol. 16, No. 2, April 2001, pp. 342–345.

Andrei, R.G., A novel fuseless capacitor bank design using conventional single bushing capacitors, *IEEE Transactions on Power Delivery*, Vol. 14, No. 3, July 1999, pp. 1124–1133.

Andrei, R.G., Kaushik, R.R., and Reinaker, R.W., Bridge capacitor bank design and operation, *IEEE Transactions on Power Delivery*, Vol. 11, No. 1, January 1996, pp. 227–233.

Bayless, R.S., Selman, J.D., Truax, D.E., and Reid, N.E., Capacitor switching and transformer transients, *IEEE Transactions on Power Delivery*, Vol. 3, No. 3, January 1988, pp. 349–357.

Bendler, J.T. and Takekoshi, T., Molecular modeling of polymers for high energy storage capacitor applications, *IEEE 35th International Power Sources Symposium*, 1992, pp. 373–376.

Bhargava, B., Khan, A.H., Imece, A.F., and DiPietre, J., Effectiveness of pre-insertion inductor for mitigating remote overvoltages due to shunt capacitor energization, *IEEE Transactions on Power Delivery*, Vol. 8, No. 3, July 1993.

Bolduc, L., Bouchard B., and Beaulieu, G., Capacitive divider substation, *IEEE Transactions on Power Delivery*, Vol. 12, No. 3, July 1997, pp. 1202–1209.

Brunello, G. Kasztenny, B., and Wester, C., Shunt capacitor bank fundamentals and protection, *Proceedings of the 2003 Conference for Protective Relay Engineers*, Texas A&M University, College Station, April 8–10, 2003, 15 pp.

Carlisle, J.C. and El-Keib, A.A., A graph search algorithm for optimal placement of fixed and switched capacitors on radial distribution systems, *IEEE Transactions on Power Delivery*, Vol. 15, No. 1, January 2000, pp. 423–428.

Choe, Y.M., Gho, J.S., Hok, H.S., Choe, G.H., and Shin, W.S., A new instantaneous output control method for inverter arc welding machine, *Power Electronics Specialist Conference*, Vol. 1, July 1999, pp. 521–526.

CIRGE Working Group 13.04, Capacitive Current Switching — State of the Art, Électra No. 155, August 1994, 33 pp.

Clerici, G., The supercapacitor, *Tecnologie Elettriche*, Vol. 20, No. 6, July/August 1993, pp. 64–67.

Cox, M.D. and Guan, H.H., Vibration and audible noise of capacitors subjected to non-sinusoidal waveforms, *IEEE Transactions on Power Delivery*, Vol. 9, No. 2, April 1994, pp. 856–862.

Delfanti, M., Granelli, G.P., Marannino, P., and Montagna, M., Optimal capacitor placement using deterministic and genetic algorithms, *IEEE Transactions on Power Systems*, Vol. 15, No. 3, August 2000, pp. 1041–1046.

Duff, W.B., Jr., Economic AC capacitors, *IEEE Power Engineering Review*, Vol. 22, No. 1, January 2002, 4 pp.

El-Amin, I.M., Duffuaa, S.O., and Bawah, A.U., Optimal shunt compensators at nonsinusoidal busbars, *IEEE Transactions on Power Systems*, Vol. 10, No. 2, May 1995, pp. 716–723.

Fenner, G.E., Considerations for specifying 69 kV shunt capacitor banks, *Transmission and Distribution World*, Vol. 44, September 1992, pp. 39–44.

Ganatra, L.M., Mysore, P.G., Mustaphi, K.K., Mulawarman, A., Mork, B., and Gopalkumar, G., Application of reclosing schemes in the presence of capacitor bank ringdown, *American Power Conference*, Vol. 61, 1999, pp. 967–972.

Girgis, A.A., Fallon, C.M., Rubino, J.C., and Catoe, R.C., Harmonics and transient overvoltages due to capacitor switching, *IEEE Transactions on Industry Applications*, Vol. 29, No. 6, November/December 1993, pp. 1184–1188.

Henault, P., Recent developments in capacitor switching transient reduction, Edison Electric Institute, Meeting No. 8, Arizona, March 30, 1999, 12 pp.

Horton, R., Fender, T.W., Harry, C.A., and Gross, C.A., Unbalance Protection of fuseless split-wye grounded capacitor banks, *IEEE Transactions on Power Delivery*, Vol. 17, No. 3, July 2002, pp. 698–701.

Huang, Y.C., Yang, H.T., and Huang, C.L., Solving the capacitor placement problem in a radial distribution system using tabusearch approach, *IEEE Transactions on Power Systems*, Vol. 11, No. 4, November 1996, pp. 1868–1873.

Kempler, M.J., Arness, R.M., Rostamkolai, N., and Clouse, S.L., Development of the EMTP model of IPL transmission system and pre-specification studies for design and installation of a 138 kV capacitor bank, *American Power Conference*, Vol. 61, 1999, pp. 97–102.

Kennedy, B., Optimize placement of in-plant power-factor correction capacitors, *Electrical World*, Vol. 209, October 1995.

Levitin, G., Kalyuzhny, A., Shenkman, A., and Chertkov, M., Optimal capacitor allocation in distribution systems using a genetic algorithm and a fast energy loss computation technique, *IEEE Transactions on Power Delivery*, Vol. 15, No. 2, April 2000, pp. 623–628.

McCoy, C.E. and Floryancic, B.L., Characteristics and measurement of capacitor switching at medium voltage distribution level, *IEEE Transactions on Industry Applications*, Vol. 30, No. 6, November/December 1994, pp. 1480–1489.

Mendis, S.R., Bishop, M.T., McCall, J.C., and Hurst, W.M., Overcurrent protection of capacitors applied on industrial distribution systems, *IEEE Transactions on Industry Applications*, Vol. 29, No. 3, May/June 1993, pp. 541–547.

Mendis, S.R., Bishop, M.T., McCall, J.C., and Hurst, W.M., Capacitor overcurrent protection for industrial distribution systems, *IEEE Industry Applications Magazine*, Vol. 2, No. 3, May/June 1996, pp. 20–27.

Mikhail, S.S. and McGranaghan, M.F., Evaluation of switching concerns associated with 345 kV shunt capacitor applications, *IEEE Transactions on Power Delivery*, Vol. 1, No. 2, April 1986, pp. 231–240.

Montanari, G.C. and Fabiani, D., The effect of non-sinusoidal voltage on intrinsic aging of cable and capacitor insulating materials, *IEEE Transactions on Dielectrics and Electrical Insulation*, Vol. 6, No. 6, December 1999, pp. 798–802.

Natarajan, R., Bucci, R.M., Juarez, M.A., and Contreas, R.U., Determination of flicker effects due to an electric arc furnace, *Proceedings of the 1994 American Power Conference*, Vol. 56, April 1994, pp. 1830–1835.

Natarajan, R., Hale, E., Ashmore, S., and Larson, C., A 230 kV power factor correction installation taking into account the low voltage filters, *Proceedings of the 1999 American Power Conference*, Vol. 61, April 1999, pp. 686–691.

Natarajan, R., Martin, D., and Sanjayan, D., Analysis of induced voltages in 230 kV substation control cables, *Proceedings of the 2000 American Power Conference*, Vol. 62, April 2000, pp. 122–126.

Natarajan, R. and Misra, V.K., Starting transient current of induction motors without and with terminal capacitors, *IEEE Transactions on Energy Conversion*, Vol. EC-6, No. 1, March 1991, pp. 134–139.

Natarajan, R., Tomlinson, J.N., and Misra, V.K., Performance of a switched capacitor starter suitable for mining application, *Mining Technology*, Vol. 73, September 1991, pp. 247–250.

Nedwick, P., Mistr, A.F., and Croasdale, E.B., Reactive management a key to survival in the 1990s, *IEEE Transactions on Power Systems*, Vol. 10, No. 2, May 1995, pp. 1036–1043.

Ng, N., Salama, M.A., and Chikhani, A.Y., Capacitor allocation by approximate reasoning: Fuzzy capacitor placement, *IEEE Transactions on Power Delivery*, Vol. 15, No. 1, January 2000, pp. 393–398.

Ng, N., Salama, M.A., and Chikhani, A.Y., Classification of capacitor allocation techniques, *IEEE Transactions on Power Delivery*, Vol. 15, No. 1, January 2000, pp. 387–392.

Okabe, S., Koto, M., Muraoka, T., Suganuma, K., and Tkahashi, K., Techniques for diagnosing deterioration of oil impregnated paper film power capacitors, *IEEE Transactions on Power Delivery*, Vol. 12, No. 4, October 1997, pp. 1751–1759.

Phillips, K.J., Conducting a power factor study, *Consulting-Specifying Engineer*, July 1994, pp. 54–58.

Porter, G.A. and McCall, J.C., *Application and Protection Considerations in Applying Distribution Capacitors*, Pennsylvania Electrical Association, 1990.

Riberro, P.F., Johnson, B.K., Crow, M.L., Arsoy, A., and Liu, Y., Energy storage systems for advanced power applications, *Proceedings of the IEEE*, Vol. 89, No. 12, December 2001, pp. 1744–1756.

Sadek, K., Pereira, M., Brandt, D.P., Gole, A.M., and Daneshpooy, A., Capacitor commutated converter circuit configurations for DC transmission, *IEEE Transactions on Power Delivery*, Vol. 13, No. 4, October 1998, pp. 1257–1264.

Sévigny, R., Ménard, S., Rajotte, C., and McVey, M., Capacitor measurement in the substation environment: A new approach, *Proceedings of the IEEE 9th International Conference on Transmission and Distribution*, October 2000, pp. 299–305.

Sochuliakova, D., Niebur, D., Nwankpa, C.O., Fischi, R., and Richardson, D., Identification of capacitor position in a radial system, *IEEE Transactions on Power Delivery*, Vol. 14, No. 4, October 1999, pp. 1368–1373.

Taylor, C.W. and Leuven, A.L.V., CAPS: Improving power system stability using the time-overvoltage capability of large shunt capacitor banks, *IEEE Transactions on Power Delivery*, Vol. 11, No. 2, April 1996, pp. 783–792.

Trammel, R. and McCarthy, K.D., Capacitor control gives voltage a lift, *Transmission & Distribution World*, Vol. 51, August 1999, 64 pp.

Wang, J.C., Chiang, H.D., Miu, K.N., and Darling, G., Capacitor placement and real time control in large scale unbalanced distribution systems: Loss reduction formula, problem formulation, solution methodology and mathematical justification, *IEEE Transactions on Power Delivery*, Vol. 12, No. 2, April 1997, pp. 953–958.

Witte, J.F., DeCesaro, F.P., and Mendis, S.R., Damaging long-term overvoltages on industrial capacitor banks due to transformer energization inrush currents, *IEEE Transactions on Industry Applications*, Vol. 30, No. 4, July/August 1994, pp. 1107–1115.

Yehling, T.O., Power systems design for a large arc furnace shop, *Proceedings of the IEEE Industry Applications Society Conference Meeting*, October 2–6, 1977, 6 pp.

## Power Capacitors for Harmonic Filters

Bonner, J.A., Hurst, W.M., Rocamora, R.G., Sharp, M.R., Dudley, R.F., and Twiss, J.A., Selecting ratings for capacitors and reactors in applications involving multiple single-tuned filters, *IEEE Transactions on Power Delivery*, Vol. 10, No. 1, January 1995, pp. 547–555.

Currence, E.J., Plizga, J.E., and Nelson, H.N., Harmonic resonance at a medium sized industrial plant, *IEEE Transactions on Industry Applications*, Vol. 31, No. 4, July/August 1995, pp. 682–690.

Day, A.L. and Mahmoud, A.A., Methods of evaluation of harmonic levels in industrial plant distribution systems, *IEEE Transactions on Industry Applications*, Vol. 23, No. 3, May/June 1987, pp. 498–503.

Golhelf, N. and Leuald, A., Harmonic filter for industrial applications, 16th International Conference and Exhibition, IEE Conference Publication No. 482, June 2001.

Gonzalez, D.A. and McCall, J.C., Design of filters to reduce harmonic distortion in industrial power systems, *IEEE Transactions on Industry Applications*, Vol. 23, No. 3, May/June 1987, pp. 504–511.

Harder, J., AC filter arrester application, *IEEE Transactions on Power Delivery*, Vol. 11, No. 3, July 1996, pp. 1355–1360.

Hammond, P.W., A harmonic filter installation to reduce voltage distortion from static power converters, *IEEE Transactions on Industry Applications*, Vol. 24, No. 1, January/February 1980, pp. 53–58.

Kawann, C. and Emanuel, A.E., Passive shunt harmonic filters for low and medium voltage: A cost comparison, *IEEE Transactions on Power Systems*, Vol. 11, No. 4, November 1996, pp. 1825–1831.

Makram, E.B., Subramaniam, E.V., Girgis, A.A., and Cotoe, R.C., Harmonic filter design using actual recorded data, *IEEE Transactions on Industry Applications*, Vol. 29, No. 6, November/December 1993, pp. 1176–1183.

McGranaghan, F. and Mueller, D., Designing harmonic filters for adjustable speed drives to comply with IEEE-519 harmonic limits, *IEEE Transactions on Industry Applications*, Vol. 15, No. 2, March/April 1999, pp. 312–318.

Merhej, S.J. and Nichols, W.H., Harmonic filtering for the offshore industry, *IEEE Transactions on Industry Applications*, Vol. 30, No. 3, May/June 1994, pp. 533–542.

Natarajan, R., Nall, A., and Ingram, D., A harmonic filter installation to improve the power factor and reduce harmonic distortion from multiple converters, *Proceedings of the 1999 American Power Conference*, Vol. 61, April 1999, pp. 680–685.

Stratford, R.P., Rectifier harmonics in power systems, *IEEE Transactions on Industry Applications*, Vol. 16, No. 2, March/April 1980, pp. 271–276.

Sueker, K.H., Hummel, S.D., and Argent, R.D., Power factor correction and harmonic mitigation in a thyristor controlled glass melter, *IEEE Transactions on Industry Applications*, Vol. 25, No. 6, November/December 1989, pp. 972–975.

Willoughby, R.D., Harmonic filters key to plant reliability, *The Line*, October 1996, 7 pp.

Wu, C.J., Chiang, J.C., Liao, C.J., Yang, J.S. and Guo, T.Y., Investigation and mitigation of harmonic amplification problems caused by single tuned filters, *IEEE Transactions on Power Delivery*, Vol. 13, No. 3, July 1998, pp. 800–806.

## Series Capacitors

Angquist, L., Lundin, B., and Samuelsson, J., Power oscillation damping using controlled reactive power compensation—A comparison between series and shunt approaches, *IEEE Transactions on Power Systems*, Vol. 8, No. 2, May 1993, pp. 687–700.

Barcus, J.M. Miske, S., Jr., Samuelsson, J., and Sévigny, R., Considerations for the application of series capacitors to radial power distribution circuits, *IEEE Transactions on Power Delivery*, Vol. 16, No. 2, April 2001, pp. 306–318.

Brown, D., DeMaster, J., Hite, D., Johnson, B., Parson, K., and Overbye, T.J., Method to improve transmission system voltage stability, *Proceedings of IEEE/PES Transmission and Distribution Conference*, 10–15 April 1994, Chicago, pp. 73–78.

Chang, J. and Chow, J.H., Time-optimal series capacitor control for damping interarea modes in interconnected power systems, *IEEE Transactions on Power Systems*, Vol. 12, No. 1, February 1997, 215 pp.

Chang, J. and Chow, J.H., Time-optimal control of power systems requiring multiple switching of series capacitors, *IEEE Transactions on Power Systems*, Vol. 13, No. 2, May 1998, pp. 367–373.

El-Marsafawy, M. and El-Emary, A., A study of the effects of power system parameters and operating conditions variation on sub-synchronous resonance, *Modeling, Simulation and Control*, Vol. 41, No. 2, 1992, pp. 1–14.

Faried, S.O. and Aboreshaid, S., Stochastic evaluation of voltage sags in series capacitor compensated radial distribution systems, *IEEE Transactions on Power Delivery*, Vol. 18, No. 3, July 2003, pp. 744–750.

Fecho, T.R., Lockwood, S.P., Sachdeva, M.K., and Wang, B.B., AEPs Kanawha River 345 kV series capacitors installation—Planning and operating considerations, *Proceedings of the American Power Conference*, 1993, pp. 712–717.

Godart, T.F., Imece, A.F., McIver, J.C., and Chebli, E.A., Feasibility of thyristor controlled series capacitor for distribution substation enhancements, *Proceedings of IEEE/PES Transmission and Distribution Conference*, April 1994, Chicago, 678 pp.

Gonzalez, R., The Manitoba–Minnesota transmission upgrade project, *Transmission and Distribution*, Vol. 44, No. 5, May 1992, pp. 66–71.

Hausler, F. and Craig, B., PacifiCorp enhances system with two 500 kV substations, *Transmission and Distribution*, Vol. 45, No. 3, March 1993, pp. 28–32.

Helbing, S.G. and Karady, G.G., Investigations of an advanced form of series compensation, *IEEE Transactions on Power Delivery*, Vol. 9, No. 2, April 1994, pp. 939–947.

IEEE Working Group, Readers' guide to sub-synchronous resonance, *IEEE Transactions on Power Systems*, Vol. 7, No. 1, February 1992, pp. 150–157.

Iliceto, F., Gatta, F.M., Cinieri, E., and Asan, G., TRVs across circuit breakers of series compensated lines—status with present technology and analysis for the Turkish 420 kV grid, *IEEE Transactions on Power Delivery*, Vol. 7, No. 2, April 1992, pp. 757–766.

Jalali, S.G., Lasseter, R.H., and Dobson, I., Dynamic response of a thyristor controlled switched capacitor, *IEEE Transactions on Power Delivery*, Vol. 9, No. 3, July 1994, pp. 1609–1615.

Johnson, R.K., Torgerson, D.R., Renz, K., Thumm, G., and Weiss, S., Thyristor control gives flexibility in series compensated transmission, *Power Technology International*, 1993, pp. 99–103.

Karady, G.G., Ortmeyer, T.H., Pilvelait, B.R., and Maratukulam, D., Continuously regulated series capacitor, *IEEE Transactions on Power Delivery*, Vol. 8, No. 3, July 1993, pp. 1348–1355.

Kazachkov, Y., Fundamentals of series capacitor commutated HVDC terminal, *IEEE Transactions on Power Delivery*, Vol. 13, No. 4, October 1998, pp. 1157–1161.

Kingston, R., Holmberg, N., Kotek, J., and Baghzouz, Y., Series capacitor placement on transmission lines with slightly distorted currents, *Proceedings of IEEE/PES Transmission and Distribution Conference*, 10–15 April 1994, Chicago, pp. 221–226.

Kong, A., Predicting life from analysis of field aged 500 kV series capacitors, *IEEE Transactions on Power Systems*, Vol. 12, No. 3, July 1997, pp. 1374–1378.

Lambert, J., Phadke, A.G., and McNabb, D., Accurate voltage phasor measurement in a series compensated network, *IEEE Transactions on Power Delivery*, Vol. 9, No. 1, January 1994, pp. 501–509.

Lee, G.E. and Goldsworthy, D.L., Equipment and protection for BPAs new intertia series capacitors, *Proceedings of the American Power Conference*, April 1992, Chicago, pp. 664–670.

Lee, G.E. and Goldsworthy, D.L., BPAs Pacific AC Intertie Series Capacitors: Experience, Equipment and Protection, *IEEE Transactions on Power Delivery*, Vol. 11, No. 1, January 1996, pp. 253–259.

Lee, S. and Liu, C.C., Damping torsional oscillations using a SIMO static VAR controller, *IEEE Proceedings C: Generation, Transmission and Distribution*, Vol. 140, No. 6, November 1993, pp. 462–468.

Lord, R., Lambert, J., Hebert, Y., Chaine, R., Riffon, P., and Parent, J., Technical considerations concerning the use of magnetic potential transformers on MOV series-compensated transmission network, *IEEE Transactions on Power Systems*, Vol. 9, No. 4, November 1994, pp. 1901–1907.

Lundin, R., Pankowski, D.T., and Targerson, D.R., Online and in control, *Transmission and Distribution*, Vol. 45, No. 3, March 1993, pp. 14–16.

McDonald, D.J., Urbanek, J., and Damsky, B.L., Modeling and testing of a thyristor for thyristor controlled series compensation (TCSC), *IEEE Transactions on Power Delivery*, Vol. 9, No. 1, January 1994, pp. 352–359.

Morgan, L., Barcus, J.M., and Ihara, S., Distribution series capacitor with high-energy varistor protection, *IEEE Transactions on Power Delivery*, Vol. 8, No. 3, July 1993, pp. 1413–1419.

Morgan, L. and Ihara, S., Distribution feeder modification to service both sensitive loads and large drives, *IEEE Transactions on Power Delivery*, Vol. 7, No. 2, April 1992, pp. 883–887.

Natarajan, R., A solid state power factor controller for continuous miner applications, *Mineral Resources Engineering*, Vol. 2, No. 3, 1989, pp. 239–248.

Nejad, M.M. and Ortmeyer, T.I., GTO Thyristor controlled series capacitor switch performance, *IEEE Transactions on Power Delivery*, Vol. 13, No. 2, April 1998, pp. 615–621.

Nguyen, C.T., Lord, R., and Do, X.D., Modeling MOV-protected series capacitors for short-circuit studies, *IEEE Transactions on Power Delivery*, Vol. 8, No. 1, January 1993, pp. 448–453.

Noroozian, M. and Anderson, G., Damping of power system oscillations by use of controllable components, *IEEE Transactions on Power Delivery*, Vol. 9, No. 4, October 1994, pp. 2046–2054.

Nyati, S., Wegner, C.A., Delmerico, R.W., Piwko, R.J., Baker, D.H., and Edris, A., Effectiveness of thyristor controlled series capacitor in enhancing power system dynamics: An analog simulator study, *IEEE Transactions on Power Delivery*, Vol. 9, No. 2, April 1994, pp. 1018–1027.

Okamoto, H., Yokoyama, A., and Sekine, Y., Stabilizing control of variable impedance power systems: Applications to variable series capacitor systems, *Electrical Engineering in Japan*, Vol. 113, No. 4, June 1993, pp. 89–100.

Paula, G., Modified feeder serves diverse loads, *Electrical World*, Vol. 206, No. 10, October 1992, pp. 58–60.

Pereira, M., Renz, K., and Unterlass, F., Digital protection schemes of advanced series compensators, 2nd International Conference on Advances in Power Systems Control, 7–10 December 1993, pp. 592–600.

Pereira, M., Renz, K., and Unterlass, F., Assessing the FACTS at Kayenta, *Modern Power Systems*, Vol. 13, No. 12, December 1993, pp. 35–37.

Pillai, G.N., Ghosh, A., and Joshi, A., Torsional interaction studies on a power system compensated by SSSC and fixed capacitor, *IEEE Transactions on Power Delivery*, Vol. 18, No. 3, July 2003, pp. 988–993.

Prabhakara, F.S., Hannett, L.N., Ringlee, R.I., and Ponder, J.Z., Geomagnetic effects modeling for the PJM interconnection system. II. Geomagnetically induced current study results, *IEEE Transactions on Power Systems*, Vol. 7, No. 2, May 1992, pp. 565–571.

Rajaraman, R., Alvarado, F., Maniaci, A., Camfield, R., and Jalali, S., Determination of location and amount of series compensation to increase power transfer capability, *IEEE Transactions on Power Systems*, Vol. 13, No. 2, May 1998, pp. 294–300.

Rajkumar, V. and Mohler, R.R., Bilinear generalized predictive control using thyristor controlled series capacitor, *IEEE Transactions on Power Systems*, Vol. 9, No. 4, November 1994, pp. 1987–1993.

Reason, J., Thyristor handles direct capacitor bank switching, *Electrical World*, Vol. 206, No. 6, June 1992, pp. 55–60.

Rent, K., Thumm, G., and Weiss, S., Advanced series compensator enhances flexibility for AC transmission systems, *2nd International Conference on Advances in Power Systems Control*, Vol. 2, December 1993, pp. 584–591.

Series Capacitor Bank Protection Working Group, Summary of the IEEE Special Publication on Series Capacitor Bank Protection, *IEEE Transactions on Power Delivery*, Vol. 14, No. 4, October 1999, pp. 1295–1297.

Song, Y.H., Fault detection technique for controllable series capacitor unit in EHV transmission systems using neural networks, *Proceedings IPEMC 94. First International Power Electronics and Motion Control Conference*, 27–30 June 1994, China, pp. 1073–1076.

Swann, G.E., Larsen, E.V., and Piwko, R.J., Major benefits from thyristor controlled series capacitors, *Power Technology International*, 1993, pp. 109–112.

Wang, Y., Mohler, R.R., Spee, R., and Mittelstadt, W., Variable structure facts controllers for power system transient stability, *IEEE Transactions on Power Systems*, Vol. 7, No. 1, February 1992, pp. 307–313.

Woodford, D.A., Solving the ferroresonance problem when compensating a DC converter station with a series capacitor, *IEEE Transactions on Power Systems*, Vol. 11, No. 3, August 1996, pp. 1325–1331.

Wilheim, M.R. and Torgerson, D., Phase controlled reactor bucks series capacitor, *Electrical World*, Vol. 206, No. 6, June 1992, pp. 60–70.

Xuan, Q.Y. and Johns, A.T., Digital simulation of series-compensated EHV transmission systems, *IEEE Colloquium on Simulation of Power Systems*, Digest No. 221, IEE, London, UK, 1992.

## Books

Arrilliga, J., Bradley, D.A., and Bodger, P.S., *Power System Harmonics*, John Wiley & Sons, New York, 1985.

Bloomquist, W.C., *Capacitors for Industry*, John Wiley & Sons, New York, 1950.

Burke, J.J., *Power Distribution Engineering*, Marcel Dekker, Inc., New York, 1994.

Heydt, G.T., *Electric Power Quality*, Stars in a Circle Publications, Second Edition, Scottsdale, Arizona, 1999.

Longland, T., Hunt, T.W., and Brecknell, W.A., *Power Capacitor Handbook*, Butterworth Heinemann, April 1984.

Marbury, R.E., *Power Capacitors*, McGraw-Hill Book Company, Inc., New York, 1949.

Natarajan, R., *Computer-Aided Power System Analysis*, Marcel Dekker, New York, April 2002.

Willis, H.L., *Power Distribution Planning Reference Book*, Marcel Dekker, New York, 1997.

## Power System Capacitor Standards

ANSI/IEEE Standard 18, IEEE Standard for Shunt Capacitors, 2002.

ANSI/IEEE Standard C37.99, IEEE Guide for Protection of Shunt Banks, 1980.

IEEE Standard 824, IEEE Standard for Series Capacitors in Power Systems, 1994.

IEEE Standard 1036, IEEE Guide for Application of Shunt Capacitors, 1992.

# INDEX

Milton Keynes UK
Ingram Content Group UK Ltd.
UKHW020005071024
449327UK00031B/2664

9 780367 393168